The Spacefaring Earth

This engaging survey of the Space Age links science and technology with politics and popular culture, war and peace, and crises and controversies. It examines the history of spaceflight as a mirror of human thought and action across the Earth.

The volume encompasses the new astronomy and sciences of the modern era, the early dreamers and pioneers after 1903, the national competitions of the First World War, the rocket states that prepared for the Second World War, the rivalries and "space race" of the Cold War between the US and USSR, as well as more recent developments including the Space Shuttle, the International Space Station, national space programs, orbital technologies, transhumanism, and military and commercial ventures in space. It also stresses the importance of geography in the geopolitics of spaceflight competition and in the nature of the planetary biosphere. Taking a chronological approach to lived human experience and threshold achievements, the chapters show how these themes have been reflected in literature, art, music, film, and our new digital worlds.

This book is essential reading for students of the history of the Space Age, as well as an excellent companion to courses on twentieth-century science and technology, the Cold War, and American history.

Michael G. Smith is Professor of History at Purdue University. He is the author of *Rockets and Revolution: A Cultural History of Early Spaceflight* (2014) and *The Rocket Lab: Maurice Zucrow, Purdue University and America's Race to Space* (2023).

The Spacefaring Earth

A History of the Space Age

Michael G. Smith

Routledge
Taylor & Francis Group
NEW YORK AND LONDON

Designed cover image: "Orbits through Space," representing "a closed elliptical orbit, a parabolic orbit, and on the outside, an open, hyperbolic orbit." All are "characteristic of the start of an interplanetary flight." Source: *Purdue Alumnus*, Volume 48, Number 2 (1 November 1960): 2. Courtesy of Purdue University Libraries, Karnes Archives and Special Collections.

First published 2025
by Routledge
605 Third Avenue, New York, NY 10158

and by Routledge
4 Park Square, Milton Park, Abingdon, Oxon, OX14 4RN

Routledge is an imprint of the Taylor & Francis Group, an informa business

Library of Congress Cataloging-in-Publication Data
Names: Smith, Michael G., 1960– author.
Title: The spacefaring Earth : a history of the space age / Michael G. Smith.
Description: New York, NY : Routledge, 2025. | Includes bibliographical references and index.
Identifiers: LCCN 2024025915 (print) | LCCN 2024025916 (ebook) |
ISBN 9781032799025 (hbk) | ISBN 9781032780719 (pbk) |
ISBN 9781003494409 (ebk) | ISBN 9781040185094 (Adobe PDF) |
ISBN 9781040185117 (Epub)
Subjects: LCSH: Astronautics—History.
Classification: LCC TL788.5 .S5837 2025 (print) | LCC TL788.5 (ebook) |
DDC 629.409—dc23/eng/20240822
LC record available at https://lccn.loc.gov/2024025915
LC ebook record available at https://lccn.loc.gov/2024025916

ISBN: 978-1-032-79902-5 (hbk)
ISBN: 978-1-032-78071-9 (pbk)
ISBN: 978-1-003-49440-9 (ebk)

DOI: 10.4324/9781003494409

Typeset in Times New Roman
by codeMantra

For our sister, Rita Ann Smith, and brother, James Adelio Smith

Contents

Figures

Tables

Preface

This textbook offers a survey for a semester-based course on the history of the Space Age, offered regularly at Purdue University since 1998. Each chapter offers several "boxes," where I define important issues and terms. Each offers a concluding commentary called, "Vertical Horizons: Above and Below Worlds." Each offers a list of Select Bibliography, some of which are annotated at the end of each list. I sometimes note important authors in the text, whose works are listed at the end of the chapter or in earlier chapters. I have collected select foreign-language sources in an appendix.

I only offer a few web addresses throughout. Otherwise, as you follow the story, please use the names and titles to search the internet for relevant images and music. They will add ballast and beat to the history. All money values refer to the time period under discussion; and readers can use an internet converter to see what the value means in currency today. Also, in transliterating Russian words, I have usually followed the Library of Congress system.

For assistance and advice over the years, I offer my thanks to Melinda Zook, Max Campbell, Timothy Olin, Arslan Jumaniyazov, Meredith Stukey, Caitlin Fendley, Bert Chapman, Roger Strater, and W. Rhodes Lacy. I thank the late Vernard Foley for his insights. I also thank all of my students over the years.

The advantage of this textbook, which I have used online since 2015, has been its openness to constant improvement in the free discussion of the classroom. Students have shared corrections and advice about the facts and ideas. They have contributed websites, movie clips, articles, books, and their own insights into the story. My advice to science and engineering students has been to adapt and refine their scientific-technological explanations for the humanities classroom. To quote Joseph Shea, manager of the Apollo Spacecraft Program Office (1963–1967): "If you understand it, you can make me understand it." This has been our standard of expectation.

At times, the writings below may seem certain and authoritative. But as history, these retellings of stories about people and events and ideas are as fallible and contingent as the lived past itself. Science teaches us cause and effect and offers us reproducible experiments and verifiable explanations. History only teaches us to understand the past, always from material and documentary evidence. This is a past limited by the knotted variables in time and space as they happened, and by our mostly imperfect powers of historical observation, reflection, and interpretation. All events are equivocal, all of us ambiguous, then and now. Nonetheless, as the following epigraph advises, we will share these facts and truths, as best as we can reconstruct them, together.

"De manera, que si el uno tiene ciencia, dar al que no la tiene, si honores, si riquezas; y así el otro al otro." Ignacio de Loyola, *Ejercicios Espirituales* #231 (1548)

Introduction

Human beings conquered the near realms of space only a few short decades after they first conquered the air. The Wright brothers made their famous first flight into the air on 17 December 1903. How long did it take us to breach the realm of space? Twenty-three years if we count the launch of the world's first liquid-propellant rocket in 1926. Almost forty years if we count the launch of the first V-2 missile in 1942. Fifty-eight years if we count the first human being in space in 1961. All of these events happened within one lifetime.

Of these significant dates, 1926 ought to count the most. That was the year when Robert H. Goddard flew the world's first liquid-propellant rocket. It matches best with the Wright brothers' first flight of a heavier-than-air craft. In terms of meticulous preparation, the Wrights spent about four years testing their various models of kites, gliders, and airplanes (from 1899). Goddard had been working on his rocket engines for about five years (from 1921), though he had really begun preliminary work in 1909.

Each event was rather modest in effect. The Wright Flyer first flew only about 120 feet in 12 seconds at a 10 feet altitude. Goddard's rocket flew about 184 feet in 2.5 seconds at a 41 feet altitude. They were both the same awkward, tentative steps. Each resulted in a long delay. The Wright brothers only first flew in public in 1908: Orville at Fort Myer, Virginia, and Wilbur at Le Mans, France. Goddard never once actually launched one of his rockets for public display. He did not publicly report his 1926 flight until ten years later. Unlike the Wright brothers' first flight, which did indeed fly in the air, Goddard's first successful launch of a liquid-propelled rocket only marked the way into outer space, and never actually breached it.

Both of these crafts flew by means of reaction propulsion, embodied in Sir Isaac Newton's third law of motion (1687): "To every action there is always opposed an equal reaction: or the mutual actions of two bodies upon each other are always equal, and directed to contrary parts." For propulsion, the Wright brothers re-invented the propeller. To power their Flyer, they built a light and efficient internal-combustion engine. To navigate the plane, they designed rudders and elevators, along with wing warping, connected by frames and control

DOI: 10.4324/9781003494409-1

wires. The brothers invested most of their initial time in learning and testing the aerodynamic principles of the wing, the box-kite "two-surface machine," able to lift a heavier-than-air machine by the geometric plane, the higher air pressure beneath helping to raise it against the lower pressure above. They had inherited this principle from the theories and experiments of scientists like Daniel Bernoulli and Sir George Cayley, between the early 1700s and early 1800s.

Robert Goddard did not need a wing. He built his rather simple cylindrical device based on the jet propulsion of the rocket. Its provenance went back a bit further, to the ancient Romans and the Chinese of the Sung era, with their rudimentary powder and solid-fuel rockets. In his many experiments, Goddard reconfigured them for future spaceflight, to launch by way of liquid propellants into the vacuum of space. Rocket power is truly unique in the arsenal of modern machines. It does not need land, water, or the air upon which to move. It needs only the propellants and engine that give it thrust, and the craft that gives it form. Unlike the internal-combustion engine that moves by wheels or blades or propellers, it is a combustion engine that moves itself, from the force applied to the rocket by the exploding gases exiting the engine and nozzle.

It seems so simple in these terms. Yet we will recount the many years when the spaceflight pioneers struggled with the varied and complex factors in this unique kind of machine. They had to use higher mathematics to calculate thrust-to-weight ratios and mass ratios, indicating the power of their rockets to accelerate to Earth orbit or beyond. They had to experiment with the chemistry of fuels and propellants (solid or liquid) to provide the right explosive force for thrust; and with the metallurgy of fashioning rocket parts to withstand the extreme heat of the combustion. They had to design engines to mix liquid propellants (oxidizers and fuels) properly. They had to plot guidance and control to steer and target the rocket. They had to design structures to protect their payloads. Rocket power demanded an intricate marriage between ballistics, chemistry, and engineering; eventually joined to celestial mechanics, astrophysics, and human physiology.

The Wright's flight and Goddard's launch were not isolated events. News of the Wrights spread to the media thanks to Orville's telegram to his father on 17 December, soon reported on the press wires. Early news about Goddard's rocket was disseminated even more widely due to a Smithsonian Institution press release. With these reports, enthusiasts around the world began to compete. In answer to the Wrights, Alberto Santos Dumont flew several documented "hops" in his box-kite airplane between August and November of 1906; as did Robert Esnault-Pelterie in 1908, flying an innovative monoplane with an air-cooled engine. In answer to Goddard's news, the Germans Max Valier and other publicists began to design rudimentary rockets for future spaceflight. The Russians, Konstantin Tsiolkovsky and Fridrikh Tsander, did the same. These enthusiasts for rocket power and space exploration were also aviation pioneers. The recent achievements of flight emboldened their own ambitions to reach beyond the atmosphere, into the near realms of space. This was the beginning of what we later called "aerospace."

In all of these ways, the history of the Space Age is really a history about people making science and technology, and about the science and technology partially remaking people. In this textbook, when we say science we are really thinking of it in relation to technology, sometimes informing it, often being informed by it. How physicists and chemists created the new knowledge by which the first pioneers elaborated models of early liquid-propellant rockets. How engineers turned these rockets into short-range and intercontinental ballistic missiles. How physicists transformed Einstein's new theories into the atomic and hydrogen bombs deployed upon these missiles. These are some of the machines of aerospace.

Science and technology are but two dimensions of history. They are rarely neutral, except perhaps in the textbook or on the blackboard. In real life, they are empowered and compromised by the political imperatives, earthbound strategies, and cultural practices of their times. Machines are never simply "what" they are. They are also a "who" and a "where," a "why" and a "how." They are implicated in human beings and our history. There are yet other dimensions. Politics intervened through a variety of different regimes: dictatorships or democracies, and all the variants in between, each with varied styles of governance, sometimes with similar purposes but usually by dramatically different means. Geography also intervened, for we are always people of the earth, embedded in locations, invested in the power and profits of place, fixed on near and far horizons of regional and global influence. Culture intervened in the writers, artists and publicists who transformed spaceflight into entertainment and sport, as public spectacle and media coverage; or into public relations and opinion polls; and even into public dissent, through jokes and songs, television commercials and various protests.

Building on these interpretive foundations, this study offers a comparative approach to the Space Age, integrating rocket designs, military rivalries, geopolitical imperatives, cultural and social factors, human exploration, cosmology, apocalyptic anxieties, and science fiction. I also offer asides on a variety of hoaxes and deceptions. They are relevant too. The Space Age was not always exactly what it seemed.

We will also study several important dynamics at work within these various dimensions of history. One dynamic is between hope and despair. The story of the Space Age involves human beings liberating and fulfilling themselves through machines of their own design, yet human beings also discovering themselves in one way or another alienated by the very machines they have made. Science and technology also often depended on both imagination and drudgery, on trial and error, accident and failure, peace and war. Politics and culture often intervened to camouflage terror and war in the romance of flight and the prestige authority of science. The world's two first and most famous rockets, the V-2 and R-7, were ballistic missiles designed by dictatorial states to kill and maim their enemies. The world's first two intercontinental ballistic missiles, the R-7 and Atlas, which carried the first nuclear payloads, were also the very same rockets that took the first humans to orbit.

Hope came in the form of utopian science fiction, "astronautics" (our reach for the stars), the hero astronaut, the Apollo program to the Moon. Despair came in the form of dystopias, artillery ballistics, the intercontinental ballistic missile, and mutually assured destruction. This dynamic also raises the question of how our values shape our technologies, and how our technologies might help shape our values anew: either to reinforce the worst in us or temper the best. In these terms, it is relevant to note that one of the most celebrated and remembered flights was not Apollo 11 which made it to the Moon, but Apollo 13 which did not. In its failure and near catastrophe, it has captivated audiences with the vicarious thrill of a real near-death experience.

A related dynamic is between liberation and oppression, one that seems to plague all of human history, with especially disastrous consequences in the modern era. Karl Marx and Friedrich Engels perhaps defined it best in *The Communist Manifesto* (1848) with their survey of the contradictions between capitalism and communism. From their perspective, capitalism bestowed remarkable gifts upon humanity, liberating human energies through mechanization, the division of labor, and the management techniques and associated labor of the factory system. Yet capitalism also created the moral and mental oppression of wage labor and commodity exchange, the oppressive conditions of factory floors and worker ghettoes, along with a vast super-structural network of what they called ideological oppression in religion, law, education, and the media. In the resulting class system, the few (the bourgeoisie) came to dominate and oppress the many (the proletariat), necessitating a workers' revolution to restore justice and equality to a world gone very wrong. These communist propositions inspired two of the major authoritarian rocket states of the twentieth century: the Union of Soviet Socialist Republics (USSR or Soviet Union) and the People's Republic of China (PRC).

Thinkers like Friedrich Hayek, Karl Popper, Eric Voegelin, and Jacques Ellul have updated this dynamic for our times and for the capitalist and democratic countries, identifying science and technology themselves as modern ideologies of both liberation and oppression. They liberate us in obvious ways, through their theories and engineering, by giving us power over nature and ourselves. Yet power and domination over nature exacts a price: the continuing domination of states over people, centers over locales, myths over reason, methods over choices. Take, for example, the relatively few technocrats of the state-capitalist and command-communist economies of the twentieth century. They shaped the destructive capacities of the First and Second World Wars and the Cold War. They counted and controlled human beings through their techniques and machines, locking people into the oppression of visible things and unseen institutions and practices.

Both democratic and authoritarian practices intertwined in the competing political systems of the twentieth century. This dynamic raises the issue of the compatibility between liberal democracy and technocracy. How are we to reconcile free elections, representative government, and the mostly free market with the increasingly complex techniques and controls of the surveillance state and big technology, their drives for order and profit? There is also the paradox between benefits and burdens. Poorer and backward states, usually dictatorships, often enjoyed the benefits of the existing advanced technologies that they could steal or buy. Richer and more advanced states, usually democracies, bore the burden of adopting some of the methods and means of the dictatorships in order to compete against them.

Another dynamic follows the analytical distinction between the "what and how" of capabilities and the "why and how" of perceptions. The modern age is split between the hard facts and visible achievements of science and technology, and the softer fictions of political action and its public reception. Capabilities and perceptions mesh and intertwine. Which influenced the course of the Cold War most? It is impossible to answer one way or the other. Their influences were reciprocal and mutually reinforcing (or canceling). As each side prepared its capabilities of machines and techniques for war, this provoked perceptions in the enemy, in both secret meetings and the public media, that war would inevitably come, forcing them to redouble their own more and better capabilities, driving the cycle further and more intensely. Every advantage or disadvantage escalated the conflict. Rational purposes and intentions often turned into both rational and irrational fears. Signals were sent and received, and often misperceived. Signals were written and read, and sometimes misread. Perceptions sometimes became misperceptions.

Take the strange case of the many Central Intelligence Agency (CIA) National Intelligence Estimates of Soviet military or space capabilities between 1945 and 1991. We will explore this case study in more depth later. But for now, please note how one tendency in the record was for analysts to regularly overestimate Soviet capabilities, leading to a political and media perception that the US was falling behind, as, for example, in the many "gaps" of the early Cold War: the nuclear gap, bomber gap, missile gap, space gap, or science gap. There was an element of bureaucratic self-preservation in these estimates and overestimates, multiplied many times over by the politicians and publicists who benefited from them. To define and publicize them was to secure one's place in the system. But this was also the more responsible and cautious approach: to overestimate the enemy and safeguard the future rather than underestimate the enemy and put one's country at risk. The Cold War was all about avoiding war by preparing for it, and in preparing for war we seemed to come ever closer to it. What a strange paradox between capabilities and perceptions.

The overarching dynamic in this story has to do with rise and decline. It is embedded in Newton's law of action and reaction, a study of the mechanical relationship between cause and effect, one thing leading to another. We will see it again in the Second Law of Thermodynamics, that "energy decreases and entropy increases." Nothing is quite stable in this world of ours. The Bolshevik writer, Evgeny Zamyatin, played upon this law in his dystopian novel, *We*, centered on rocket power and the conquest of space, featuring the perfect order of his One State devolving into chaos. We will see it in the Apollo 13 mission, a case study of the human advance upon and retreat from the Moon, success turned into failure and back again. We will also see it in the popular science survey, by Peter Ward and Donald Brownlee, *The Life and Death of Planet Earth* (2004), featuring the possible origins and ends of our planet, and its mortal place in the big universe. Human beings cannot escape this flow of oppositions in nature and humanity, one key to understanding the history of our spacefaring Earth, so far.

In all these ways, the history of the Space Age takes us backward, back through time. We cannot talk about space without talking about time. Our evolutionary paradigms make it so, demanding that we address the origins and ends of the Earth and its humanity, and some of the phases and stages and eras in between. As for origins, we shall introduce several theories that mark the very beginning of life on Earth, for example, in the form of "seeds" that come from outer space. As for ends, we shall meet a number of important thinkers and philosophers who see our human future and salvation in outer space. Einstein's theories of relativity also make it so, as for example his theories on space-time, that space travel approaching or at the speed of light might even compress time.

Space and Time

From ancient times, human beings have often combined the notions of space and time. In his scheme of categories, Aristotle defined time as the change in the position of a thing either before or after. From Plato, he also inherited the impact of time's decay on the visible lives of persons and the material lives of regimes. Space was bound to time's disorder. The word space in the original Latin (*spatium*) meant both space and time: an encompassed area of space, and a measure or stretch of time, as in a distance or interval. For Christian believers, according to the theology of the soul, time is a realm of earth, a place of consciousness and free will, sin and redemption, in the course of a human life, the realm of the secular in preparation for eternity. Eternity and infinity, origins and ends, belong to God, above time. As Alberto Melucci has surveyed, human ways of perceiving time have also been embedded in the things of physical space, as in the burning of a candle, or the motion of sand through an hourglass, or the changing shadows of a sundial. This combined spatial-temporal meaning lasted into the early modern era (the 1400s) when, thanks to the new sciences of the Renaissance and later Enlightenment, we began to delineate and elaborate more precisely between space and time. Still, our temporal metaphors of the Space Age remain beholden to physical space, as in the countdowns of the hands on a mechanical clock; the use of terms like leaps and gaps, breakthroughs and watersheds, thresholds and milestones, crash programs and Moonshots; and even Neil Armstrong's, "one small step."

This very term, Space Age, is fraught with problems. In using it, we are really defining ourselves in a relatively short amount of time, really the era after the launch of the first artificial satellite, Sputnik (1957), humanity's first significant step beyond Earth, at least to orbit. We have been spacefaring people ever since. Yet other ages also compete for our attention. We are still embroiled in an Atomic Age, marked by the first human-made release of atomic energy, in the atomic bombs of 1945, later adapted into nuclear reactors for peaceful civilian use. We are also in an Information Age, which we variably mark from the first computers and transistors, or from the first digital and genetic technologies. Of all these ages, the Information Age seems to have made its most dramatic impact on popular culture and everyday life. According to a 2010 comment by Eric Schmidt, chief executive officer at Google, "there were 5 exabytes of information created between the dawn of civilization through 2003, but that much information is now created every 2 days, and the pace is increasing." Though the exact numbers are open to dispute, he made a legitimate point: that we truly live in an Information Age, or perhaps an Information "overload" Age. This textbook is my small contribution.

These are the three recent ages of our times. Each is contained within an even greater era: the millennia since the last Ice Age, about 11,700 years ago, what the International Union of Geological Scientists calls in neutral terms, the Holocene Era, for the "most recent." Paul Crutzen and his enthusiasts have renamed it, in rather charged terms, as the Anthropocene Era, for the "new man," spotlighting us, the human beings who have transformed the globe's natural environment, mostly for bad: through the agricultural and industrial revolutions, with all their attendant deforestations, mass extinctions, blights, pollution, and climate change. Yet the Anthropocene has also witnessed the Green Revolution, which has fed millions of people, and along with medical and social advances, has vastly improved their lives.

These ages and revolutions relate to one machine or another. Machines dominate our lives, both old and new ones. Think about how the automobile takes us everywhere, transforms our environment with the new architecture of roads and highways, shopping malls and parking lots. Think about how trucks and railroads intersect our landscapes, or how airplanes crisscross the skies in predictable contrail patterns. These are all relatively old technologies, but they still surround us. The romance of the Space Age centers on the newer and rarer machines by which we aspire to higher worlds: rockets, interplanetary probes, spaceplanes, spacecraft, space stations. Unlike most other machines in the normal world, these are essential to life in outer space. Without the rocket and spacecraft and all their attendant mechanisms in environmental controls and spacesuits, the human being would perish in space, the most forbidding of environments. The machine is essential to us there, the human being clothed in the machine and its processes.

Part of the romance of the Space Age is in its rivalries too, in the drama of what we call the "space race" between the US and the USSR after 1957. The very phrase may seem a rather trivial way to describe such a rich and complex history, especially since it was related to several more races: the nuclear race and the missile race. But it remains a vital idiom and reflection of the times. Perhaps even too much. The media today is filled with warnings and predictions in the very metaphors of the Space Age: of a new "space race for 5G" (fifth generation cellular technology), or of a new "Cold War" between the US and China for military satellites, or of the need for a "Moonshot" to confront the challenges of AI (artificial intelligence). Knowing some of the historical facts behind these metaphors will help us be a little wiser today as we read the news.

In the space race, both the USSR and US enjoyed significant gains along the way. The USSR's breakthroughs are beyond dispute. They include these firsts: intercontinental ballistic missile (ICBM); satellite to orbit; dog in orbit; man in space; woman in space; spacewalk; unpiloted orbit of the Moon with safe return; spacecraft landings on the Moon and Venus; and a series of long duration flights in space. The USSR also launched much bigger payloads early on, at least until 1964. Its legacies endure in the way the National Aeronautics and Space Administration (NASA) trains astronauts for long-term space flight. Or in how the international consortium configured the modularity of the International Space Station.

The US also had its share of firsts, if rather less spectacular than the Soviet kind, as, for example, the X-1 to X-15 aerospace programs. The Gemini program firsts included: ten piloted flights in 20 months; astronaut-controlled precision maneuvers; rendezvous of two craft; docking of a craft with a propulsion stage; use of a propulsion stage to move a craft to a higher orbit; flights of a week or more; the longer duration spacewalks; controlled re-entry and precision landings; and the first artificial gravity tethering tests. Beyond these, we ought to mention the significant achievements of Apollo, in landing on and exploring the Moon. Or take the Space Shuttle: in launching and servicing the Hubble telescope, among many other space-based platforms; in important space laboratory experiments; in secret military maneuvers; and in raising and building the components of the International Space Station, also an American achievement if we count the costs, borne largely by the US Treasury.

Given these alternating advances and retreats, who really won the race to space? The answer depends upon how we more precisely frame it. Neil Armstrong said it best, in an address he delivered on the 40th anniversary of Apollo 11. The "space race," he cautioned, was not a simple race but a series of "heats," something like a marathon or triathlon. Granted, at the beginning, the Soviets won quote "heat after heat." Yet America kept pace, leading to what Armstrong called quote "the apex of the space race," his own Apollo.

I take the question historically, understanding the race as a long and challenging ordeal of technological feats and spectacular firsts and sophisticated achievements. In any case, each of these "race" metaphors tells a truth about the space race: that it was essentially a kind of sport, one of the most visible public displays of rivalry between the US and USSR in the Cold War. Popular culture determined the nature of Russian and American competition in space as much as technological innovation and governmental policy. Space technologies were a matter of public consumption in both the commercial capitalism of the US and the command communism of the USSR. Each side attempted and achieved highly publicized Olympian feats. With one crucial difference: American feats, announced and covered live, carried public relations risks; the Russian ones were secret until successful.

Along with covering the fascination with races and firsts, which at times encompass a dizzying chronicle of prideful national achievements, this survey will also focus on thresholds, the relatively simultaneous moments of shared discovery and achievement. Firsts are selfish. Thresholds are shared. The threshold is an entrance point to a new place in time, a moment when people changed, when they created new techniques and pathways and routines, and transformed their mindsets and cultures.

The rivalries between the US and USSR play a significant role in this story. Rivalry presumes some similarity. Both sides envisioned rocket propulsion and space travel along parallel tracks. They built similar kinds of space shuttles and orbital stations. They calculated the same kind of probabilities of extraterrestrial life. They experienced hauntingly similar space disasters. They each defined themselves in the reflection of the other. In the 1920s, the USSR was already inspired by Robert Goddard's experiments, driven to "catch and surpass" the tremendous potentials of American technology and market capitalism. By the 1960s, as Walter McDougall has argued, the US was so desperate for spaceflight recognition that it began to adopt some of the enemy's very own models of authoritarian and technocratic statism.

None of the similarities mean equivalences. Focusing on science and technology should not divorce spaceflight history from the broader human element, especially the common desires and rights of all people for freedom, dignity, and prosperity.

The differences between the US and USSR were decisive. Americans understood space as a function of their industrial optimism, the vast achievements of their market capitalism and national wealth, mobilizing the project-contract system, and the power of state funding and private enterprise, to serve a national purpose. The Soviets received spaceflight as a function of their historical pessimism, their drive to reverse Russia's age-old backwardness and poverty, their fear of European capitalism, their reliance on the command-administrative system and authoritarian communism. These contrasting mentalities informed how each side designed hardware: the US with high costs and sophistication; the USSR more for cheaper mass production and durability. There were advantages to both.

The space race pitted a democratic, free-market US against a one-party, dictatorial, command-economy USSR. Democracies are not perfect. They are fallible and sometimes corrupt. With their free markets, they are embroiled in networks of production and consumption. But they have an advantage, as Sylvia Kraemer has argued, in the capacity for transparency and deliberation, for truth and renewal. They do not make claims, like dictatorships, of infallibility and inevitability. The model of democratic pluralism of the US eventually won out over the bankrupt totalitarianism of the USSR, which collapsed in 1991. The same dynamic plays out in the current rivalry between the US and the PRC. True popular sovereignty and human dignity matter.

In all of these cases, the history of the Space Age always returns us to Earth. Aimed upwards, the artifacts and events of the Space Age bring us back to the ground. This insight is hardly original. It is at the center of the leading comparative histories by Walter McDougall and William Burrows, by T.A. Heppenheimer and Mike Gruntman. Space is a mirror, a mirror of our complex and conflicted nature as human beings. It is not yet a place where we permanently belong, but a place where we project ourselves, fight our earthly battles, find areas of conflict and collaboration, and plot victories or defeats.

The achievements of the Space Age both in and after the twentieth century have always served rather modest, realistic, earthly needs. Keplerian and Newtonian theories of planetary motion and gravity, hardly revolutionary paradigms by contemporary standards, have governed much of our planning: from the drafts for rocket propulsion vehicles, to the construct of lunar orbit rendezvous, to the trajectories for gravity-assisted interplanetary probes. Politicians conceived of rockets and missiles as weapons in the arsenal of strategic air power and bombing. Satellites served the practical needs of military surveillance and economic profit. Engineers and physicians designed orbital space stations so that spacefarers could someday return home, to recover from the debilitating effects of microgravity.

These lessons are properly humbling. We have yet to travel and adapt (evolve) in a sustained way beyond the confines of orbit. The stunning Apollo 8 photograph of "Earthrise," suspended in the black void of outer space, is indeed a perfect metaphor for the Space Age. The living earth, not dead space beyond, has always been the central character in this very human of dramas. To divorce the history of spaceflight from the affairs of humanity on earth, for all their good and ill, means to objectify and privilege that history in a vacuum, like outer space itself. The lyricist and singer, Nick Drake, captured some of this truth in one of the Apollo program's culminating years, in his album, *Pink Moon* (1972), whose front cover featured an American rocket to the Moon, and which included the song "From the Morning," a celebration of this earth of day and night, of the rising sun and moon, and of the rising and knowing of the human being most of all.

Some of this story has been for our human good, as for example in the stunning changes in our everyday lives brought by smartphones and laptop computers. Think of the achievements of orbiting satellites, in the global positioning that navigates our cars, or announces the arrival of the next bus. Or in the web-based maps that plot our quickest travel routes and show us what our homes look like from space. Or on the weather channels that monitor fronts and temperatures with mind-boggling accuracy. Or on Sirius satellite radio that broadcasts shock talk shows (maybe not so good). Take the interplanetary probes and space spectroscopes and telescopes that have mapped and measured the nearby planets, Venus and Mars. As we shall discuss in Part III, these probes and tools discovered important truths about these planets, but they have also discovered a dramatic truth about us: that Venus and Mars are dead Earths. Thanks to what we know about them, we understand our own Earth as fragile and mortal. Thanks to how we use and abuse the new tools of information technology, we reveal our own fragility as well.

This model of an Earth-centered approach has an intellectual foundation: in phenomenology, a philosophy focused on the many nuances of our experience and consciousness, of our perceptions of nature and ourselves. A quote from the French philosopher, Maurice Merleau-Ponty, who was writing in conversation with the German philosophers, Max Scheler and Edmund Husserl, best expresses what we will be doing here.

> The Earth is the matrix of our time as it is of our space. Every constructed notion of time presupposes our proto-history as physical beings bound by consciousness to a single world. Every evocation of possible worlds refers to a way of seeing our own world. Every possibility is a variant of our reality, an effective possibility of reality.

I will take this quote as our standard for this survey. It tells us that our imaginations are always earthbound. We are nothing without this spacefaring Earth, without the ground and gravity upon which we move: the soil upon which we walk and talk, perceive and think, upon which we create the very natural sciences, and their attendant tools and technologies. As fragile human beings in the real world, we are bound to our bodies and to the planet, to our environments and cultures, worlds of fact expressed through art and science and always language. This is the phenomenological sense of the human being as the sole agent to rise above and reconnoiter the whole cosmos, as Scheler offered; of the Earth as our "primitive home-place" and "ark of the world," in Husserl's terms. Thus, to imagine other inhabited planets is but to imagine other humanities and other Earths.

Our goals here are historical. As historians, we are setting out to map persons and events and ideas through time. We will share the same humbling and experiential point of view. For my older readers, two-thirds of this book, at least from the 1950s, is their own lived experience. Thus, we will begin this course not from our imagination, or science fiction, or spaceflight alone, but from the real world, our first order of experience upon this earth. This means the science and technology that defines and transforms the world, the politics that often governs it, and the popular culture that sometimes gives its consent.

All of these terms refer to the essential spatial and temporal horizons of our lives. We have witnessed them in the vistas of spacious deserts or prairies, in sunsets or in starry skies, and in the moments when our visual field collapses into what seems like an infinite line or expanse. These kinds of distances take us to a threshold, motivate us to wonder and imagine. In the plotlines of the spaceflight revolution, these lived horizons become vertical horizons, aiming us upwards.

Select Bibliography

Bainbridge, William Sims. *The Spaceflight Revolution: A Sociological Study*. New York: Wiley, 1976.

Burrows, William. *This New Ocean: The Story of the First Space Age*. New York: Random House, 1998.

Chapman, Bert. *Space Warfare and Defense: A Historical Encyclopedia and Research Guide*. Santa Barbara: ABC-CLIO, 2008.

Deudney, David. *Dark Skies: Space Expansionism, Planetary Geopolitics, and the Ends of Humanity*. New York: Oxford University Press, 2020.

Ellul, Jacques. *The Technological Society*. New York: Vintage, 1964.

Gruntman, Mike, *Blazing the Trail: The Early History of Spacecraft and Rocketry*. Reston: AIAA, 2004.

Harrison, Albert, *Spacefaring: The Human Dimension*. Berkeley: University of California Press, 2001.

Hayek Friedrich. *The Constitution of Liberty*. Chicago: University of Chicago Press, 1960.

Heppenheimer, T.A. *Countdown: A History of Spaceflight*. New York: Wiley, 1999.

Husserl, Edmund. "Foundational Investigations of the Phenomenological Origin of the Spatiality of Nature" (1934), in P. McCormick and F. Elliston, ed. *Husserl: Shorter Works*. South Bend: University of Notre Dame Press, 1981: 222–223.

Johnson, Stephen B. et al., ed. *Space Exploration and Humanity: A Historical Encyclopedia*, in two volumes. Santa Barbara: ABC-CLIO, 2010.

Kempshall, Matthew. "Space, Time, and the World," in *The Later Middle Ages*, Isabella Lazzarini, ed.. Oxford: Oxford University Press, 2021: 137–168.

Kraemer, Sylvia. *Science and Technology Policy in the United States: Open Systems in Action*. New Brunswick: Rutgers University Press, 2006.

Launius, Roger. *The Smithsonian History of Space Exploration: From the Ancient World to the Extraterrestrial Future*. Washington, DC: Smithsonian, 2018.

Logsdon, John. *The Penguin Book of Outer Space Exploration: NASA and the Incredible Story of Human Spaceflight*. New York: Penguin, 2018.

McDougall, Walter. *The Heavens and the Earth: A Political History of the Space Race*. New York: Basic Books, 1985.

Melucci, Alberto. *The Playing Self: Person and Meaning in the Planetary Society*. Cambridge: Cambridge University Press, 1996.

Merleau-Ponty, Maurice. *Signs*. Evanston: Northwestern University Press, 1960.

Mumford, Lewis. "Authoritarian and Democratic Technics," *Technology and Culture* 5/1 (Winter 1964): 1–8.

Neufeld, Michael J., *Spaceflight: A Concise History*. Cambridge: MIT Press, 2018.

Popper, Karl. *The Open Society and its Enemies*. London: Routledge, 1945.

Scheler, Max. *The Human Place in the Cosmos*. Trans. Manfred Frings. Evanston: Northwestern University Press, 2008.
Sparrow, Giles. *Spaceflight: The Complete Story from Sputnik to Curiosity*. London: DK Books, 2019.
Spitzmiller, Ted. *The History of Human Space Flight*. Gainesville: University of Florida Press, 2017.
Voegelin, Eric, *The New Science of Politics: An Introduction*. Chicago: University of Chicago Press, 1952.

For vital sources and descriptions, see the following web pages:

Sergei Khlynin's *Epizody Kosmonavtiki*, 2024. http://epizodsspace.no-ip.org/
Mark Wade's *Encyclopedia Astronautica*, 2019. www.astronautix.com/index.html
Anatoly Zak's *Russian Space Web*, 2024. www.russianspaceweb.com/

Part I

The Spaceflight Visionaries

1 Parabola

Picturing the Cosmos

This first chapter takes us to the crossroads of new scientific theories about our planet and universe: between the old cosmology of the ancient and medieval eras; and the new physics, astronomy, and biology that took shape between the fifteenth and nineteenth centuries, in the eras of the Renaissance, Enlightenment, and scientific revolution. These may seem rather distant origins for the new theories of rocketry and spaceflight at the turn of the twentieth century. But they are essential. Writers of science fiction turned some of these theories into plotlines for their stories about spaceflight, part real and part fantasy, stories that inspired several new generations of experimenters and engineers. The common thread was one of motion and ascent, by people and planetary bodies, moving through time, space, and mind.

Throughout Part I, which is dedicated to the rocket pioneers, our overarching theme will be the sign of the spaceflight **parabola**, the shape of an arc or arrow that represents the trajectory of an object moving at high velocities through outer space. At the beginning of the scientific revolution, a favorite sign of the parabola was the comet, that planetary body sometimes seen from Earth as streaking through the heavens and signifying disaster. The Great Comet of 1577 was just such a case, though this was also the beginning of the scientific explanation about comets, thanks to Tyco Brahe's observations and theories.

In the modern era, the public received comets as portents of doom. Halley's Comet, which appeared in Europe's skies in 1910, brought all kinds of media panic that its tail would send poisonous clouds of gas to kill thousands upon Earth. At the same time, the spaceflight pioneers, both the creators of rocket science and the writers of science fiction, began to entertain the human being in some kind of spacecraft, especially in a rocket, as vying with the comets. As comets had once signified some kind of cataclysm coming to Earth from the heavens, now we began to see the parabolic rocket as a new artificial comet coming from Earth to conquer the heavens.

In these ways, we human beings demarcated the old against new. We turned the scientific revolution and its parabolic trajectories into dramatic turning points in history. As Herbert Butterfield once wrote, "transforming the whole diagram

DOI: 10.4324/9781003494409-3

of the physical universe and the very texture of human life itself," that revolution "looms so large as the real origin both of the modern world and of the modern mentality." Yet nothing in history ever happens quite that way, especially intellectual history. In terms of the scientific revolution, there is no exact turn, but more of a plodding, gradual shift. The spaceflight pioneers shared European intellectual history as a common heritage. They shared a distinctly modern cosmology and astronomy, a fascination with the moon and sun, planets and stars, in relation to each other and to the Earth and us. They shared the scientific and cultural paradigms that made for the "spaceflight revolution" of the twentieth century and beyond.

Between the Old and New Cosmologies

The old cosmology, however old, was a remarkable intellectual edifice that lasted over 1,500 years. It had its origins in the principles summarized by the Greek philosopher, Aristotle, between 384–322 BCE (Before the Common Era). His simple physics held that the elements of earth and water fell back toward it, while air and fire fell away. Matter was centered on Earth, the center of the universe. The higher celestial spheres, made of pure and incorruptible ether (after the philosopher, Plato), were truly out of this world.

The Greek scholar, Claudius Ptolemy, who lived in the 200s CE (Common Era), completed remarkable studies of both the heavens and the earth, along Aristotelian lines. As an astronomer, he revived the ancient conception of Earth at the center of the universe, around which revolved the sun, moon, and stars in "uniform, circular orbits." But given a number of rather minor anomalies in observation and timing, he positioned the Earth just off center and added a number of epicycles, small circles around which the planets moved as they moved in larger circles around Earth. As a geographer, he also helped to measure and size the Earth as a gridded globe of latitudes and longitudes, creating one of the first and most influential sets of atlases of the known world: that of Europe, Africa, and Asia. It was not perfect, nor complete, but when it was republished in 1482 it became the model by which the Mediterranean explorers set off to round the cape of Africa, find a shorter route to the east Indies, and eventually "discovered" all of the Americas and two oceans.

Astronomy was inseparable from geography, and outer space of a fabric with *terra firma*. From pre-modern times, we watched the stars and planets to predict the seasons and map our own movements on the land.

This Ptolemaic system lasted into the medieval era, as astronomers plotted the elegant cycles and epicycles of the planets. Scholasticism, a method of argumentation primarily associated with Saint Thomas Aquinas (1225–1274 CE), and one that dominated the universities during the late medieval era, added Scriptural elements to define the perfect spheres of the rounding planets and stars, as well as the heavenly spheres of the angels. As Aquinas wrote, "The world is composed of the whole of its matter. For it is not possible for there to be another Earth than this one, since every Earth would naturally be carried to this central one, whatever it was." Scholastic theologians grafted Christian ideas onto classical cosmology and metaphysics, including the superiority of astral substances, the incorruptibility of the heavenly bodies, and the redeemable inferiority and corruptibility of us.

Ancient and medieval cosmology defined the Earth at rest, but it still saw the planets and stars as moving. The heavens were in motion, moved by a merciful and loving creator God and his angels and hosts. Sometimes the early cosmologists even imagined us moving to the heavens, to worlds perhaps inhabited by other created beings. Lucian of Samosata's *True History* (from the 200s CE), a satire on the travel adventures of the ancient Greeks, took his imaginary travelers to witness a solar-lunar war and several kinds of alien life and aerial beings. We can find similar interplanetary adventures in folktales, like the Islamic world's, *One Thousand and One Arabian Nights*, and Japan's *The Tale of the Bamboo Cutter* (both from the 900s CE). The "plurality of worlds" thesis preceded modern science. We have only rarely imagined ourselves either totally at rest or alone in the universe.

Two images reveal this truth in their picturing and empowering of the human person. Both show a mighty being in form and nobility, expressing the ancient and medieval concept of microcosm-macrocosm, on how each one of us is a model of the whole universe, as one complete mind and body, spirit and matter. The one, by the mystical theologian (and Doctor of the Church), Hildegard of Bingen, encloses us in Christian creation and symbolism. Her "Second Vision," from *The Book of Divine Works* (1173), centers the human being in a Christ-like pose, arms extended, embedded in a weave of terrestrial lines, surrounded by symbolic circles of air and fire and the movements of the planets. The other, by the artist, polymath, and engineer, Leonardo da Vinci, is one of the most reproduced and narrated of all time. His "Vitruvian Man" (1490), a study of the symmetrical proportions of the human body in its relations (through circle and square) to the greater universe, frees us as a fuller body in motion.

The most dramatic event in the rather slow transformation between the medieval and modern conception of the universe was the publication of Nicholas Copernicus's, *On the Revolution of the Heavenly Bodies* in 1543. As turning points go in history, this was rather equivocal, coming at the end of Copernicus's life, and followed by a century or more of controversy, critique, and neglect. He cleansed the Ptolemaic system of some of its disorder, yet retained his complex epicycles, all with the aim of joining mathematical calculation and purer Platonic forms. But it was the beginning of a new science about the universe and humanity's place within it: a sun-centered or heliocentric universe, around which the Earth and planets moved with observable and verifiable regularity and precision.

Copernicus did not write in an intellectual or cultural vacuum. His treatise came at a point of some culmination in the many years of ferment and debate within the Roman Catholic Church over the nature of the human being, Earth, and the heavens. He also wrote at the very beginning of the Age of Reformation and the religious wars that followed, as Martin Luther and John Calvin and their varied successors challenged the authority and dogmas of the Church. Luther and some of his followers were as adamantly against Copernicus's sun-centered universe as leading Catholic clerics and scholars. But in time the Protestant states became more tolerant and supportive of the new sciences, one of the reasons they could flourish in a freer market of ideas in a freer Europe, if rent by religious and civil wars into the mid-1600s.

European explorers and conquerors, meanwhile, broke out of their small peninsula of Eurasia and set out upon the oceans to navigate, map, and subdue new lands, rename them for Europe. They followed the sun, moon and stars on these travels, joining above worlds and below worlds to track paths through unknown seas and lands. They used a variety of instruments: the magnetic compass to point north; the sextant, to measure angles between a celestial body and the horizon; the astrolabe, to define latitude by measuring the altitude of the sun or star. In all these ways, our forbearers literally envisioned their paths on Earth through a new optics of the skies.

The sun-centered universe laid down a marker for the "giants" of astronomy and physics to come: Tyco Brahe, Johannes Kepler, Galileo Galilei, Isaac Newton. These creative thinkers made the new sciences possible with new tools of observation (telescopes), new ways of experimentation, and new ways of calculation (higher mathematics and eventually calculus). They were innovators of eyes and mind and pen. As Galileo wrote, "the grand book of Nature … is written in the language of mathematics." Tyco Brahe, one of the founders of the new astronomy, made a remarkable discovery with his observations and conclusions about the Great Comet of 1577. He theorized that the heavens were not a perfect set of crystal spheres, but a realm of nature filled with regularities and irregularities, including parabolic comets that could pass right through these distant realms, all demanding fuller observation and measurement.

Building upon the precise observations and findings of Brahe's observatory, the work of Johannes Kepler (1571–1630) confirmed the Copernican thesis with his mathematical calculations. He summarized them in his three laws of planetary motion: that the planets are physical objects that move through space in elliptical orbits; at varying speeds in relation to their distance from the sun; and by verifiable and predictable mathematical laws and natural causes. Or as he proclaimed in the "Introduction" to his *New Astronomy* (1609), "I prove philosophically not only that the Earth is round, not only that it is inhabited all the way around at the antipodes, not only that it is contemptibly small, but also that it is carried along among the stars." Kepler introduced a planet in motion, the human imagination stirring upon a spacefaring Earth.

Kepler's work is also a telling example of some of the dimensions and dynamics covered in this survey. He dedicated the book to a political patron, Rudolph II, the Holy Roman Emperor. He wrote it for the elite reading public of his day and for posterity, with all due honor to his predecessors Copernicus and Brahe, among others. And he wrote the work, like his later, *The Harmonies of the World* (1619), with a near-mystical fascination with the power of mathematics to order and explain nature. Kepler also wrote a piece of clever science fiction, his *Dream / Somnium* (1634), a demon's story of a launch and journey through space to a cratered and inhabited moon, is filled with astronomical detail and the perspective of Earth from space.

Galileo Galilei (1564–1642) took the readers of his *The Starry Messenger* (1610) to the moon as well. This time thanks to one of the world's first telescopes, which he adapted from a Dutch model, to view the lunar surface not as a perfect glassy plane according to the old cosmology, but as a landscape of mountains and valleys "not unlike the face of the Earth," almost like another Earth. Bohemia, to be exact. He redefined the lights of the night sky as traveling stars. He found the moons orbiting Jupiter, just as it and the nearer planets revolve around the sun. In his *Dialogue Concerning the Two Chief World Systems* (1632), he confirmed the Copernican worldview. Galileo was also a visionary physicist. With his experiments, observations, and precise mathematical recordings of the behavior of falling bodies, he redefined motion. Not by any inherent Aristotelian heaviness of matter, but by the relationship of matter to time and distance, all quantifiable and verifiable.

Modern Science: Newton, Darwin, Einstein

These foundations for a new science of the universe, and the Earth within it, found their great culmination in the work of Isaac Newton, *Mathematical Principles of Natural Philosophy* (1687). Newton united Kepler's explanations about the movements of the planets, and Galileo's explanations about the motions of bodies on Earth, into one superb mathematical edifice for the whole universe. It was governed by universal gravitation and the three laws of motion: I. that "an object will stay at rest or move with a constant velocity, unless it is acted upon by an external force" (inertia); II. that the change of a body's velocity is proportional to the force acting upon it; and III. "to every action there is always opposed an equal reaction: or the mutual actions of two bodies upon each other are always equal, and directed to contrary parts." He contributed a perfect metaphor for his age of Enlightenment: the notion of a benevolent God overseeing all of nature made like a "great clock," in which energy is never lost or wasted, perfect in its motions and symmetries … except for us. Gravitation holds everything in place; planets always return to the start of their orbits.

Lost in Translation

Like the other great works of the scientific revolution, Newton wrote his masterpiece in Latin, as *Philosophiæ Naturalis Principia Mathematica*. Compare his third law, offered in English just above, and in the original Latin here: *Lex III: Actioni contrariam semper et æqualem esse reactionem: sive corporum duorum actiones in se mutuo semper esse æquales et in partes contrarias dirigi*. Everything we know about these works comes through laborious and somewhat treacherous translations. It is interesting to note that the translator of Kepler's *Astronomia Nova* took longer to finish his translation into English than Kepler did to write the original. Yet something, however nuanced, was invariably lost in the translation. Many of the works that follow come to us this way, especially from French, German, and Russian sources. As a result, I will offer translations and definitions below with care.

Newton's universe moved by its own precise laws of gravitation and motion in space. Two Enlightenment thinkers gave it a new kind of motion through time. The "Nebular Hypothesis," advanced by Immanuel Kant in 1775 and Pierre de Laplace in 1796, held that our solar system first formed from a nebula of gasses, constantly rotating and contracting and condensing, making for the formation of the planets, first the outer and later the inner planets, in an evolutionary sequence. The planets formed and developed in a kind of grand historical line, meaning Mars was older and Venus younger, a common theme in science and science fiction. Our near neighbors, in other words, were screens upon which we might visualize our futures and pasts. In the late 1700s, explorers like Captain James Cook even set around the world on a quest for the best vantage point to view the "transit of Venus," its revealing path across the face of the sun. Here was a superlative moment when humanity's search of the heavens intersected with navigation and exploration here on Earth.

The search for evolutionary origins on Earth proved far more elusive, as Paolo Rossi has surveyed. All through the Enlightenment of the sixteenth and seventeenth centuries, philosophers like René Descartes and G.W. Leibniz attempted to make sense of the patterns of order and chaos, necessity and accident, that they found in nature and history. Their challenge was to move beyond the Mosaic law and biblical conceptions, discover truths in the real world, and still encapsulate some meaning in God's heavenly design. From philosophy, Descartes detached our minds from matter, set apart and adrift from nature. From science, beginning with James Hutton's, *The Theory of the Earth* (1795), scholars searched for the natural laws of geology to explain and date new origins of the Earth, map its time scales in the solar system, and its mutable and diverse species, as well as our dominion over it all, as bearers of language and civilization.

The revolutionary tract of Charles Darwin, *On the Origin of Species* (1859) offered the decisive theory. Like Cook and the others in search of Venus, Darwin moved about our own globe near and far before he ever plotted his novel vision for physical and biological evolutionary change. In the voyages of the *Beagle*, he traveled along the coasts of South America, all the way to the Galapagos Islands, collecting samples of plants and animals for further study. Study he did, in order to extrapolate from the variety and complexity of his specimens. He deduced the grand theory of the organic evolution of the species, that all was struggle and change and flux in the world: in chance mutations, natural selection, and the survival of the fittest. In the everyday struggle to survive, some specimens of plants and animals do not, some do. Those that do tend to pass on their superior and survivable features to future generations, sometimes even to create whole new species.

Darwin's ideas expressed a zeal for marking time in its origins and ends. In the many popularizations of his thought, including the sophisticated varieties like those of Ernst Haeckel, evolution ascends, and species like us constantly improve and adapt, adapt and improve, moving from lower to higher life forms. This was a "vitalism," so T. H. Huxley lectured in 1868, that began with the very origins of life on Earth. Life began within the complex chemical reactions that created the original "protoplasm" of part-inorganic, part-organic matter. It was a vitalism that continued, as H.G. Wells wrote in *The Discovery of the Future* (1902), in our "great will to live struggling out of the intertidal slime," to rise by spiral curves of "long ascent" to seize the land, to "master the air," and even someday to reach outer space.

The theory of evolution soon gave rise to a popular socio-political movement: eugenics. This was a term coined by Francis Galton in 1883, who sought to turn Darwin's theory into actual practice, to improve the health and breeding of the human race by methods both positive (public health and natalism) and negative (segregation, sterilization, abortion). The topic may seem far removed from rocketry and spaceflight. But both evolution and eugenics were intimately connected to the origins of the Space Age, as their patrons came to see all the new theories, tools, and technologies as means to advance and elevate the human race, and even adapt our bodies for a new life in space.

Although his work has less immediate relevance for the early history of rocketry and spaceflight, we ought to mention the writings of Albert Einstein, summarized in his Special Theory of Relativity (1905) and General Theory of Relativity (1915). Einstein confirmed the speed of light as a constant throughout the universe. He unified the concepts of mass and energy. These were no longer the separate and stable features of Newton. Einstein found energy within mass through his famous formula: $E = mc^2$, that energy is equal to mass times the speed of light squared. It was a formula later confirmed in the massive power of the atomic bomb, the very same energy we see in the power projected by the stars. He also introduced the concept of space-time, redefining our understanding of gravity, seeing objects with mass as creating a curve in space-time, much as a heavy ball creates an indentation or cavity when placed on a soft pillow. Stars like the sun create space-time distortions that influence the planets of lesser mass around it, creating the curvatures of space-time (orbits) through which they travel. Motion thus changes time and space, makes them relative. Increases in velocity will even slow time down, a favorite theme in the space science fiction to come.

Einstein's theories shared the scientific stage with a variety of new physicists, namely Henri Poincare and Paul Langevin, on relativity in general; Ernest Rutherford on the parts and relations within the atom; Max Planck on quantum theory and atomic and sub-atomic processes; Neils Bohr on light waves and particles. Their theories remained in some dispute among scientists, and somewhat obscure to the public, for decades to come. Yet they led the way to the coming technologies of transistors and integrated circuits and microprocessors, tools that have become essential to spaceflight operations in our own day. They all contributed to new explanations and understandings of the infinitesimal and the infinite, the behaviors of the smallest things around and within us (physiology), and the behavior of the immensities and pluralities of the universe beyond us (astrophysics). By the 1920s, popular science in Europe and the Americas began to celebrate these new microcosms and macrocosms of existence: how the smallest of worlds, the relations within atoms, were intricate systems of attraction and repulsion just like the largest of worlds, the solar system and galaxy, with the possibilities of millions or billions more of these. far beyond our own.

Time Travel

Space travel has often been a case study in time travel. By the terms of the Kant–Laplace theory and the Big Bang theory, the whole universe, its galaxies and its solar systems, are actually maps of past time more than of current space. The planets in our own solar system, for example, formed and developed from first to last. To explore them is to visit a moment in evolutionary time quite apart from our own. In the Perry Rhodan stories of the 1960s, and S.M. Stirling's novel, *The Sky People* (2006), human explorers find dinosaurs, neanderthals, and primitive species on a younger Venus. Conversely, we associate older Mars with an advanced extraterrestrial race, as in the classic example, H.G. Wells's *War of the Worlds* (1898), wherein the evolved invaders from the Red Planet deploy their advanced technology to destroy humanity but end up destroying themselves.

In the most radical measure, according to Einstein's Theory of Relativity, time will move slower for spacefarers traveling at the speed of light than for those who remain on Earth. In essence, spacecraft might become time machines, speeding it up or slowing it down, bending time, and transporting humans into the future or the past. The film, *Planet of the Apes* (1968) held this premise. Astronauts crash on a strange and desolate planet, where humans are mute and ruled over by tyrannical apes. In the final scene, the hero comes upon a ruined Statue of Liberty and realizes he has been on Earth all along. Humans have destroyed their civilization by atomic warfare and apes have risen in their place. *Interstellar* (2014) portrayed astronauts traveling through a wormhole in their quest to locate a new planet for an imperiled humanity, thereby experiencing time dilation, a fascinating if imaginary portal between present and future.

We might best describe all the achievements of the old and new sciences, from Aristotle to Aquinas, from da Vinci to Einstein, in one word: magnitude. The term has a powerful metaphorical complexity. It can mean greatness in rank or size or influence, as in the magnitude of the heavens in either the old or new cosmologies. But it also has an astronomical meaning, as when we speak of the magnitude of the degree of brightness of a star or planet. It has a geological meaning, as in the total amount of energy measured by an earthquake. And it has a mathematical and engineering meaning: a number expressing the ratio of two quantities; and "the volume of a sphere" or "length of a vector." Orders of magnitude help us count by leaps and bounds: by tens and hundreds, thousands and ten thousands, millions and billions. All of these kinds of magnitudes can diminish and increase us, frighten and ennoble us. They show us how small we are in the universe, and how large we can become. They express the leaps that humanity made from the Copernican to the Keplerian, from the Newtonian to Einsteinian universe. These notions of magnitude are also at the heart of the most famous words of the Space Age: "one small step for a man, one giant leap for mankind."

The Plurality of Worlds

By the last decades of the nineteenth century, in Europe and the US, the new sciences of human explanation aroused the reading public's fascination with astronomy and the planets. Science gave shape to a renovated thesis on the "plurality of worlds," that the universe is inhabited by many more worlds than our own. One of its most influential advocates was the French astronomer, Camille Flammarion (1842–1925). His life was not all glamour. An astronomer at the French Bureau of Longitudes and Paris Observatory, he spent countless hours observing the planets and stars, helping to synchronize astronomical and Earth time, also amassing a catalog of thousands of double stars. Yet he also had the rare ability to turn the recent discoveries of astronomy into a new public consciousness about our place in the universe. He was a founder of the French Astronomical Society (1877) and the journal, *Astronomy*. Beginning in his early 20s and lasting into old age, he wrote a series of popular and profitable works: *The Plurality of Inhabited Worlds* (1862); *Real and Imaginary Worlds* (1864); *The Wonders of the Heavens* (1871); and *Popular Astronomy* (1894). Flammarion's gift was to convey, in colorful language and mystical terms, that the universe not only moves but is inhabited, lives and dies, and breathes a kind of life all its own.

Cosmos and Creation

Flammarion was one of the most successful of the "pluralists," but he was not alone. The naturalist, Alexander von Humboldt, in his five-volume study, *Cosmos: A Sketch of a Physical Description of the Universe* (1845–1862), celebrated how "the aspect of external nature, as it presents itself in its generality to thoughtful contemplation, is that of unity in diversity, and of connection, resemblance and order, among created things most dissimilar in their form – one fair harmonious whole." Earth and its humanity were part of a beautiful, orderly, oceanic universe. Writers and poets put us in motion within it. At the start of his classic novel, *Moby Dick* (1851), Herman Melville celebrated the natural human attractions to our far horizons, or what he called our "loomings," our wild and narcissistic desires "to go to sea," the "mystical vibration" of seeing ourselves in "in all rivers and oceans." Walt Whitman meditated upon all these wonders and wanderings in his poem, "On the Beach at Night Alone," from the "Sea-Drift" cycle of *Leaves of Grass* (1855):

> A vast similitude interlocks all,
> All spheres, grown, ungrown, small, large, suns, moons, planets,
> All distances of place however wide,
> All distances of time, all inanimate forms,
> All souls, all living bodies though they be ever so different, or in different worlds,
> All gaseous, watery, vegetable, mineral processes, the fishes, the brutes,
> All nations, colors, barbarisms, civilizations, languages,
> All identities that have existed or may exist on this globe, or any globe,
> All lives and deaths, all of the past, present, future,
> This vast similitude spans them, and always has spann'd,
> And shall forever span them and compactly hold and enclose them.

Works like Richard Proctor's, *Other Worlds than Ours* (1870), and Max Wilhelm Meyer's, *The Making of the World* (1906), helped to spread these notions of diversity and plurality, observation and exploration. Space was a new frontier, or at least the last frontier remaining for Earth. Space was a new ocean, a frontier of frontiers. Space was alive and within reach. Technology was helping us to see and explain it, thanks to improved telescopes, the camera, the photometer (to measure the intensity of light and other optical properties), and the spectroscope (an optical device to represent different chemical elements). We began to see and think about the planets as at one with us, especially a living and perhaps even civilized Mars, our neighbor in the solar system.

Another set of technologies informed our new astronomical consciousness of the late 1800s: the newly mechanized Earth. We could not look out at the planets without drawing analogies to them from our own. This was a moment when, as never before, people began to understand the world as one place. The Industrial Revolution was uniting the world by steamship lines, telegraph networks, undersea cables, the Suez and Panama canals, and transcontinental railroads. New and integrating scientific discoveries become inventions: the telephone, phonograph, light bulb, x-ray, and motion picture. Thanks to lighter-than-air balloons, the new heavier-than-air flying machines, and aerial photography, we were seeing the Earth in altogether new ways from the skies, along with all the straight lines and circles of the engineering achievements of our advanced civilization. People were reading about all these developments in an ever-growing international network of print media (newspapers, magazines, mass-edition books). There were pitfalls, as when in the late summer of 1835, the *New York Sun* published a series of articles on the discovery of a civilization on the Moon. It was a hoax, fabricated by the author, Richard Adams Locke, though presented as truth.

Many of these machines and engineering feats were powered in one way or another by the new steam engine, the power plant that transformed fuel and heat into motion and work. To explain this wondrous machine, and the conservation and transformation of energy more broadly, thinkers like Nicolas Sadi Carnot and William Thomson (aka Lord Kelvin) devised the new Laws of Thermodynamics, centered on the Second Law. It held that the energy of a closed system is bound to decrease and decay, as entropy. This notion of entropy both complemented and complicated the new sciences of biology and physics, as well as philosophy and politics. Building on the ancient Greek concepts of cosmos and chaos, order and anarchy, publicists like Henry Adams found entropy in the cyclical course of history: the rule of life and death, category and contingency, ascent and decline. As part of nature and life, human beings were bound to rise and fall.

Scholars applied energy and entropy, on the Earth's own natural and historical cycles, to the universe beyond. We made Mars conform to us and our senses, if as an elder Earth. As telescopes and spectroscopes discovered new features and facts about the near planets, astronomers colorized Mars, painted it with green oceans or skies, red soils or vegetation. Astronomers anthropomorphized the planet, finding "canals" upon it, naming its locations and moons from ancient mythology and world history. Camille Flammarion published his two-volume work, *The Planet Mars and its Living Conditions* (1892 and 1909), summarizing that the planet was alive with vegetation. Perhaps the canals were the great feats of intelligent engineering, markers of an older and superior Martian civilization that was experiencing decline.

The invention of global communications like the telegraph, wireless telegraph and radio, inspired our longing to talk with Mars, what the magazine *Scientific American* termed "interplanetary communication." The French Academy of Sciences announced the Guzman Prize, a sum of 100,000 francs for "the person of whatever nation who will find the means within the next ten years of communicating with a star (planet or otherwise) and of receiving a response." Scientists vied for the prize, and for the world's media attention. William Henry Pickering, of the Lick Observatory, claimed he saw light signals coming from Mars in 1901. He and others proposed sending signals back, through vast arrays of reflecting mirrors, or even huge geometric figures written upon the ground, or by way of the new electromagnetic technologies of Guglielmo Marconi and Nikola Tesla. Global and interplanetary communication also helped inspire a new linguistics, the search for a planetary language. English and French, along with the artificial language, Esperanto, were favorites.

These new ventures in the "plurality" thesis went to some of their farthest and most popular extremes with the writings of the Harvard-trained mathematician, Percival Lowell (1855–1916), patron of the Lowell Observatory at Flagstaff, Arizona. He was a world traveler. Educated for a time in France, he also worked for his family's cotton textile mills, a premier international business of the day; and he traveled widely in East Asia (Japan and Korea) and was fluent in Japanese. His acclaimed lecture tour in the spring of 1910 took him to the Royal Institute of London and the Astronomical Society of Paris.

From his observations of both Mars and Earth as planets, Lowell drew a series of ever-stunning conclusions about a likely civilization there. He published these findings in the press and in a series of popular books, whose titles rose in a crescendo of expectation: *Mars* (1895), *Mars and Its Canals* (1906), *Mars as the Abode of Life* (1908). Lowell had observed the changes of the seasons upon the planet, its hues of light and dark, which he read as "lines" of water moving from the poles toward the equator. He surmised that the planet was in crisis, engineering vast canals, "thoroughly Martian in their mathematical precision," to transfer water to the cities and farms. He called all this "rational deduction" and "deciphering," a new science of "planetology," with the added pathos of a Martian civilization struggling to survive upon a dying planet. Here was a prophesy of our own terrestrial death. All this amazed a growing reading public, by now versed in Darwinian evolution and the Nebular Hypothesis.

Not every scientist agreed with Lowell. Critics dissented in the popular press, interspersed with his claims, fueling drama and intrigue. Leading astronomers, like Edward Emerson Barnard, saw no canals, nor did George Ellery Hale, or Asaph Hall. The eminent Darwinist, Alfred Russel Wallace, in his *Is Mars Habitable? A Critical Examination of Professor Percival Lowell's Book* (1907), demolished Lowell's supposed proofs by revealing how the canals made no sense given Martian planetary and atmospheric conditions. Lowell was just seeing the same kinds of climates and shapes that existed naturally here on Earth.

Others questioned Lowell's eyesight. The astronomer, A.E. Douglass, one of Lowell's own assistants, pointed out that human vision through telescopes was prone to all kinds of optical illusions. In 1903, E.W. Maunder and J.E. Evans made a sociological study of middle-school boys (maybe not the most trustworthy of experimental groups). They found that the boys, on viewing Martian planetary smudges from a distance, easily turned them into discrete lines. Within a few years, better photographic evidence also confirmed that the Lowellian lines or canals really were smudges. More recently, several astronomers have argued that, given Lowell's observations of Venus, and the narrow aperture of his telescope, he created a kind of ophthalmoscope, one that did not picture Venus but deceived him, mirroring the circulatory system in his own eye.

From Science Facts to Science Fiction

All of these scientific tracts and philosophical speculations bring us to the great works of science fiction from the late nineteenth century. These were the ones that inspired the pioneers and enthusiasts of space travel. Their stories were colorful and cinematic. They took readers to outer space, much like the older extraordinary adventures, if now with more scientific and technological precision. Authors often set these stories in the real world, sometimes even prefaced as a recently discovered historical manuscript, or news story, or personal testimony.

Jules Verne nearly single-handedly created the genre. Verne was famous for a number of fun adventures: flights around the world by balloon, voyages to the bottom of the sea, trips to the core of the Earth. His foray into outer space, *From the Earth to the Moon* (1865), was a best seller. Verne's story had an edge, a satire of American bravado and militarism in the Civil War, turning its refinement of the canon and battlefield ballistics into new instruments of peace; and turning manifest destiny from the intercontinental to the interplanetary. It also had some science. Though he launched his piloted lunar shell rather implausibly by canon, Verne still applied some basic calculations for velocity, and inserted realistic detail about launch and descent conditions, gravity forces, and various physical and astronomical facts as then known.

Verne's story captivated reading audiences young and old. As we will see in the next chapter, each of the three leading rocketry and spaceflight pioneers read and loved him in their youth, so they testified. His story was inspiring, likely because it introduced astronomy and engineering into the tale and made it more real. He and his successors also transformed the plurality of worlds into a plurality of crafts to take us there. Verne's own sequel, *Around the Moon* (1870), featured the spacecraft as a kind of rocket. The trend was to employ more and more advanced and sophisticated technologies to lure readers into the real possibilities of space travel. For a long time, "anti-gravity" was a favorite trick of the trade. It was easy because it was self-explanatory. It was also a reaction power of a kind, a prelude to rocket power, what writers often called "repulsive energy."

The trend was also to reveal the spacecraft as an earthly habitat, as a new roving home for humanity. Plotlines transformed Verne's all-male crew into men and women both, microcosms of our Earthly existence, filled with the romance of making baby earthlings, and civilization, anew in outer space. George Griffith's, *A Honeymoon in Space* (1901) built his whole plot around the love between the Earl of Redgrave and his wife, newlywed spacefarers through the plural evolutions of the solar system. One of Verne's most colorful successors was John Jacob Astor: millionaire heir, inventor, veteran of the Spanish–American War, builder of the Waldorf-Astoria hotel in New York City, and victim of the Titanic disaster. Quite a resume. Astor's *A Journey in Other Worlds* (1894), set in a future of a fully engineered Earth, applied the anti-gravity force, "apergy," to launch an American conquest of Saturn and Jupiter. Friedrich Mader's, *Distant Worlds* (1911) took his readers in a spherical anti-gravity ship on a colorful journey to variously evolved planets, featuring an international crew of professors and engineers, several female companions, and two Darwin-inspired pet chimpanzees, Dick and Bobs.

Verne's most illustrious and industrious successor was the great writer and socialist, H. G. Wells, and his *War of the Worlds* (1898). Technology is important in the story. The Martians are outfitted with jet propulsion and flying ships, armored fighting machines, death rays and poisonous gasses. But biology, from Well's university training, took up center stage, according to Darwin's survival of the fittest. When the Martian ships first arrived at Earth, they laid waste to all the advances of human civilization and its concentrated urban masses. Vampires of a sort, they have come to feast on earthlings, their new planetary quarry. Yet they collapse, in the end, in one of the great surprises of modern literature, by the power of the tiny and lowly bacteria. Humans are immune to them. The Martians are not, so vulnerable under their impressive armor and machinery. Perhaps more than any other story of the twentieth century, *War of the Worlds* had staying power: translated and republished into dozens of multi-lingual editions; adapted into a famous radio broadcast of 1938, a film adaptation of 1953, a popular Steven Spielberg film of 2000, and a less successful television series in 2020. Another version may be due soon.

Martian invasion novels, and related "coming war" novels, became a trend. Garrett Serviss's *Edison's Conquest of Mars* (1898) turned the Wells story backward, in a pirated version, with Thomas Edison coopting Martian and human technology to create a fleet of anti-gravity ships (with death rays) to travel to Mars and take our revenge. In a dramatic turn of events, we become the invaders this time. Edgar Rice Burroughs, famous for creating *Tarzan of the Apes* (1914), built a career with his Mars serials, centered on a *Princess of Mars* (1917), in which John Carter, a Confederate soldier from the American Civil War, is thrust straight into Percival Lowell's Mars, even using one of his maps. He battles alien creatures, falls in love, flies anti-gravity ships, adapts to Martian conditions and politics, falls in love again, and brings human smarts to the struggle against scarcity and depravity. In orchestral music, Gustav Holst dedicated one part of his seven-movement suite, *The Planets* (1916), to the planet "Mars" (also the Roman god of war), filled with the riveting sounds of attack and counterattack, via a crescendo of percussion, brass, and strings. The piece premiered at a time when Earth itself was embroiled in the catastrophe of the First World War.

Clever entrepreneurs also turned some of these stories into amusement rides, part science lesson and part circus fun, as, for example, Frederic Thompson's, "A Trip to the Moon," at the Pan-American Exposition in Buffalo, New York (1901). The paying customers (a hefty 50 cents) left the world of the exposition for the ride's otherworldliness of dark and shadow. They sat in a cigar-shaped craft with wings, mounted on rails and simulating a rough flight, transported through space to a fairy-tale-style moonscape of dancing maidens and an electric fountain. Electricity, the light bulb, and motion pictures were among the highlights of the Exposition's celebration of technological progress. Inventor Thomas Edison took the ride. President William McKinley took it too, though he was also gunned down at the fair, while greeting the public, shot by the anarchist, Leon Czolgosz. So much for progress.

Thompson's ride was such a profit-making hit that its patrons moved it to Coney Island in 1903 as "Luna Park," part of a vast amusement park built around the theme of the ride, at a total cost of $700,000. Thompson also operated the Hippodrome theater in downtown Manhattan, whose show, "A Yankee Circus on Mars," featured 64 "diving girls" entertaining the king of that newsworthy planet. The theme of these attractions was also the subject of one of the first science-fiction films: Georges Méliès's *Voyage to the Moon* (1902), a replaying of Verne's inaugural story, now with fairy-tale elements and dancing girls. A drawing from the movie, as a backdrop to France landing on the moon, showed a comet trail sweeping across the sky, symbol of the new human parabola of spaceflight.

Méliès's interplanetary shell took on the recognizable "cigar" shape of the flying machines that now filled the print media of the day, in the photographs of new rigid-dirigible airships. Writers adapted them to imaginary science fiction. People even began to see these lighter-than-air boats, as alien ships, flying through the air by the late 1890s. Case Gilson of California claimed to have seen "a great black cigar with a fishlike tail" flying at tremendous speeds toward San Francisco. His was but one of the first in a series of Unidentified Flying Object (UFO) sightings that later peaked in the Cold War.

These stories and diversions made their mark on popular culture. Some of the impressions were light-hearted, as in the case of James Thornton's vaudeville tune, *My Sweetheart's the Man in the Moon* (1892), a classic romantic tale about lovers communicating in the heavens, only now by telephone and balloon. Other impressions were more serious, as in the case of Thomas Edison's famous quote, "the impossibilities of yesterday are the commonplaces of today." It became one of the enduring myths in the popular literature, as if by some modern alchemy we were transforming our literary dreams into scientific-technological realities. The early rocket pioneers, among them Konstantin Tsiolkovsky and Robert Goddard, repeated this catchy slogan, that an era of science fiction was giving way to an era of science fact. It remains a common chapter heading for our own histories of spaceflight: "From Dream to Reality" or "From Legends to Science."

Does science ever really happen this way? Does imagination come before reason, or are they not already constituent and reciprocal forces? I argue that we get our imagination and dreams from the actual and potential in the world around us. Science fiction is not a leap into the unreal and impossible but is already a measure of the real and possible. Science fiction depends on science facts, such as we know them. Popular science in its day was a strange combination of both science and story. The rocket pioneers were dedicated readers of popular science in newspapers and magazines, a science filled with facts and techniques, but also with all the hyperbole of drama and spectacle.

It is also but a short step from hyperbole to hubris, as when we talk of our dreams transforming into reality, or of our imaginations wishing technological progress into existence, or of our power to conquer space and time. The mathematician and physicist, Blaise Pascal, also one of the founders of the scientific method, understood these dangers, cautioning in his *Thoughts / Pensées* (1662) against exuberant belief in the plurality of worlds or in the coming conquest of the cosmos.

> I must not seek my dignity from space, but from the sovereignty of my own thought. There is no advantage in possessing more worlds. By space the universe encompasses and swallows me up like an atom. By thought I comprehend the world … The entire universe need not arm itself to crush us. A vapor, a drop of water, suffices to kill us. But if the universe were to crush us, we would still be more noble that that which killed us, because we know that we die and the universe knows nothing of this. All our dignity consists, then, in thought. By it we must elevate ourselves.

The thoughtful human being was the measure of most things, large and small.

Alfred North Whitehead said something similar in the twentieth century that, without us, "Nature is a dull affair, soundless, scentless, colorless; merely the hurrying of material, endless and meaningless." Nature offers little without the human perception and intuition that give it meaning. We are the ones writing and reading this story, after all.

Vertical Horizons: Above and Below Worlds

This box is a special feature at the end of each chapter, offering glimpses into the ways human beings distinguish our outer boundaries of imagination and exploration, mark our far horizons as either above worlds or as below worlds, sometimes as heavens and hells, but always with humanity hovering somewhere in between.

From the *Book of Genesis* (11: 2–9), we have the powerful tale of the Tower of Babel, originating in a story about a failed competition by humanity against God, much like that of Adam and Eve. After the Great Flood, so the story goes, the peoples of the Earth had one language. At Babylon, they sought to build a great tower of brick to reach to the heavens in order to "make a name for ourselves." But when God saw what they were doing, he decided that "as one people speaking the same language … nothing they plan to do will be impossible for them." "And so the Lord confused their language and scattered them over the face of the whole Earth," answering hubris with humility. "Above" and "below" intersect so dramatically in the story: the people building a stairway to heaven from the deep clay and stone of the Earth; God above punishing them with the curse of distance and plural languages. Pieter Bruegel the Elder memorialized the scene in his canvas, *The Tower of Babel* (1563).

In the ancient era, around 128 CE, Roman architects and engineers built the Pantheon, not as a spiral tower but as a dome, a temple that sits as a kind of vault of the heavens, framed by tons of concrete, decorated with marble, granite, and porphyry. It remains the best existing structure of old Rome, amid a modern city filled with the crush of busy streets, government buildings, apartment blocks, and plenty of other ruins. The Pantheon was dedicated to the Roman gods, derived from the tales of the Greeks, whose images filled the various platforms and niches. Zeus (Jupiter) ruled over the earth from the heights of Mount Olympus, and Apollo as the sun god, from which they shaped human destinies. Some deities lurked about in the depths of the underworld (Hades or Hell, the land of the dead) or in the fathoms of the sea, like Poseidon (Neptune). The varied circles and squares of the Pantheon's interior probably featured the planets too, a model for later planetariums. The "eye" or oculus above represented the eyes of the gods looking down at humanity. It opened the whole structure to the skies and weather; small drains in the floors channeled out accumulating rainwater.

In the late medieval era, Dante Alighieri transformed these parameters into one of the great classics of world literature, his *Divine Comedy* in three parts (1308–1321), a tale of his own journey from hell through purgatory and up to heaven. Not before or since has any author traced the depths of depravity and punishment, and the heights of salvation, in such colorful and compelling terms. The poem is filled with images both terrifying and blissful, which evoke a wide range of emotions: fear, pity, disgust, amusement, awe and even joy. In the Basilica of Florence, Santa Maria del Fiore, Domenico di Michelino (1465) painted a fresco of Dante's vision, in the shape of the Tower of Babel, a spiral showing the punishing path of sinners from Purgatory to Heaven. He included the Duomo (Dome) of the Basilica, an engineering marvel of the time to rival the Pantheon, if now to give glory to God … and Florence … and Filippo Brunelleschi, who engineered it.

As our survey continues, please look for these images of the spiral and dome in the words and pictures of the history of spaceflight. Think about how we cultivate and adapt them in new ways, as layered ziggurats and geometric planes that reach for the heights, that mimic the envelopes of air and gravity that enclose us. They do not move, but they may still prefigure the moving envelopes of the aircraft and spacecraft to come. We always gaze upon or take ourselves to realms beyond.

Select Bibliography

Adams, Henry, *The Degradation of the Democratic Dogma*. New York: Macmillan, 1920.

Armytage, W.H.G., *Yesterday's Tomorrows*. London: Routledge, 1968.

Atkins, Peter. *Galileo's Finger: The Ten Great Ideas of Science*. New York: Oxford, 2003

Bailey, J.O. *Pilgrims through Space and Time*. New York: Argus Books, 1947.

Banerjee, Anindita. *We Soviet People: Science Fiction and the Making of Russian Modernity*. Middletown: Wesleyan University Press, 2012.

Blumenberg, Hans. *The Genesis of the Copernican World*. Trans. Robert M. Wallace. Cambridge: MIT Press, 1987.

Butterfield, Herbert. *The Origins of Modern Science*. New York: Macmillan, 1965.

Clair, Jean, ed. *Cosmos: From Romanticism to the Avant-garde*. Munich: Prestel, 1999.

Crowe, Michael. *The Extraterrestrial Life Debate*. Mineola: Dover, 1999.

Dick, Steven J. *Plurality of Worlds: The Origins of the Extraterrestrial Life Debate from Democritus to Kant*. New York. Cambridge University Press, 1982.

Dick, Steven J. *The Biological Universe: The Twentieth Century Extraterrestrial Life Debate and the Limits of Science*. New York: Cambridge University Press, 1996.

Guthke, Karl. *The Last Frontier: Imagining Other Worlds, from the Copernican Revolution to Modern Science Fiction*. Ithaca: Cornell University Press, 1990.

Humboldt, Alexander von. *Cosmos: A Sketch of a Physical Description of the Universe*, in five volumes. Baltimore: Johns Hopkins University Press, 1997 [1845–1862].

Jacob, Margaret C. *The Cultural Meaning of the Scientific Revolution*. New York: Knopf, 1988.

Kepler, Johannes. *New Astronomy* [1609], trans. William H. Donahue. Cambridge: Cambridge University Press, 1992.

Kragh, Helga. *Conceptions of Cosmos: From Myths to the Accelerating Universe*. New York: Oxford University Press, 2007.

MacDonald, William L. *The Pantheon: Design, Meaning, and Progeny*. Cambridge: Harvard University Press, 1976.

Markley, Robert. *Dying Planet: Mars in Science and the Imagination*. Durham: Duke University Press, 2005.

Nicolson, Marjorie Hope. *Voyages to the Moon*. New York: Macmillan, 1948.

Prigogine, Ilya, and Isabelle Stengers. *Order out of Chaos. Man's New Dialogue with Nature*. New York: Bantam Books, 1984.

Rossi, Paolo. *The Dark Abyss of Time: The History of the Earth and the History of Nations from Hooke to Vico*. Chicago: University of Chicago Press, 1984.

Solomon, Matthew. *Fantastic Voyages of the Cinematic Imagination: Georges Méliès's Trip to the Moon*. Albany: State University of New York Press, 2011.

Trotta, Roberto. *Starborn: How the Stars Made Us*. New York: Basic, 2023.

Wachhorst, Wyn. "Kepler's Children: The Dream of Spaceflight," *The Yale Review* (April 1996): 112–131.

Wells, H.G. *The Discovery of the Future: A Discourse delivered to the Royal Institution on January 24, 1902*. London: Fisher Unwin, 1902.

Whitehead, Alfred North. *The Concept of Nature*. Cambridge: Cambridge University Press, 1920.

Wulf, Andrea. *The Transit of Venus: The Race to Measure the Heavens*. New York: Knopf, 2013.

On the new sense of globalism in the nineteenth and twentieth centuries, in the literary imagination and in real technologies, see the following books. Rosalind Williams, *The Triumph of Human Empire: Verne, Morris, and Stevenson at the End of the World*. Chicago: University of Chicago Press, 2013; David Hochfelder, *The Telegraph in America, 1883–1920*. Baltimore: Johns Hopkins University Press, 2014; Daniel Headrick, *The Invisible Weapon: Telecommunications and International Politics, 1851–1945*. New York: Oxford University Press, 2014. For the grand historical sweep, see Joyce Chaplin, *Round about the Earth: Circumnavigation from Magellan to Orbit*. New York: Simon and Schuster, 2013.

2 Rocket

The Space Pioneers

The history of the Space Age, as a study of science and technology, begins with the people who designed the first mathematical theories and interplanetary machines for spaceflight. It is a story of eccentrics and innovators, people who stood at the peripheries of their scientific fields but who eventually brought their fields straight back to them. They were a small cadre of pioneers who were raised in the astronomical consciousness of the late 1800s, inspired by the optical tools that brought the planets and stars closer, enthused by the industrial machinery and engineering projects here on Earth ready for adaptation to space travel. A diverse group, they were united by a fascination with aviation technology. In their day they represented the core values and national purposes of their home countries, values and purposes that positioned them and their enthusiasts in an ever-intensifying competition. This was, already in 1903, something of a global race toward space.

These national rivalries were part of a pattern of geopolitical competition originally called the "Great Game," meaning British and Russian leveraging for domination in Central Asia, in places like Iran and Afghanistan, where their empires met. The grand theorist, Halford Mackinder, translated that regional rivalry onto a global context, dividing the world into the vast Eurasian land mass of the "pivot" (Russia), and a series of inner "crescents" (Europe, the Middle East, South and East Asia) and outer ones (Great Britain, the Americas, and Australia). These categories gave new meaning to the older cultural oppositions between the East and the West. Mackinder raised technology as the key new variable in the competition of the pivot and crescents for global influence. Technology was a pivot all its own, providing Russia with railroad "mobility" across land, and Great Britain with steamship "mobility" across oceans.

The rocket pioneers that I introduce in this chapter were all supremely geographic thinkers, inspired by new advances in aerial technologies, in the dirigible after 1884 and airplanes after 1903. Here were new mediums for global mobility and power. Their foundational theories in rocket science, and engineering designs for rockets, were to serve their homelands with a new aerial reach: to far horizons on Earth and into the cosmos beyond.

DOI: 10.4324/9781003494409-4

Their race toward space was also personal. The pioneers each laid explicit claim to being the first to conceive the technical possibilities of spaceflight: Konstantin Tsiolkovsky in a manuscript of 1883, Robert Esnault-Pelterie in his predictions of 1908, Robert Goddard in his notebooks of 1909, Hermann Oberth in his early designs of 1909. There is no reason to dispute these firsts. They all ring true. Yet it was the public moment, the words and formulas in print, which counted most. Thanks to his celebrity, Goddard tended to seize the spotlight. He received the majority share of media attention around the world, often at their expense. Each felt threatened or insulted by the others, though at times they also reached out in friendly ways.

For all these rivalries, the pioneers and their enthusiasts also believed that they were part of a radically new shared moment: to create the machines to free humanity from Earth and to discover plural worlds. They saw outer space as a site of liberation from Earthly gravity and cares. Space was a place where Newton's laws of motion and Darwin's laws of evolution and Einstein's theories were to be finally tested and proven. It was a place to conquer the curse of entropy. The pioneers saw, in both science and fiction, worlds of possibilities, not fantasies. They also saw outer space as a realm of national purpose, even a future battlefield for their own nation's victory. Their boldness of vision ended in rather meager results, at least in terms of the sizes and altitudes of their rockets. But these were still the foundations for the later achievements of the Space Age.

The Russian Empire and Tsiolkovsky

The story begins in a surprising place, the Russian Empire, a country well behind its Western European neighbors in terms of industrial development, a manufacturing base, free-market incentives, civil society, and standards of living. Russia enjoyed a premier Academy of Sciences in the capital, St. Petersburg, thanks to the eighteenth-century founder of both, Tsar Emperor Peter I (often called "Great"), whose radical array of state initiatives set out to Westernize and

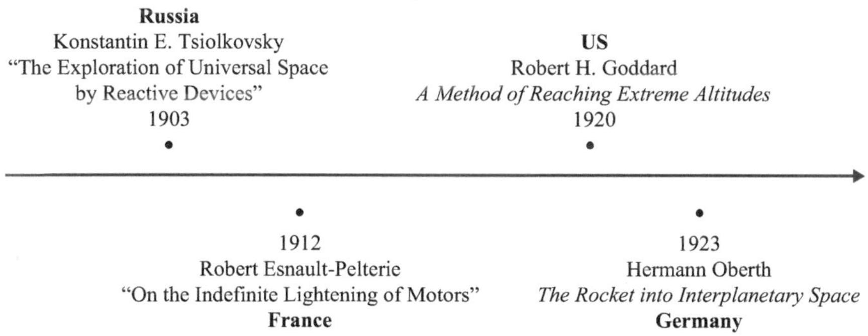

Figure 2.1 A timeline of intellectual discovery. Author.

modernize the country into a "great power." Peter and his successors brought German science to the country, Dutch shipbuilding, German and English military innovations, Italian artists and architects. They invited European financial and engineering experts to build mines and early manufacturing and industrial plants, to fashion transport networks for the export of grain, and to build the Trans-Siberian railroad.

These were all coercive, state-sponsored directives, not free entrepreneurial initiatives. This was a state capitalism that depended on foreign money and talent to build the economy and infrastructure. State patronage of science and technology, of industry and the military, was a function of the country's very backwardness and poverty. Yet it was also a means to cancel that very backwardness and mobilize its forces for Westernization. State power became a recurring substitute for the goals themselves, caught in a web of fear and longing for the West, for its machines and techniques. The geopolitical gambit saw both its successes and failures. The Petrine Empire under the Romanovs created and fortified the armies that won over the Napoleonic forces in 1814, but that also lost the Crimean War in 1856; that won over the Ottoman state in 1878, but that also began the First World War with a reckless mobilization, and ultimately lost that war in 1917.

Russia found strength in the font of its native genius, creative thinkers like the chemist Dmitrii Mendeleev and physiologist Ivan Pavlov, in the many great artists, writers, and composers of the nineteenth century. However, it also failed to democratize and fortify a mass system of education, or craft a viable civil society. Russia freed the enslaved serfs by 1861. Though it also mismanaged the liberation and "redemption" with cumbersome and corrupt state bureaucracies, and a communal land system in the villages burdened by inefficiency, and too little land for too many people. These paradoxes became myths in the Russian national mentality of backwardness: how far the country had come, in leaps and bounds, turning disadvantage to advantage, yet how far it had yet to go.

Konstantin Tsiolkovsky (1857–1935) was born into this conflicted empire, at the edge of a forestry preserve in the Ryazan province, southeast of Moscow. He suffered a bout of scarlet fever as a boy, which left him nearly deaf. At the age of 16, he spent several years of impoverished study in Moscow, largely self-taught. He became a schoolteacher, settling after 1892 in Kaluga, where the later Soviet state eventually honored him with the State Museum of Cosmonautics. Tsiolkovsky published innovative work on chemistry and aeronautics in Russia's leading journals, but he was never a credentialed academic or a part of normal science. He was the classic prodigy and outsider, a "sufferer" for science, as he often put it.

During his years of study in Moscow, Tsiolkovsky was probably inspired by the ideas of the librarian and philosopher, Nikolai Fedorov, author of a radical program for human transformation, what he called the "philosophy of the common task." Fedorov believed that Western science and engineering had empowered humanity with the tools to cultivate the whole Earth, and even explore and civilize the vastness of outer space. The reach and might of the Russian Empire, across the Eurasian landmass, into the Arctic, and Siberia, and Central Asia, also proved that this could be done. He framed his "task" within a grand mystical and paternalist religious project: to celebrate God the Father's creation, and God the Son's resurrection, with a new fraternal army of cosmic engineers. For Fedorov, humanity's compulsory task was to reassemble all of the disparate atoms of physically dead human beings spread throughout the universe, to bring them back to fuller life, and to achieve immortality for the rest of us.

Tsiolkovsky shared some of these ideals, though as a materialist and non-believer. He even wrote a series of booklets and essays surveying his own occult vision of a pan-psychic universe, the home of all-feeling and thinking atoms, from which humanity came, and to which it would return. He also advocated the eugenic purification of the human race: euthanizing the disabled and breeding more perfect human beings for cosmic ends.

Tsiolkovsky was very much a provincial. He read widely, including his favorite venues of basic textbooks, encyclopedias, science journals, and magazine and newspaper articles. But he only ever read them in his native Russian, and without the full discipline and debate of academic exchange. He wrote a manuscript, "Free Space" in 1883, in which he discussed the physics and mechanics of microgravity, one of his passions, and of human adaptation to it, predicting some kind of reactive power as the best means to reach and navigate in outer space. He also wrote two well-received science-fiction novels on this same theme: *On the Moon* (1893) and *Dreams of Earth and Sky* (1895), lighthearted stories about the physics of space voyages and weightlessness.

Two important technical works influenced his most original writing: Alexander Fedorov's, *New Principle of Aerial Flight* (1896), which offered general engineering designs for jet and rocket power; and Ivan Meshcherskii's classic, *Dynamics of a Point of Variable Mass* (1897), which defined the calculus necessary to enable rocket power. Inspired by their designs and proofs, Tsiolkovsky wrote the world's first significant tract on rocketry and spaceflight: "The Exploration of Universal Space by Reactive Devices," published in its first part in the Russian journal, *Science Review*, in May of 1903. Here he named his own Russian rocket, not Verne's canon shell, as the best means to reach outer space.

Tsiolkovsky's key innovation in this tract was the rocket equation, the calculus of the thrust-to-weight ratio, from which he derived the crucial parabolic or escape velocity to leave Earth orbit: 11.2 kilometers or 6.95 miles per second. For the correct thrust and velocity to launch and fly, the rocket needed to carry just enough weight of the right propellants. As to these, he recommended liquid oxygen as the premier oxidizer, and liquid hydrogen as the best fuel, for the explosive force to propel the rocket craft and payload. He also discussed the multiple-staged rocket, one to carry the needed propellants in stages that, once spent, could be jettisoned to save weight and provide a more positive ratio.

Cosmonautics and Astronautics

Tsiolkovsky's "The Exploration of Universal Space by Reactive Devices," published in three parts in 1903, in 1911–1912, and in 1914, confirmed the possibilities of exploring the universe by rocket. He also wrote a summary edition of the whole series, entitled "The Rocket in Cosmic Space" (1924), now with an accent on the "cosmic," to confirm his priority in the field, as against recent American and German claims. Tsiolkovsky's use of the phrase, "cosmic space" (*kosmische raum*), soon became a mainstay of the new international idiom of rocketry, an honor Hermann Oberth recognized by coining of the term *cosmonautics*, meaning rocket travel to the cosmos or universe. The term *astronautics* had a later provenance. In 1927, a committee in the French Astronomical Society, including Robert Esnault-Pelterie and the physicist Paul Langevin, invented the term (*astronautique*) to mean rocket science and travel to "the stars."

The Russian aviation establishment eventually received Tsiolkovsky's ideas with enthusiasm. Seeing their country as behind the US and France in the new field of aeronautics, scientists and engineers valued Tsiolkovsky's innovative theories as a distinct Russian advantage, a way to leap forward in flight. They disseminated his articles, and articles about him, in Russia's popular-science and engineering press, including Iakov Perelman's book, *Interplanetary Voyages* (1915), the world's first scientific survey of rocketry and spaceflight, devoted mostly to astronomy, but featuring Tsiolkovsky's space rockets, and later republished with updates on his and foreign developments into the 1930s.

The Russian Revolution of 1917 reinforced this dynamic of competition with the West. Tsar Nicholas II abdicated power to an initially liberal Provisional Government. Vladimir Lenin and his Bolshevik Party overthrew that government in a violent coup in October of 1917, and fought a brutal four-year civil war to maintain power. The human costs were high, counting at least 5 million dead, resulting in a one-party dictatorship ruled by executions and terror, labor armies and commodity rationing, grain confiscations and artificial famines. The Bolsheviks called this "war communism," renaming themselves as the Communist Party in 1918, and their new state the Union of Soviet Socialist Republics (USSR) in 1922.

As Marxist ideologues ruled over one-sixth of the planet's land mass, they also inherited the contradiction of imperial Russia: a powerful state ideology and apparatus governing over a relatively backward society and economy. They now transformed the empire to serve the struggle of communism against capitalism. Soviet Russia, so isolated in a hostile capitalist world, needed to "catch and surpass" the science and technology of the West in order to destroy it. The irony was that the US helped save this regime, through the efforts of Herbert Hoover's

American Relief Agency. Several hundred American administrators, financed by $10 million from the US Congress, and $90 million in other donations, fed some 10 million Russians between 1921 and 1923.

Lenin's party-state regime was a combination of dictatorship, political repression, and what it called techno-utopian "Americanism." The regime suppressed its political enemies, socialists and conservatives alike, in imprisonment and executions, concentration camps and forced exile. It abolished freedom of the press, of speech, of assembly, and of religion. Yet the regime also built and celebrated American-style electrification projects and hydroelectric dams; and created the State Planning Agency (Gosplan) to manage the country's finances and economy. It educated tens of thousands of young engineers. "Americanization" meant applying the assembly line approaches of Henry Ford, as well as the time-body-motion studies of Frederick W. Taylor, to industry and culture. Initially designed to raise profits for capitalism, the Soviets adapted "Fordism" and "Taylorism" to raise productivity for communism.

In order to compete better with capitalism, the engineer Nikolai Kondratiev wrote the original work, *The Major Economic Cycles* (1925), defining what we now call the "Kondratiev Wave," usually associated with technological innovation, like the steam engine, or electrification, in the advanced industrial economies. It featured the spiral "S" curve of an economic cycle lasting approximately every 50 years. The cycle rose slowly at first, then rapidly, reaching a plateau and then falling into crisis or depression. Knowing the cycles would help Soviet Russia calibrate its state economic policies and technological investments in relation to those of Europe and the US.

American methods in science and technology were all the rage among the communists. So were rockets and spaceflight, an enterprise that Soviet elites considered their own, thanks in part to Tsiolkovsky. His rockets had real military potential, for aerial and artillery warfare, and practical-symbolic value, to reach space someday and prove communism's inherent superiority.

The Soviet Air Force was primed for Tsiolkovsky's visions. As its hymn, "Ever Higher / *Vse Vyshe*" (1921) proclaimed, "We were born to make fairy tales come true, / Conquer all distance and space." In the fall of 1923, the Air Force Academy invited Tsiolkovsky to offer lectures, to the delight of the students. In the spring of 1924, several of the cadets, together with interested professionals and workers from around Moscow, founded and later enlarged the world's first rocket society, the "Society for the Study of Space Travel" (in Russian, *Obshchestvo izucheniia mezhplanetnykh soobshchenii* or OIMS). It only lasted a year, but Soviet state leaders and publicists took an interest, promoting rocketry and spaceflight theories in the top party-state newspapers and magazines. Fridrikh Tsander joined the Society, an aviation engineer who had developed Tsiolkovsky's ideas into his own designs for rocket planes to ascend into outer space.

In the spring of 1924, partially in response to news reports out of the US about Robert Goddard's own space-bound rocket, the Soviet regime premiered something quite extraordinary: the first media image portraying "Rocket Trajectories Launched by Different Speeds." It was a threshold moment, an image of the three lines of advance by rocket beyond Earth. These included the orbital circle of 8 kilometers a second; the parabolic arc to escape Earth's gravity at 11 kilometers a second; and the hyperbola at more than 11 kilometers a second. The image appeared in the Communist Party newspaper, *Truth / Pravda* and its journal, *Young Guard / Molodaia gvardiia* (April–May 1924). A version of this image is on the front cover of the book you are now reading. The leading proletarian poet, Vladimir Mayakovsky, penned some delightful lines to accompany the dream: "Have some fun, / Shooting the heavens by rocket. / Chart your own course, / Take a spin along its parabola."

The Soviet regime also promoted a wider movement in Russian "cosmism," taking shape in literature, poetry, and art. For example, Tsiolkovsky published his utopian novel, *Beyond the Earth* (1918), the first in world literature to send humanity to space in a true liquid-propellant, multistage rocket, its crew bound to enjoy the liberating effects of microgravity and space travel. He also wrote visionary cosmist manifestoes through the 1920s, discussing the gradual conquest of the atmosphere by jet propulsion, and of outer space by rocket and orbital stations, with eventual human migration to the asteroid belt, planets, galaxy, and beyond. He understood that we would eventually have to flee our own dying sun. He called all this his rocket's ascending "spiral" through space. This was also Lenin's favorite image to describe his historical philosophy of dialectical materialism, humanity's successive path to communism, no matter the human cost. Revolutionary progress aimed upward and forward.

In poetry, a group of "proletarian" and "cosmist" writers turned to both technological and mystical themes, promising to outstrip American achievements, re-engineer the Earth for communism, set off for the planets, conquer the craters of the Moon and the canals of Mars. As M. P. Gerasimov wrote, "Dressed by the winged dawn, / We will boldly fly to the skies / Like a fuming comet / We will cut through the Milky Way … / We the cosmic millions …" A small school of Russian revolutionary anarchists even promoted a radical program of "interplanetarism and immortalism," with the goal of flying into space and conquering death.

In art, painters and graphic artists drew from pre-revolutionary trends, namely Kazimir Malevich's "Suprematism," to turn his sharp-colored geometric shapes into stunning landscapes of aerial cities and interplanetary crafts. The artist El Lissitzky turned this style upon us, in his *New Man* (1920), re-creating the human form as a parabola or shooting star, a new Vitruvian person retooled for outer space. Vladimir Tatlin's Monument to the Third International, a model premiering in Petrograd and Moscow in late 1920, was a Soviet spiral toward outer space. To be made of metal and glass, the telescopic structure was designed to rotate in time with the heavenly bodies: the cube (legislative halls) at the bottom once a year; the pyramid (executive offices) at the center once a month; and the cylinder and hemisphere (propaganda and communications rooms) at the top once a day and hour. Like the Tower of Babel of old, it was dedicated to communism's utopian reach.

One of the most successful of these Soviet cultural ventures was Aleksei Tolstoi's ever-popular novel, *Aelita* (1923), also adapted to the screen as Iakov Protozanov's *Aelita: Queen of Mars* (1924). The story was part Jules Verne on space travel, part H.G. Wells on the dying civilization of Mars. The story featured an Einsteinian time warp and strands of occult and cosmist philosophies. It also presented a fictional version of Konstantin Tsiolkovsky, in the main character and rocket engineer, Mstislav Los', who launched his rocket from an impoverished and broken Soviet Russia (after the civil war), to an advanced and futuristic Mars, there to meet the star of erotic attraction, Aelita.

The great exception to these visionary enterprises was Evgeny Zamyatin's classic science-fiction novel, *We* (1922), the archetypical dystopian adventure. Zamyatin wrote a finely threaded parody of Soviet Americanism and cosmism. His modernistic and oppressive "One State" focused all its energies on building an interplanetary rocket, of the Tsiolkovsky style, but at the expense of human freedom and dignity. The regime reduced people to numbers, dictated all facets of their lives, brainwashed them from childhood, and tortured them to keep order and happiness. Zamyatin instead celebrated human freedom, the revolutionary energy of free choice and will against the entropy of dictatorship and technique. Banned in the USSR as an insult to the Soviet regime, Zamyatin's novel was published in an influential English edition by 1924, a model story that warned about humanity's ambitions, pretensions, and drives to self-destruction.

These real and imagined trends in spaceflight culture peaked in 1927 with another set of Soviet firsts, the "First World Exhibition of Interplanetary Machines and Apparatuses," held in Kyiv, Ukraine and later moved to Moscow. In honor of the tenth anniversary of the Soviet state, the exhibit highlighted all the world's great pioneers and their rockets, including those of Tsiolkovsky, Robert Esnault-Pelterie, Robert Goddard, and Hermann Oberth. The next year, one of the country's leading aviation engineers and rocket enthusiasts, Nikolai Rynin, published the world's first comprehensive encyclopedia of spaceflight, *Interplanetary Flight and Communication* (1928–1932, in nine volumes).

French Aviation and Esnault-Pelterie

By every indication of the historical record, Tsiolkovsky was first in rocketry and spaceflight theories. Without denying his priority, we also need to admit that he was very much a man of his time. His debut essay in 1903, the very year that the Wright brothers conquered heavier-than-air flight, was more than a coincidence. Inventors and experimenters in Europe and the Americas were competing to be first to fly an airplane, gearing for new and rapid advances in aviation. Tsiolkovsky was part of this wave. In the context of the race for dirigibles (powered, navigable balloons), he even proposed an innovative all-metal version for Russia to compete with the German Zeppelins.

Tsiolkovsky was not alone in advocating rocketry and spaceflight ideas. They were already a currency in Europe and the Americas. The rocket was on people's minds, in popular-science articles and science-fiction novels. In the spring of 1903, the French popular magazine, *La revue*, publicized the coming promise of "interplanetary communication" and travel. Humanity was about to send wireless signals to Mars, or even build spaceships to get there. Not artillery shells as Verne entertained, but as true liquid-propellant rockets, applying the reactive and "staged" method, whose explosive force came "from within" and achieved the correct "takeoff velocity." Several popular novels also showcased liquid-propellant rockets for outer space: Hans Ganswindt's plan for a reaction-style "interplanetary ship" in 1891; Georges Le Faure and Henri de Graffigny's, *The Extraordinary Adventures of a Russian Savant* (1888–1896); John Munro's *A Trip to Venus* (1897); and L.B. Afanasiev's *Journey to Mars* (1901).

The premier advocate for rocketry and spaceflight in Europe was Robert Esnault-Pelterie. A pioneer aviator and engineer, he was one of the inventors of the aileron, the movable portion of a wing to control the plane's roll. He also invented a new kind of radial engine, and the control column (joystick) to operate and navigate the airplane in flight. In 1908, he also predicted the coming replacement of the airplane propeller with either jet (air-breathing) or rocket (liquid-propellant) power. The French popular-science magazine, *I Know All*, published the idea in a prominent article, including a drawing of Esnault-Pelterie's streamlined jet-propulsion craft, prototype of a space-bound rocket. The year 1908 saw a dramatic turn in aviation, what with the debut of the new Wright Flyer on the world stage and all the accompanying French aerial advances. Publicists predicted that the airplane would transform history, unite the world as never before, join continent to continent in a truly global civilization. They named the leading aviators as "supermen," and foresaw a future world language to match their new world technologies, even the coming of interplanetary communications and travel from this footstool Earth.

In February of 1912, while in Russia to negotiate the sale of French airplanes, Esnault-Pelterie gave a paper before the Imperial All-Russian Aerial Club. Its title, "Considerations of the Results of the Indefinite Lightening of Motors," addressed a contentious issue in aviation, how to create lighter and stronger motors to help drive airplanes faster and higher. His use of the term "indefinite" implied something far more, how liquid-propellant rockets, by the exhaust velocity that "lightened" their engines, might even fly "to infinity."

In this paper, Esnault-Pelterie made the stunning claim that human beings had at least the theoretical and technological means for full-fledged space travel, especially to the Moon, Mars and Venus. Like Tsiolkovsky some ten years before, he defined a mathematical equation for the parabolic or escape velocity by rocket, what he called "self-propulsion." He discussed the precise means of propulsion and navigation to the planets. He promised that human beings could adapt to the extreme heat and cold and zero gravity of outer space with new environments of their own making. The only real problem was the thrust-to-weight ratio. He did not believe that we then had the proper sources of energy in either solid or liquid form to provide the necessary thrust. But he was an optimist, and atomic energy was one solution. Humanity would become a new "Halley's comet" and achieve "fantastic" velocities into interplanetary space.

Esnault-Pelterie gave this lecture again in Paris in November of 1912, before the French Academy of Physics; and published a summary of the talk in the March 1913 issue of the French *Journal of Theoretical and Applied Physics*. His authority was beyond doubt. So was his pride. He had learned about Tsiolkovsky's priority work while in Russia, but he never recognized that until 1930.

The proximity of these two innovators raises the important issue of simultaneous or parallel moments of discovery in history. Did the invention of space rockets happen this way? Simultaneous discovery is actually quite common. The heroic image of lone inventors, the "singletons" as Robert Merton once called them, is mostly false. "Multiples" are the rule, those who arrive at similar conclusions and insights, based on a common set of texts and contexts, all at around the same time. Scientists discovered oxygen, invented calculus, and conceived of evolution in this way. Alexander Graham Bell and Elisha Gray invented the telephone simultaneously in 1876. Research scientists at AT&T Bell Laboratories and the Westinghouse Laboratory invented the transistor in 1948 in a parallel moment of discovery and application. The list goes on.

Do these simultaneous or parallel moments of discovery and invention matter? They most certainly do, especially to the people doing the discovering and inventing, who prefer their own singularity, of course. At stake is the prestige factor of being first, given how societies celebrate scientists; as well as the profit motive, no little thing given the money people can make on their publications and even more on the mass production of their inventions. Just look at the examples of Thomas Edison or Henry Ford. Yet we ought to take care in assigning superiority and advantage. Simultaneous discovery reveals how we are all people of our times, of threshold moments. Context is always bigger than any one of us, though any one of us can also always rise within our context to contribute the new and the better.

America's Republic of Technology and Goddard

The US had significant advantages over the world's industrial economies. This "republic of technology," as Thomas Hughes has called it, enjoyed the stability of democratic governance and the prosperity of a free-market economy and private enterprise. These core American values were not without their strains, as, for example in political corruption, business scandals, civil-rights abuses, and labor unrest. But by the first decades of the twentieth century, the US had become the world's first industrial and manufacturing giant: in the railroads that linked the country, the skyscrapers and bridges of Manhattan, the textile industries of New England and new steel industries of the Midwest, and the corporations of Bell Telephone and General Electric, Dupont and Standard Oil, to name a few.

The government played its role in these advances: in the bureaucratic and military networks that marshaled American money and talent to help win the First World War, in the rebuilding of the navy and army air forces afterward, in turning the US into the world's financial powerhouse and occasional guarantor of European peace and order. In the new field of aviation, the government founded the National Advisory Committee on Aeronautics (NACA) in 1915, with facilities at Langley Airfield in Norfolk, Virginia, to compete with some of the rather stunning achievements in Europe during the war. For the next 40 years, NACA presided over some of the world's best aeronautics advances: aerodynamics testing, new wing (airfoil) designs, and streamlined cowled engines.

Robert H. Goddard (1882–1945), America's premier rocketry pioneer, shared in these confidences about the machine. His heroes were the great masters of American invention and commercial capitalism: Thomas Edison, Henry Ford, Frederick Taylor, and Harvey Firestone. These were makers of machines and builders of systems. Goddard read about their exploits and achievements in his favorite newspapers and magazines, including *Scientific American*, *Popular Science*, and *Popular Astronomy*. He patterned his own career of trial and experimentation, publication and commercial patenting, on their models.

From his youth, Goddard was also a fan of science fiction, especially *War of the Worlds* by H.G. Wells, which he probably first read in its pirated version by Garrett Serviss, the one about *Edison's Conquest of Mars*. At the impressionable age of 17, he even turned these stories into a mystical vision of his own, experienced in the cherry tree of his backyard, which he had climbed to cut off some dead limbs. He dreamed about transforming the ladder, saw and hatchet into some great gravity-defying machine. It was not a vision of the Martians attacking Earth, but of him fighting gravity and launching off into interplanetary space to conquer them.

The date was 19 October 1899, what he forever celebrated as his "Anniversary Day," the moment he began to dream of "the possibility of ascending to Mars." The tree became his boyhood friend and his fond adulthood memory, by which he measured his career progress and mission. When the tree collapsed in a storm, he wrote in his diary, "must go on alone." Mars became his lifelong nemesis in a noble pursuit of defending the mortal danger it posed against Earth, or even of saving humanity from Earth's very own mortality as a planetary body.

This youthful exuberance, together with his dedication to methodological research, defined Goddard's career path. He spent most of his life (47 years, from 1882 to 1929) in or near Worcester, Massachusetts. At 21 years old, he was a graduate of South High School (1904), then received a B.Sc. in Physics from Worcester Polytechnic Institute, and a Ph.D. from Clark University (1911), where he served on-and-off as a Physics professor into the early 1930s. He was 30 years old when he made his first trip away from Worcester, to Washington, DC, on an overnight trip in 1911. He married 12 years later at the age of 42, by which time he had cultivated a taste for cigars and a stiff drink.

As a consummate physicist and engineer, Goddard began his "applied" career at Clark, where he experimented with what he called the various "efficiencies" of solid-fuel rockets, the combustive force (or thrust) packed in their various combinations of explosives. In the academic year beginning in the fall of 1912, he traveled to Princeton University to continue his rocket experiments at the renowned Palmer Physical Laboratory. These few months and their immediate aftermath offer us a curious moment of intersecting lives. While there, Goddard likely met or learned about the work of Robert Williams Wood, Princeton professor of Physics, who had been working on reaction power since the late 1880s and was co-author of *The Man Who Rocked the Earth* (1915), a science-fiction novel that featured a rocket into space, perhaps inspired by his contact with Goddard. Here was a fascinating overlap between science and fiction.

An even more significant intersection of lives happened after Goddard became ill in the winter of 1912–1913, returning home to Worcester in March to cure his tuberculosis. He was bedridden for several months, and subject to a slow regimen of exercise and walks over the summer and fall of 1913 and into the next year. He only resumed his career, as a Physics instructor at Clark University, in the fall of 1914. The period of recuperation was actually one of the most intense intellectual times in Goddard's life. He started the process of patenting his rockets. He drafted an article on "The Navigation of Interplanetary Space" to save humanity from a dying sun. He prepared lectures and presentations on these topics, what he eventually even called going "an infinite distance after an infinite time." One puzzle remains. Did Goddard read Esnault-Pelterie's article on rockets and interplanetary flight in the *Journal of Theoretical and Applies Physics*? It was soon available at Clark University's library and was one of the premier academic journals in the world. Could he, a research professor of Physics, have missed it? The likely scenario is that he read it and retooled his professional work motivated by the article, even though he later claimed to have known about Esnault-Pelterie only after 1922. Esnault-Pelterie likely motivated Goddard.

In 1917, Goddard received his first major career advancement with a grant from the Smithsonian Institution, for $5,000, sponsored partly by the director, Charles G. Abbot, who was also a leading astronomer and a fierce promoter of the plurality of worlds theory. Goddard used it to refine his research work into "A Method of Reaching Extreme Altitudes" (1919), published by the Smithsonian. Most of the paper calculated the thrust-to-weight ratios and exhaust velocities for a two-staged, solid-fuel rocket to study the upper atmosphere. The only dramatic part was on the last pages, where Goddard discussed the possibility of sending a rocket to the moon, filled with enough "flash powder" to create a small explosion and signal on the lunar surface, visible from Earth. Humanity would finally achieve a measure of its own interplanetary communication.

Thanks to a Smithsonian media release and widespread newspaper coverage of Goddard's "moon rocket," he became one of America's most famous scientists, with continuing coverage and teasing promise from Goddard over the next few decades. American audiences were eager and expectant.

Goddard made partly good on his promise, as he turned from solid-fuel to liquid-propellant rockets, which by his calculations had the requisite explosive force for space travel. By September of 1921 he was testing engines with a combination of liquid oxygen and gasoline, and by November 1923 built a prototype. On 16 March 1926, Goddard made the successful launch of the world's first liquid-propellant rocket (liquid oxygen, which had to be maintained at a very cold minus 297 degrees Fahrenheit, and gasoline). This was a culmination of about 20 years of his theoretical work, including 12 years of practical experiments. He described it launching at "express-train speed" and "almost magical" without roar or harsh flame.

In this notebook entries, Goddard revealed his sense of self and history. Here in his liquid-propellant rocket was the foundation for a new technology of mass production and profit, something like the Model T Ford. He launched several more liquid-propellant rockets into late spring; and began refinements for larger rockets (with improved fuel pumps and regenerative cooling) and for improved launch towers. These continued to the summer of 1929, though the tests never reached more than several hundred feet high or 60 miles per hour. These thresholds were only the very beginning of his rocket trials, as he was about to set off for Roswell, New Mexico and the hope of greater gains and fame.

German Reactionary Modernism and Oberth

The German Empire offered one of the most promising grounds for the introduction of rocketry and spaceflight. In the nineteenth century, its universities were among the most admired models in higher education, its students trained in seminars and laboratories, interacting with professors in a dynamic learning environment. German science, thanks to the mathematics and physics faculties at Gottingen and Heidelberg, was the best in the world. Yet German culture was also the site of what Jeffrey Herf has called a "reactionary modernism," a political and cultural obsession with the machine as a measure of national worth and accomplishment. It was both a romance and an abuse of the machine. It was an irrational obsession with rational science and its technologies so as to fulfill Germany's national destiny and to destroy its enemies, what ideologues defined as English commercialism, French republicanism, and "Jewish cosmopolitanism."

This complex was inspired, in part, by Oswald Spengler's *The Decline of the West* (1918), damning European culture in general for its bourgeois corruptions and conceits, yet celebrating Germany's Faustian desire to seize the future and fill the world with its national dynamism. He symbolized this desire in the shape of the Gothic arch and in the uplift and projection of artillery guns and airplanes. Germans found their pride and confidence in machines and engineering projects: industry, railroads, Zeppelin airships, Opel cars, autobahns, and Volkswagens. This trend culminated, as we shall see in the next chapter, in the V-2 ballistic missile, one of the "wonder weapons" of the Second World War.

Germany first proved its scientific-technical and industrial might, with stunning new machines and methods, in the First World War. This was a new kind of total war, with weapons of mass destruction and defense like flamethrowers, tanks, armored vehicles, and barbed wire. German industry made 18 different types of poison gas, including chlorine and mustard gas, which tore through uniforms to burn the skin, or blinded and asphyxiated their victims. The Germans organized meticulous "Special Brigades" for their battlefield gas attacks, sometimes demanding several days of preparation: in the storage, transport, readying, and launch. This new kind of weapon required the kind of care and handling we will later see with the V-2 missile. Poison gas also demanded a new kind of delivery system: improved rockets for distance and accuracy, and to leave no trace of point of origin, diminishing the potential of any counterattack. One kind of weapon led to another.

One of the most prescient of Germany's new weapons, at least in terms of future rocketry developments, was the "Paris Gun," sometimes mistakenly referred to as "Big Bertha." The German military introduced it rather late in the war, and without any significant military advantage, but with a terrifying media and legacy effect. Between 23 March and 9 August of 1918, three of these massive guns sent about 351 shells against Paris, Germany's essential military target throughout the war, killing some 256 and wounding about 620 others. Mounted on a central pivot and trailing wheels, its barrel set at a 50-degree angle, the gun sent a 228-pound shell about 80 miles in three minutes, at an altitude of 25 miles (40 kilometers). As Mark Wade has noted, "this was the highest altitude attained by a man-made object until the first successful V-2 flight test on October 3, 1942." The Paris Gun was a predecessor of the V-2, and a stunning proof, along with the strategic reach of military aviation, that the coming war would involve destruction and terror from above.

The Edge of Space

The shell of the Paris Gun only ever approached the edge of space, a realm for which we do not have a universally accepted boundary. According to the International Aeronautics Federation, space begins at 62 miles (100 kilometers) based on ascent from Earth, applying Theodore von Karman's calculations related to the absence of air and the loss of normal aerodynamic lift. NASA has applied several different standards. NASA Mission Control measured space by descent back to Earth, at 76 miles (122 kilometers), known as "Entry Interface," as, for example, when the Space Shuttle began to experience atmospheric drag. NASA also recognizes 50 miles (80 kilometers) as the designation point at which aviators become "astronauts." This specific boundary actually originated in an article, "How High Up?" in *Scientific American* (July 1926). It was a threshold image, a map of advance into outer space by landmarks both natural and human. Devoted to Goddard's rocket, the article featured a cross-section map of the heights of the Earth (Mount Everest and Pike's Peak) and the heights of the air (cirrus and stratus clouds). Outer space (the realm of falling meteors) began at about 50 miles. In between Earth and space were the ascending machines of human invention: kites, balloons, airplanes, and the shell of the Paris Gun. The map awaited one thing: Goddard's rocket.

These achievements were markers of German national power. They promoted a public romance of the machine. Hermann Oberth (1894–1989), our last of the spaceflight pioneers, came of age at this moment, a youthful devotee of popular science, science fiction, and space travel. A German ethnic born in the Austro-Hungarian Empire, in Transylvania, what later became part of independent Romania, he was something of a loner and outsider, and ended up as a schoolteacher for much of his career, rather like Tsiolkovsky in these regards. He attempted to break into the German academic establishment, first with medical studies at Munich, though interrupted by the First World War. During the war, he served as an orderly, and even medicated himself with anesthetics and narcotics to simulate the feelings of weightlessness.

After the war, he began doctoral studies in physics at the University of Heidelberg, where he presented his mentors with a thesis about rocketry and spaceflight. They rejected it, on the grounds that it was not yet science, though the scientific principles and calculus were all sound. Oberth then paid his own funds to have it published, as *The Rocket into Planetary Space* (1923). A creative and comprehensive work, Oberth argued that humanity was ready, with the right engineering designs, to construct a machine to break free of Earth's gravity and travel through the vacuum of interplanetary space. That machine was a rocket. He offered precise exhaust velocities and liquid-propellant combinations to reach the upper atmosphere, Earth orbit, and even escape velocity. He proposed liquid oxygen as the oxidizer, and alcohol and water (or liquid hydrogen) as the fuels, with multiple stages. He proposed aluminum for lightness and strength as a metal, with asbestos linings for protection and cooling. This was no mere theoretical exercise. He mapped space as a place for human habitation, in the spacecraft itself and in orbital space stations, for Earth observation and communication. He even discussed space mirrors to reflect concentrated heat upon Earth, a means to save the planet from global cooling.

Oberth was the least successful of our pioneers in terms of his career path, but he was the most honest. He openly admitted his profit motive to build and manufacture rockets for mass production. He also recognized Robert Goddard's work, convincingly asserting that he had read it only after his own book was already complete.

Still, Oberth was also the most influential of the pioneers, at least in Europe, in terms of his public fame and lasting legacies. In science fiction, adventure novels like Bruno Bürgel's, *The Star of Africa* (1921), or Otto Willi Gail's, *The Shot into Infinity* (1925) and *The Stone from the Moon* (1926), centered their interplanetary adventure stories around Oberth-style (or Goddard) rockets. These rockets gave a believable, high-technology edge to otherwise fantastic and entertaining stories.

Within a few years of Oberth's book, a number of popular and scientific works answered or supplemented his mission. Max Valier's, *The Advance into Space* (1924–1927 in four editions) turned Oberth's complex mathematics into a thrilling adventure piece, filled with new terms like *rocket flight* and *spaceship*, *spaceflight* and *cosmic pilot*. Walter Hohmann's intensely mathematical, *The Attainability of Heavenly Bodies* (1925), proposed a series of possible trajectories to Earth orbit, to Mars, to Venus, and ultimately to the outer planets of the solar system. Hohmann's front cover presented a metallic human being in a Vitruvian pose, ready for spaceflight. By 1926, the engineers Franz von Hoefft and Guido von Pirquet charted similar interplanetary paths, built upon intricate networks of rockets, space stations, and piloted missions. Oberth inspired it all. He also lived the longest, long enough to witness the Apollo 11 Moon mission, one of the most remarkable "dreams come true" of the twentieth century.

Vertical Horizons: Above and Below Worlds

Just as the pioneers prepared for rocket flights into outer space, they also prepared for the "coming war" here on Earth. Following the bankrupt victories and bitter defeats of the First World War, the media (newspapers, radio, magazines, and books) regularly featured doomsday scenarios of this imaginary war, half known and half unknown, the next war that would settle the unfinished business of the last one. New maps of the atmosphere and stratosphere spotlighted outer space as a "high" battlefield to come. Military planners sometimes navigated these futuristic scenarios in their own planning and predictions.

The rocketry pioneers were wise to all of this. They understood the military implications of rocket planes and moon rockets. Tsiolkovsky once predicted that they would eventually replace artillery as modes of lethal long-distance bombing. His student, Fridrikh Tsander, promoted the military implications of the rocket as a new means of artillery and high-altitude flight. His lectures at Moscow State University in October of 1924 included a major section on the "Military Significance of Interplanetary Flights," surveying the role of "long-range rockets," chemical warheads, speed and distance limits, surveillance, and even global positioning through satellites.

From his laboratory at Boulogne-sur-Seine, Esnault-Pelterie focused on the military applications of rockets after 1928, financed with the modest support of the Ministry of War. He promised a medium-range ballistic missile to fulfill strategic warfare doctrines, though by 1936 his experiments with fuel pumps and rocket launches were all trial and error. Between 1914 and 1940, Goddard made recurring appeals to the US military, requesting funds to develop the rocket. He worked directly with the Army on a "trench rocket," a hand-held "bazooka" type weapon for short-range battlefield use; and with the Navy on solid-fuel rockets for anti-submarine warfare. Oberth proposed a liquid-propellant ballistic missile to the German imperial army in 1917. He later called this a "long-distance rocket" in an appeal to Adolph Hitler in 1939. In his *Ways to Spaceflight* and several lectures, combining fascination and trepidation, he foresaw "death-dealing rockets" and a "rain of rockets," filled with poison gas enough to "exterminate entire populations."

All of these proposals highlight the curious place of liquid-propellant rocketry in national and international life. Each of these rocket pioneers attempted, at one point or another, to convince the military of the war-making applications of their designs and inventions. Each time they were heard, and sometimes promoted and funded. But these initiatives and partnerships did not develop into any lasting or formidable state rocketry programs, at least not yet.

Select Bibliography

Andrews, James. *Red Cosmos: K.E. Tsiolkovsky, Grandfather of Soviet Rocketry*. College Station: Texas A&M University Press, 2009.

Cheng, John. *Astounding Wonder: Imagining Science and Science Fiction in Interwar America*. Philadelphia: University of Pennsylvania Press, 2012.

Durant, Frederick, and George James, ed. *First Steps toward Space*. Washington, DC: Smithsonian, 1968.

Fritzsche, Peter. *A Nation of Fliers: German Aviation and the Popular Imagination*. Cambridge: Harvard University Press, 1992.

Goddard, Esther C., and G. Edward Pendray, ed. *The Papers of Robert H. Goddard*. New York: McGraw-Hill, 1961.

Gorin, Peter. "Rising from a Cradle: Soviet Public Perceptions of Space Flight before Sputnik," in *Reconsidering Sputnik Forty Years Since the Soviet Satellite*, Roger Launius et al., ed. New York: Routledge, 2000 [1997], 11–42.

Hagemeister, Michael. "The Conquest of Space and the Bliss of the Atoms: Konstantin Tsiolkovskii," in *Soviet Space Culture*, Eva Maurer, ed. Houndmills: Palgrave, 2012, 27–41.

Hagemeister, Michael. "Konstantin Tsiolkovskii and the Occult Roots of Soviet Space Travel," in *The New Age of Russia: Occult and Esoteric Dimensions*, M. Hagemeister, B.G. Rosenthal, and B. Menzel, ed. Munich: Verlag Otto Sagner, 2012, 135–150.

Herf, Jeffrey. *Reactionary Modernism: Technology, Culture, and Politics in Weimar and the Third Reich*. New York: Cambridge University Press, 1984.

Hughes, Thomas. *American Genesis: A Century of Invention and Technological Enthusiasm*. Chicago: University of Chicago Press, 2004.

Mackinder, Halford J. "The Geographical Pivot of History," *The Geographical Journal* 23/4 (April 1904): 421–444.

Merton, Robert. "Resistance to the Systematic Study of Multiple Discoveries in Science," *European Journal of Sociology* 4/2 (1963): 237–282.

Neufeld, Michael J. "The Reichswehr, the Rocket, and the Versailles Treaty: A Popular Myth Reexamined." *Journal of the British Interplanetary Society* 53 (2000): 163–172.

Neufeld, Michael J. "Weimar Culture and Futuristic Technology: The Rocketry and Spaceflight Fad in Germany." *Technology and Culture* 31/4 (1990): 725–752.

Patenaude, Bertrand. *The Big Show in Bololand: The American Relief Expedition to Soviet Russia in the Famine of 1921*. Stanford: Stanford University Press, 2002.

Stites, Richard. *Revolutionary Dreams: Utopian Vision and Experimental Life in the Russian Revolution*. New York: Oxford University Press, 1989.

Parts of this chapter are based on the story told in Michael G. Smith, *Rockets and Revolution: A Cultural History of Early Spaceflight*. Lincoln: University of Nebraska Press, 2014. In it, readers will also find the two images that I discuss in this chapter: the Soviet "Rocket Trajectories Launched by Different Speeds" (1924); and the Goddard-style map of Earth, air, and outer space, from "How High Up?" (1926).

3 Missile

Rocket States and War

The US had Robert Goddard, the world's most famous rocket experimenter. The USSR had Konstantin Tsiolkovsky and his popularizers, along with the world's first amateur rocketry and spaceflight society, however fleeting its existence. Weimar Germany had Hermann Oberth and a small cadre of spaceflight enthusiasts. It also had significantly more. Thanks to its core values and commitment to the renovating powers of the machine, and its unique predicament in the world, Weimar Germany also experienced the most sustained and popular wave of interest in rocketry and spaceflight. By the early 1930s, Germany became the center of a growing international movement.

What the individual pioneers had begun, thousands of enthusiasts around the world now grew and fortified in the new amateur societies. They were small in size but visible and boisterous in the forums of popular culture and the media. They joined in a common, idealistic promise: that of a united planetary effort to launch human beings into interplanetary space. Here was a short-lived, if vibrant fraternity of spaceflight enthusiasts, often driven from within by national purpose and competition, but also bridged by mutual respect, personal friendships, collaborative projects, and even a few instances of spying. This chapter surveys the rise of these societies, along with several turning points: their modest successes in making and launching rockets; their more dramatic successes in publicizing and propagandizing rocketry and spaceflight; and their most stunning achievements of all, convincing two "rocket" states, Stalin's USSR and Hitler's Germany, to fund and mobilize national energies for liquid-propellant rocketry. We will canvass these transitions from rocket societies to rocket states, and from rocket states to a world at war, a world under the threat of a new destructive weapon: the V-2 missile.

The German Societies and the Weimar Republic

In the wake of Hermann Oberth's small but creative book, *The Rocket in Planetary Space*, and his first popularizers, writers like Max Valier and Otto Willi Gail, Germany enjoyed the first successful and sustained amateur rocket society,

DOI: 10.4324/9781003494409-5

the "Society for Space Travel" (*Verein für Raumschiffahrt*, hereafter called the Verein), founded in Breslau in 1927. The Verein lasted for about five years, but in fits and starts, largely due to personality conflicts and the limiting effects of the Great Depression. Yet it had an influence and left a legacy far beyond its numbers and achievements, inspiring Soviet-Russian and American rocket groups to form on its model, and convincing both the German and Soviet-Russian militaries to secretly begin investing and supporting its methods. The Verein published an innovative magazine, *The Rocket*, devoted to a wide array of rocketry and spaceflight themes, including astronomy, interplanetary trajectories, and especially surveys of liquid-propellant rocketry. Its editor, Johannes Winkler, tutored readers on these issues, with a smart popular-science approach and inspiring essays on humanity's inevitable power: "to pierce the armor of Earth's gravity," in one of Max Valier's slogans, with the new armor of the cylindrical rocket.

This new publicity peaked in popular science and culture with two intersecting events. The first was the new edition of Oberth's 1923 book, now retooled as *Ways to Spaceflight* (1929). The book proved once again that Oberth was the most creative and comprehensive of all the pioneers. It was an expanded edition, still a mathematical survey of thrust-to-weight ratios and escape velocities, based on precise physics and thermodynamics. He also included storytelling for the general reader, dramatizing spaceflight in poetic terms. He proved that spaceflight was a new and necessary science of human endeavor, what he now called "cosmonautics." Oberth argued for multiple-staged rockets to achieve escape velocity; and discussed new kinds of metals to provide lightness yet also withstand the intense heat of launch and re-entry. He surveyed the advantages of liquid propellants; delved into inertial stabilization and guidance, as well as orbital and interplanetary pathways; reviewed life support systems for the craft and for venturing out into space in suits; and defined re-entry and recovery techniques.

The second event was the premiere of the science-fiction film, *Woman in the Moon* (1929), directed by the accomplished cinematographer, Fritz Lang, with a screenplay by his wife, Thea von Harbou, a popular writer. Lang's film company helped to finance Oberth's book, hiring him to build a prototype rocket for the movie's premiere. Its plotline followed Oberth's "Luna" rocket, a multiple-staged rocket to be launched from the Indian Ocean in the summer of 1934 with a human crew to explore the Moon. The movie was an international success. In vivid scenes, the German scientists and engineers plotted their way to the Moon, built Oberth's grand rocket, launched its crew in realistic scenes of acceleration and weightlessness, and explored the lunar surface.

The film was more sober futurism than science fiction, marrying what was realistically possible with a fun adventure story. Oberth's prototype never flew for the movie premiere, though some quick-thinking studio modelers created a miniature rocket ship that launched from a small globe and sped along a wire into the dark sky. All Oberth could muster in the summer of 1930 was a small engine, the "Cone Nozzle" (*Kegeldüse*), which created a measurable thrust, and which for a time was considered the world's first successful test of a liquid-propellant engine (Goddard did not publicize his 1926 flight until 1936).

The Verein accomplished what Oberth could not. At an abandoned military barracks outside Berlin, its members set up a "Rocket Flying Field" (*Raketenflugplatz*) to test a variety of small solid-fuel and liquid-propellant rockets between 1930 and 1932. They were really starting from scratch, to replicate what Goddard had been doing for almost 20 years, although the Verein publicized its methods and results while Goddard kept his mostly secret. Its rocketry enthusiasts learned, by trial and error, how to locate the tanks for the oxidizer and fuel, how to position and design the engine for maximum combustion and efficiency; how to handle and mix the propellants and fuels through a variety of lines, pumps and valves; how to apply light but strong metals to withstand the heat of combustion; and how to cool the superhot engine. With their simple "Minimum Rocket" (*Mirak*) series and more sophisticated "Repulsor" series, the Verein made hundreds of tests and actual launches, reaching as high as 5,000 feet altitude.

These feats, often covered in the news media, inspired a veritable rocket craze in Weimer Germany, as enthusiasts began to build and test mostly solid-fuel rocket ice sleds and bikes, rocket gliders and planes, and rocket cars. Popular-science and spaceflight writer Max Valier, backed by two industrialists, Fritz von Opel and Paul Hylandt, began to test solid-fuel rocket cars at the Avus Speedway in Berlin, reaching speeds of over 100 miles per hour with the Opel-Rak II in the spring of 1928. Valier died in an explosion of one of his liquid-fuel rocket cars in 1930, but not before his popular book, *Rocket Flight* (1930), mapped a pathway to outer space. It featured his supersonic rocket plane, rising far above the skyscrapers of our greatest cities, spanning the continents through the stratosphere, bridging the Atlantic in hours not days.

From Vienna, Eugen Sänger's *Rocket Flight Engineering* (1934) offered precise confirmation that just such a stratospheric plane was possible, thanks to his studies of the gas dynamics and internal efficiencies of rocket engines, and of the aerodynamics of supersonic flight. Hermann "Noordung" Potočnik's, *The Problem of Spaceflight* (1929), with vision and panache, actually positioned humanity living in outer space, in an avant-garde "living wheel" space station, with detailed descriptions and placements of airlocks, observation platforms, solar mirrors to collect energy, living quarters, and us.

The rest of Europe watched and read about these experiments and works in widely reported news stories. From France and Italy, rocketry and spaceflight enthusiasts retooled their own efforts and fashioned new initiatives. Robert Esnault-Pelterie revived his earlier essays in expanded book editions, mostly in answer to Oberth's work, now backed by the French Astronomical Society and its new commitment to the science of "Astronautics." At Paris, the Polish émigré, Ari Shternfeld, inspired by Tsiolkovsky and Oberth, wrote the first draft of his masterpiece, *Introduction to Cosmonautics* (1933), a sophisticated scientific discourse covering the higher mathematics of spaceflight and all manner of our human adaptations in spacecraft and spacesuits.

Still other publicists, like the Russian émigré, Alexandre Ananoff, adapted these theories for popular audiences in his *Interplanetary Navigation* (1935) and in his displays at the Paris International Exhibition (1937). He regaled the possibilities of "cosmic rockets" to explore the "beauties of infinity," the "immensities of space." Fascist Italy joined the wave of interest, sponsoring the Volta Conference of 1935 on "High Altitude Velocity in Aviation," on how to surpass the limitations of the internal combustion engine and propeller of the airplane, which had effectively reached these thresholds in altitude and speed. The conference was dedicated to a number of new studies of high-speed aerodynamics, including air-breathing jet engines; the swept-back wing for supersonic flight (in reducing drag); and the ultimate engine of all, pure rocket power.

The German military took notice of the Verein, in a series of surveillance and intelligence operations under the authority of Lt. Colonel Karl Becker, one of Germany's leading ballistics experts, known as the "colonel professor," chief of the ballistics and munitions unit within the Weapons Testing Division of the Army Ordnance Office. A veteran of the Paris Gun program and a promoter of liquid-propellant rocketry, Becker had hired a young and dynamic Captain Walter Dornberger in 1930, soon to be a Major General and the head of the V-2 missile program. As Becker's leading engineer, Dornberger put on civilian clothing and met with the Verein amateurs, taking careful notes of their work and even passing along financial support to maintain their tests. Both Becker and Dornberger, who witnessed several rocket flights, were impressed with the Verein's achievements, especially with fuel mixtures and engine injectors, but were aghast at its disorganization, lack of adequate measurements, and poor records. The enthusiasts had proven the value of liquid-propellant rocketry but were squandering its potential.

In 1932–1933, Becker and Dornberger set out to quietly dismantle the group, hiring off several of its most talented young engineers to work for them at a new liquid-propellant rocket facility, at Kummersdorf, an Army proving ground outside of Berlin. Among them were Walter Reidel, Walter Grunow, Arthur Rudolph, and a young Wernher von Braun, who was just about to graduate with a university degree in mechanical engineering. It was an interesting and impressionable group of young men, all of whom were knowledgeable about Tsiolkovsky's theories, had read about the Goddard rocket frenzy after 1920, and were passionate fans of Oberth's works.

At Kummersdorf, this small team began the methodical design and development, testing and redesign, of the "Assembly" (*Aggregat*) series of rockets, through the A-1, A-2, A-3, A-5, and especially the A-4, renamed the "V-2" missile in the Second World War. Von Braun contributed to the range of experiments and improvements, learning as he worked. Dornberger facilitated his Ph.D. studies in applied physics from the University of Berlin (1934), with a dissertation on the gas dynamics of liquid-propellant rocketry, classified as top secret. Von Braun graduated just as Adolph Hitler consolidated the new German national socialist (Nazi) regime, a menacing event for the near future of liquid-propellant rocketry.

Stalin's Command State and Rocketry

The USSR also experienced this transition from amateur society to rocket state between 1928 and 1933, although we need to remember that it was a dictatorial regime and a rising "command" economy since 1917, if radically transformed again with Joseph Stalin's massive and brutal program for collectivization and industrialization after 1928.

In its drive to "catch and surpass" the technological and industrial achievements of the West, this Stalinist command system dictated production quotas and military-industrial "tasks" (*zadachi*). It exported grain on the global market, at the high human cost of artificial famines in Ukraine and Kazakhstan, and violent collectivization throughout Russia, in order to pay for European and American engineering expertise and machinery. Stalin devised a series of "great construction sites of communism" (*velikie stroiki kommunizma*), including the White Sea and Baltic Canal, built largely by slave labor. His regime hired French, German, and American engineers to design and help build its vast industrial complexes at places like Magnitogorsk, Chelyabinsk, and Mariupol (Ukraine).

Stalin's regime even mapped the stratosphere and cosmos as its ultimate geopolitical goals, as portrayed in the film, *Circus* (1936), which featured a stratospheric rocket show, soon popularized in a chocolate candy, *Stratosfera*. The movie's popular song, "Wide is My Motherland," glorified Soviet technology and the Soviet Empire, "from Moscow to its farthest borders." Rockets had both geographic and symbolic reach.

This was no Weimar type of republic. Although remarkably different in their political systems, both Weimar Germany and Soviet Russia did share something special: a secret military alliance that lasted between 1922 and 1933. The two countries came to a public rapprochement under the Treaty of Rapallo, which recognized mutual territorial claims and economic interests. As part of this cooperation, they also negotiated a series of top-secret military protocols that allowed a defeated German military, in violation of the Treaty of Versailles, to test its chemical warfare technologies, tank systems, and advanced military aircraft in the closed spaces of the USSR. Two pariah states, one broken by the First World War, the other isolated by its communist regime, now joined forces to prepare for the "coming war."

Accompanying these policies, Lenin's government had established a Gas Dynamics Laboratory (GDL) in 1921, dedicated to the testing and development of solid-fuel rockets for the Soviet military, for use as airplane take-off boosters and missiles (surface to air, and air to ground). The future designer of Soviet rocket engines, Valentin P. Glushko, joined the Laboratory after his university graduation in 1929. In his youth, he had been a devotee of astronomy, science fiction, and all things Tsiolkovsky. Now he began a series of designs and test-stand experiments for his liquid-propellant "Experimental Rocket Motor," including some truly innovative work with electric rocket engines, and even a proposal to turn his rocket exhaust into a military death ray. Taking note of this work was Red Army Commander Mikhail Tukhachevsky, promoter of a series of military innovations, including liquid-propellant rocketry, as part of his strategy of "deep-battlefield" warfare, the ability to project a nation's military power fast and deep into an enemy's homeland so as to overwhelm and defeat it.

The USSR also formed a paramilitary organization at this time, the "Society for the Promotion of Aviation and Chemical Defense Needs," as a bridge between the educated public and the military, a means to mobilize the population for civil defense needs, aviation, and invention. Within the USSR's state-administered command economy, it was a way to create a pipeline for Soviet science and technology and to promote the flow of talent and ideas into the military establishment.

In 1931, Fridrikh Tsander and his rocketry enthusiasts established a "Group for the Study of Reactive Motion" (*Gruppa Izucheniia Reaktivnogo Dvizheniia* or GIRD) within this society, largely in response to the rocketry craze in Weimar Germany. It became the leading paramilitary rocketry group in the country, filled with idealistic young engineers for the Moscow Higher Technical School and associated state aviation firms. The military supervised it, and by the spring of 1932 retooled it as an integral part of the Directorate of Military Inventions of the Red Army. The core Moscow Group had a sister organization in Leningrad, centered around Tsiolkovsky's admirers and publicists from just before the Russian Revolution, Iakov Perelman and Nikolai Rynin, who propagandized rocketry and spaceflight issues in popular magazines and newspapers, in engineering seminars and aviation clubs.

Tsander was an engineer and designer of rare talent, though somewhat too consumed with Tsiolkovsky's ideas and with the promise of flying rocket planes and pure rockets into space. The military was patient with him at the start, though it soon replaced him with a young aviation engineer in the GIRD, Sergei P. Korolev, future Chief Designer of the Soviet space program. Under Korolev's guidance, and just after Tsander's untimely death, the GIRD launched the USSR's first hybrid gel-propellant rocket, the GIRD 09, reaching 1,300 feet altitude in 18 seconds (17 August 1933); and its first pure liquid-propellant rocket, the GIRD X (25 November 1933), powered by one of Tsander's engines.

In 1933, Stalin's government joined the Leningrad GDL (with Glushko) and the Moscow GIRD (under Korolev) into the world's first state institution devoted to rocket power: the "Reactive Scientific-Research Institute" (*Reaktivnii Nauchno-Issledovatelskii Institut* or RNII). In Germany, Becker and Dornberger had already formed their Kummersdorf unit. The RNII was something altogether new and proactive; in fact, Stalin approved it based partially on the reports of Russian émigrés who had joined the Verein, and of his secret agents who were observing German developments. The USSR, he knew, needed to be first to go farther, so he and his regime committed the RNII to the design and development of air-breathing jet engines, and solid-fuel and liquid-propellant rocket engines, for a variety of means: artillery, aircraft, and ballistic missiles.

The achievements of the RNII engineers were significant. Georgii Langemak and Valentin Glushko wrote an engineering primer, *Rockets* (1935), on the rocket as ballistic missile, filled with all the necessary calculus, physics, chemistry, metallurgy, and engineering. Langemak prepared an impressive dictionary of standardized terms. Mikhail K. Tikhonravov, one of the future stars of the Sputnik launch, advanced his initial theories and designs for interplanetary rockets while working in the RNII. Ari Shternfeld, who emigrated to the USSR from France, working as an engineer in the RNII, finally published his manuscript as a book, the Russian classic, *Introduction to Cosmonautics* (1937), another first in its breadth and depth of scope.

In April of 1934, the RNII and the Soviet Academy of Sciences sponsored the All-Union Conference on the Study of the Stratosphere, the first such conference in history devoted to investigating the potential of jet engines and pure rocket power as replacements for the internal combustion engine and propeller of the traditional airplane.

Not all was well within the Soviet bureaucracy. It was ridden with rivalries, intrigues, and chaos, thanks to the regime's command-administrative system. For example, Stalin placed the RNII under his own loyalists in the Commissariat of Heavy Industry, with other rocketry developments under his loyalists in the Commissariat of Defense. His aim was to control the ambitions of his rivals, Commander Tukhachevsky and the officers of the Red Army. In response, Tukhachevsky created his own new liquid-propellant rocket program in "Construction Bureau #7" (KB-7), which competed with the RNII and which arguably surpassed it by launching the country's first prototype ballistic missiles in 1937.

Other kinds of chaos infiltrated the system, what Mark Harrison has called an artificial "market" of competing offices and contracts, part of the Stalin regime's design to promote capitalist-style competition and innovation underneath an umbrella of unilateral control. It was often messy and inefficient. Still other conflicts stressed this system, as, for example, the struggle between conservative artillery (and aviation) officers, already invested in proven canons or planes, and the radical rocketeers, who were demanding status and financing in an already strained bureaucracy.

These tensions, along with the inhuman pace and violence of Stalinist collectivization and industrialization, drove the purges and terror after 1934. In waves of arrests, imprisonments, and executions, Stalin and his secret police disciplined and punished the command-administrative system. They first targeted industrial managers from the old regime, then wealthy and politically suspect peasants, then his party-state political opponents, then industrial and agricultural managers within the system, then real and imagined oppositionists from all walks of life. From above, the secret police widened its spotlight upon supposed "enemies of the people." From below, the people themselves followed waves of panic, accusing each other of crimes, settling old scores with accusations, finding sabotage and "wrecking" in any hint of mismanagement. Hundreds of thousands perish in executions. Millions suffered exile or imprisonment in labor camps.

In a later wave, Stalin approved the arrest and execution of Tukhachevsky, and the top and middle echelons of the military's officer corps, including the managers of the RNII and related rocketry programs. Rocketry was such a vulnerable enterprise because it was a young science and technology, prone to the normal trial and error of experimentation, of mishaps and complications, disputes, and rivalries, all of which were now named as crimes against the state. As a manager within the rocket state, Korolev was arrested, and suffered torture in Moscow's prisons, and the deprivations of the prison mines at Kolyma, Siberia. He barely escaped execution and was amnestied to contribute his engineering skills in the Second World War, when he worked at a prison laboratory run by the airplane designer, Andrei Tupolev. There he joined Glushko, his former colleague from the RNII and his future partner in the Soviet space program.

For all the disorder within this system, the rocketry lobby found support within Stalin's propaganda state, which showcased cosmism and spaceflight in dramatic ways, largely centered on the person and ideas of Konstantin Tsiolkovsky. The state published a series of science-fiction novels that took communist explorers in Tsiolkovsky rockets to colonize the asteroids, to build "new arks" of inhabited space stations, and to explore primitive life on the planet Venus. It sponsored rocketry and stratospheric-plane exhibits, seminars, and publication runs that reached hundreds of thousands of people, especially youngsters. During the May Day parades in Red Square in 1934, before Lenin's Tomb and the assembled political leadership and crowds, as dirigibles and airplanes flew overhead, Soviet radio broadcast a speech by Tsiolkovsky, promising the first "trans-atmospheric flights," when "heroes, men of courage, will inaugurate new air routes: Earth–Moon orbit, Moscow–Moon, Kaluga–Mars."

To fortify the message in a major key, the state commissioned a film on the subject: *Cosmic Flight* (1936). It was the story of yet another Tsiolkovsky-like character, academician Pavel Sedykh, director of the "All-Union Institute for Interplanetary Communications," who leads the Young Communist Marina and the Young Pioneer Andriusha on an expedition to the Moon. Tsiolkovsky served as a scientific consultant on the film, providing the director with sketches of the rocket and spacecraft, advice about the mechanics and physics of spaceflight, and about the look of the planets and stars from outer space. M.M. Gromov, a famous pilot, advised about the look of the internal cabin. K.N. Shestkovskii, an astronomer, advised about the scenes of the lunar landscape. The film was visually stunning in its use of models and in a compelling plotline: the rocket launched from a long inclining ramp just outside of Moscow, emblazoned with the name, "Joseph Stalin"; its crew on a Tsiolkovsky-inspired trek to survive acceleration, experience weightlessness, and explore the Moon. In an ultimate insult, they even explode flash powder there, as Goddard had once advised, but now in the letters: "USSR."

Goddard and the American Societies

Rocketry in the US proceeded along two tracks, one old and one new: Goddard's track, if renewed with more money and attention; and an amateur rocket society, if quite separate from his influence. Goddard's professional life changed dramatically after he met Charles Lindbergh in 1929, America's greatest aviator of the moment, who parlayed his own interest in jet and rocket power into an initial $50,000 grant for Goddard from the wealthy Guggenheim family. Relocating to the favorable weather conditions at Roswell, New Mexico, he initiated a series of experiments between 1930 and 1932, and again between 1934 and 1940, culminating in a variety of classic cylindrical-shaped rockets with tapered cones, powered by liquid oxygen and gasoline. They came in a variety of shapes and sizes, some larger or smaller, others thinner or heftier. Size and shape mattered less than performance, of course. They were at most 18 feet high and reached 9,000 feet altitude. There was the A series, the L series, and the P series, with categories and sub-categories within them, all named and numbered to fit Goddard's precise methods and meticulous recording techniques. He worked, as always, with small crews of technicians, in an air of isolation and secrecy. Yet he achieved a number of impressive and continuing firsts through these years.

Table 3.1 Robert Goddard's rocketry milestones. Author.

Robert Goddard's Rocketry Milestones
Rocket equation for high altitudes and spaceflight (1912)
Proves rocket propulsion in a vacuum
Bazooka prototype
Patent for a multi-stage rocket
Nozzle design for a rocket
Mechanical design for a multi-stage, liquid-propellant rocket
Launch of a liquid-propellant (liquid oxygen and gasoline) rocket (1926)
Combustion chamber regenerative and film cooling
Internal stabilization and guidance (gyroscopic) control
Blast vanes for guidance control by gyroscope
Gimbaled and clustered engines
Turbo pumps for feeding propellants
Parachute recovery of the rocket with instruments (payloads)
Launch of a vehicle that moved faster than the speed of sound (1935)

In all of these ways, Goddard was ahead of his time. He knew it and his public knew it, especially when the Smithsonian Institution published his *Liquid Propellant Rocket Development* (1936), a summary of his work to date, including a final confirmation of the world's first liquid-propellant rocket in 1926. Goddard was one of the most famous and best-financed scientists of the interwar era, this thanks to the hundreds of national and international news reports about him; and to the significant grants he received, in time approaching $200,000 from the Guggenheim funds. Americans were confident and the rest of the world was confident too, awaiting news of Goddard's next flight, expectant of his promises of a rocket flight to the stratosphere or to the Moon.

Yet there was a paradox to Goddard's media image. He was in the news but also in the comics, hovering somewhere between respectability and hilarity. After 1926, the science-fiction magazine, *Amazing Stories*, began to feature variations of the "Goddard rocket" on its colorful front covers and in its plotlines, often set in interplanetary space. Some of those stories were respectful and factual, with a focus on the new science of rocketry. Yet over time, readers tired of the rocket and demanded more fantastic plotlines. They wanted the entertainment and adventure of the "space opera." The "Buck Rogers" series of interplanetary adventures was born in *Amazing Stories* in 1928, soon to become a comic strip, radio show, film series and eventually a television show. In the early years, one of the central characters of Buck's band of space travelers was the indomitable Dr. Huer, creator of many of Buck's machines and gadgets, from rocket spaceships to death rays and x-ray glasses. He likely became the picture in many Americans' minds of the classic rocket scientist, bald and bespectacled, quirky and brilliant, Goddard in comic disguise.

These themes and forums spotlight what Scott Bukatman has called a spirit of "levity" in the science fiction and popular science of the day, a thrill for lightness and light-heartedness. We saw this same levity in the dancing girls at the thrill ride, "A Trip to the Moon" at the Pan-American Exhibition, or in George Melies's *Voyage to the Moon*. American audiences now saw it again, beginning in December of 1932, with the famed entertainment show, the Radio City Music Hall "Rockettes," from an earlier routine called the "Rockets," famous for their straight and tall, gravity-defying dance routines.

The deeper paradox, as Theodore von Karman once recognized, is that Goddard left few actual legacies beyond his media celebrity. His patents and publications were influential, building a set of ever-rising expectations and corresponding rivalries to match or beat him. Yet his rockets only inspired, rather than modelled, rockets to come. The American Interplanetary Society (AIS) soon set out to overtake him. Founded in New York City in 1930, the AIS was initially devoted to science fiction and space travel themes. Two of its founders, G. Edward Pendray and David Lasser, were editors from Hugo Gernsback's publishing enterprise of science-fiction and popular science magazines. They helped to popularize the growing international and American movement in such other leading magazines as *Nature* and *Popular Science*. Lasser's *The Conquest of Space* (1931) was a colorful and inspiring primer on the coming achievements of space travel. Pendray was also a reporter for the *New York Herald Tribune*, which competed with the *New York Times* for readers of rocketry and spaceflight news.

Rocket Science

The term *rocket science*, which has become something of a cliché in American parlance, had its distant origins in 1915, when Robert Goddard gave a talk on rocketry, "New Methods of Reaching High Altitudes," at the Physics Colloquium of the Worcester Polytechnic Institute. A small whirlwind of local press coverage followed, including rumors about his fantastic invention and the possibility of reaching into outer space. One reporter even called it a "veritable new science." The founder of the American Interplanetary Society, G. Edward Pendray, was apparently the first person to use the full term, "rocket science," in an article in *Popular Science* (August 1931), though devoted not to rocketry in the US but especially in Germany, and now spreading across the world, a new international science for interplanetary flight. After the Second World War, the news media disseminated several more colloquialisms, "it isn't rocket science," along with "you don't have to be a rocket scientist," which were already popular and lighthearted ways of expressing all the complexities and challenges of making rockets.

All of the news stories out of Germany soon turned the attention of the AIS to rockets. Pendray corresponded with the Verein spaceflight publicist, Willy Ley, and visited the Rocket Flying Field in August of 1931. He was astounded by the futuristic look of the place and by the vanguard launches, returning home to duplicate liquid-propellant rocketry, American style, if without Goddard, who refused his offers for collaboration. So inspired by their Verein brethren, AIS members launched a series of small liquid-propellant rockets between 1933 and 1934. These were awkward and modest affairs, launches from a Stockton, New Jersey farm, near the Delaware River, and from Great Kills and Marine Park in Staten Island. The rockets reached distances of 1,338 feet and speeds of 1,000 feet a second. Some of the first rockets were hobbled together from a cocktail shaker and a saucepan, with coming donations of liquid oxygen from the Air Reduction Company, and of aluminum from the Aluminum Corporation of America.

In 1934, the group changed its name to the American Rocket Society (ARS) in an appeal for more technical and serious-minded members. Its officers standardized a new terminology for the enterprise: "rocketry" for the science and engineering, "rocketor" for the engineer, "thrust" for the "push of a rocket," and "shot" for the rocket flight itself. In an all-out quest for seriousness and professionalism, members like John Shesta and Alfred Africano began testing rocket engine designs, showcased in the *Journal of the American Rocket Society* (later *Astronautics*). They also published their research in some of the leading US engineering and applied physics journals. Rocketry was joining mainstream science.

A number of international and national personalities helped to build upon this prestige. Robert Esnault-Pelterie offered his research to a public gathering of the AIS at the hall of the American Museum of Natural History in New York City in 1931, to great acclaim. John Q. Stewart, a Princeton University astronomer, joined the media attention with plans to build a spherical metal spacecraft, propelled by ionized hydrogen, to reach the Moon by 2050 with a 70-person crew. More modestly, by 1940, after a decade of publications and experiments, the American Rocket Society enjoyed a measure of confidence, with hundreds of members from around the country and varied professions. On its board were representatives from the chemical and metals industries, along with Dr. James Kimball of the US Weather Bureau, Dr. H.H. Shelton of the American Institute of New York, and Dr. Alexander Klemin, director of the Daniel Guggenheim School of Aeronautics of New York University.

One among these new national rocketry advocates was Theodore von Karman, a Hungarian émigré who came to the US in 1930, from the heart of German academic culture. His credentials were impeccable: a doctorate at the University of Göttingen, director of research in the Austro-Hungarian air forces during the First World War; professor and director of the Institute of Aeronautics at Aachen, Germany; expert in aerodynamics, specializing in the study of fluid streams around bodies and the forces of "drag" waves. He had also participated in the debates over liquid-propellant rockets in the German academic and popular press in the 1920s.

Von Karman became director of the Guggenheim Aeronautical Laboratory at the California Institute of Technology (GALCIT), supported by the same family that financed Goddard, and with whom he now vied for preeminence in the rocketry field. GALCIT's wind-tunnel aerodynamic studies were some of the best in the world, around which America's young aerospace industry now coalesced (the Douglas, Northrop, Hughes, Lockheed, North American, Vultee, and Consolidated corporations). In 1935, von Karman and his graduate students, inspired by Eugen Sänger's work in Austria, and by several vanguard Italian studies, began experimenting with rocket planes and engines. Local rocket enthusiasts, William Forman and John Parsons, joined the group. Parsons was also a satanist, who mixed occult rituals along with his explosive powders. Goddard would not have fit in very well with this motley group. When GALCIT's Frank Malina visited Roswell in August of 1936, the secretive Goddard politely refused cooperation.

Through all of these rocketry developments, most of the average American and European public remained attuned to spaceflight and cosmic themes in various ways, often mixing hope with an undertone of despair. Take the feature film, *Things to Come* (1936), based on a book and screenplay by H.G. Wells. The film's plotline first destroyed the Earth through a world war, reducing civilization to ashes from aerial bombings. In the end, a futuristic underground civilization arises, one promoting equality and justice, the conquest of Earth's natural resources, and even interplanetary flight (by 2036). "All the universe or nothing," exclaims the main character at the end, announcing a rocket migration into outer space, via a Verne-like launch by canon.

The two world fairs of the late 1930s also projected hope and despair. The New York World Fair (1939) spotlighted the latest scientific and technological, cultural and economic achievements. It also portrayed American cosmic values in the main attraction, the Trylon and Perisphere, an obelisk and great sphere, within which was a futuristic "Democracity." Other attractions, one designed by Raymond Loewy, took fairgoers on stratospheric rockets across the Atlantic or on farther journeys to the planets and stars. In a stark contrast, the Paris Exhibition (1937) featured two ominous pavilions from Soviet Russia and Nazi Germany, propagandizing the "progressive" values of the two dictatorial regimes, communism and national socialism, just before their pact that dismembered Poland in their joint invasions that began the Second World War.

One of the most memorable interwar, interplanetary events happened on Halloween, Sunday 30 October 1938, when Manhattan's Mercury Theater, under the direction of Orson Welles, read an adapted radio version of the classic, *War of the Worlds*, at 8 pm. Only this time the Martians invaded not London or Boston or New York City but the small American town, Grovers Mill, New Jersey. The players, broadcast coast to coast, read the story as if the events were actually taking place. Live. Waves of what the newspapers called "mass hysteria" ensued. Approximately 6 million Americans heard the show, and at least a portion of them, thousands of people, believed for a time that it was real, either because they did not hear the disclaimers, or were just plain gullible. Police reports and eyewitness testimonies confirm that people looked up in fear to the skies, or barricaded themselves in their homes, or sought out comfort in churches. Some thought that Earth really was under a Martian attack. Others believed a meteor had fallen and was spreading poison gas. Still others simply thought this was the end of the world.

This was one of those rare moments in the twentieth century, of a simultaneous experience on a mass scale, the first of several more to come: Pearl Harbor, the Cuban Missile Crisis, the assassination of John F. Kennedy, the Challenger Disaster, and the "September 11" Attack on America. The 1938 radio broadcast was a hoax, of course. But people still believed. It was also a strange omen of the Nazi blitzkriegs in the Second World War, very real, and just about to come.

The Nazi Regime and V-2 Missile

Like most of the German military forces, Karl Becker's small but efficient team at Kummersdorf probably welcomed the Nazi victories in the German parliamentary elections and Adolph Hitler's appointment as Chancellor in 1933. Hitler's

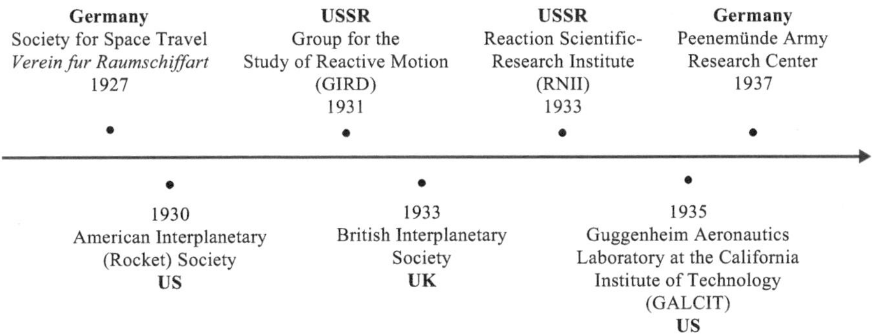

Figure 3.1 The rise of rocket societies and states. Author.

regime, fashioned as a Third Reich (1933–1945), was dedicated to the recovery and reconstruction of the German economy, then suffering from financial and industrial devastation and massive unemployment. It was also dedicated to re-militarization, aimed against the harsh terms of the Versailles treaty imposed by France and the United Kingdom. It was also dedicated to a program of territorial expansion, and eventual racial extermination in the east, against Central Europe and the vast Eurasian holdings of the USSR.

With new sources of Nazi funding, the German military focused its attention on the achievements and promises of Becker's liquid-propellant rocket as a new weapon in the "coming war." Werner von Fritsch, commander in chief of the German Army (*Wehrmacht*), financed the initial experiments with the famous "Assembly" (*Aggregat*) series at Kummersdorf. On witnessing a test-stand firing of the A-4 in March of 1936, he approved expanded funding and investment: the move to a new and improved rocket facility at Peenemünde, a peninsula and series of islands jutting out into the Baltic Sea from northern Germany. In this, von Fritsch found support from Hermann Goering, head of the German air forces (*Luftwaffe*), and Wolfram von Richthofen, patron of jet and rocket airplanes. Both of them saw huge potential for aviation in liquid-propellant rockets. In the rising military-industrial complex of Nazi Germany, Minister of Armaments, Albert Speer, approved of these investments.

At Peenemünde, as Michael Neufeld's work has surveyed, Becker's small Kummersdorf team transformed into a state-sponsored economy of scale, concentrating rocketry talents and resources (from the government, military, and academia) under one management hierarchy. Here, among a variety of high-technology projects, Germany built the world's first medium-range, liquid-propellant ballistic missile: the A-4 (V-2). Under Walter Dornberger's expert management skills, Peenemünde focused on the ultimate objective of building a missile that could carry a powerful warhead for military use and, at 47 feet long and 5 feet in diameter, also fit on the German transport and tunnel network. From this, Dornberger set a series of discrete goals.

First, was a goal to design the optimum aerodynamics of the missile, taking into account the stabilizing fins and the effects of the exhaust. Based on advanced wind-tunnel tests, engineers designed a "corset," the dynamic frame with flexible joints that allowed the missile to be controlled in its three-axis control (pitch, roll, and yaw).

Second, was a very challenging objective to ensure that the propellants would accurately mix and combust in the engine in order to meet the required thrust and velocity. Designers placed the engine, a bulbous combustion chamber with an injection system of 18 nozzles, at the bottom of the missile, fed by a lower tank of liquid oxygen as oxidizer, and an upper tank of 75% ethyl alcohol and 25% water as fuel. They solved an incessant but necessary problem of overheating with regenerative cooling (by ethyl alcohol and water). Nitrogen operated some of the valves and kept the alcohol tank pressurized. Hydrogen peroxide and sodium permanganate forced the propellants into the combustion chamber, where they ignited for the combustive force and thrust.

Third, was the essential objective of crafting lightweight but effective mechanisms to control the missile in flight: the gyroscope, basically a motorized spinning top, placed in the nose, whose preset mechanisms kept the missile on course by sensing and fixing changes in its orientation, steering it by the carbon rudders. Electronics for radio signals also helped to keep the missile on course and to shut down the engine for a ballistic return to Earth. The Peenemünde teams paid constant and methodical attention to all of these details and improvements. After the A-4's first successful flight on 3 October 1942, they made 56,000 further improvements to the missile over the next year in order to improve fuel and engine efficiency, stabilization and guidance. The end result was a masterpiece of German science and engineering … and mass terror.

The Ballistic Missile

Based on an artillery paradigm, the V-2 was a ballistic missile, meaning that it used its velocity and altitude to fall back to Earth in order to destroy its intended target. Besides its power plant, the great difference between the V-2 and more conventional artillery (or the Paris Gun) was its altitude. If the shell of the Paris Gun reached 25 miles altitude, the V-2 reached about 50 miles, close to the edge of space. At 25 seconds into its flight, the V-2's engines shut down, as it then coasted to its highest altitude in order to return to Earth in a gravity-bound ballistic trajectory. We will meet the gravity assist in Chapter 9, how a rocket craft or probe, to save on propulsion, can skip and jump along gravitational fields on their peaceful, exploratory pathways in space. The V-2 needed gravity as well, but only to destroy people and things back on this planet.

Although mostly well organized and efficient in its rocketry developments, the work of Peenemünde was subject to several incapacitating burdens. Hitler and his subordinates made unrealistic demands for deadlines or quotas, these to meet purely political or military goals, and well beyond Peenemünde's capacities. There were a series of rivalries between the various bureaucracies. The Army and Air Force vied against each other for priority funding for either their artillery or aviation needs. Speer's Ministry of Armaments sometimes adjudicated these rivalries, sometimes pursued interests of its own. In time, the infamous "Defense Corps" (*Schutzstaffel* or *SS*) intervened too. Originally a fiercely ideological paramilitary corps for the Nazi Party, under Reichsführer Heinrich Himmler the SS governed over significant dimensions of the German military and industry. It managed the concentration camp system and the racial extermination programs, but also vied for authority over vanguard technology programs, including rocketry, especially after Himmler witnessed a perfect A-4 flight, with accurate targeting, in the summer of 1943.

Wernher von Braun knew Himmler from that visit, as he knew Goering and other Nazi elites from their visits and contacts. He also personally met at least four times with Hitler, who visited Kummersdorf in March of 1939, and who was briefed by Dornberger and von Braun in July 1943, after which Hitler awarded von Braun with the War Service Cross and the title, Research Professor. Von Braun remembered Hitler as a "godless" being, but also charismatic and "brilliant." Hitler's Nazi regime had indeed been generous to the young man, only 25 years old in 1937 but already a technical director and chief engineer at Peenemünde, rewarded with a position in the Luftwaffe and the use of a Messerschmitt 109 airplane for his personal use. His Nazi Party number (from 1937) was 5,738,692 and his SS number was 185,068; he was admitted to the SS in 1940 and through this he advanced to Second Lieutenant (*Untersturmführer*), First Lieutenant (*Obersturmführer*), Captain (*Hauptsturmführer*), all the way by 1943 to Major (*Sturmbannführer*). With all of his duties and posts came the spoils of money and girlfriends; and the rare moments of levity too, as when he and his fellow engineers painted a risqué logo of the "Woman in the Moon" on an A-4 missile, or when they organized a costume party, "Mardi Gras on Mars," in February of 1938.

Von Braun's positions, though centered on his unique managerial talents, sometimes placed intense stress on his professional and personal life, often pulled in different directions by either his loyalties to the Army, or Air Force, or SS. On 15 March 1944, the SS arrested and jailed von Braun for several weeks, with two others, Klaus Reidel and Helmut Gröttrup, on charges of sabotage, ostensibly because they valued the goal of creating a rocket for space travel over the state goal of using the rocket as a weapon of war. But the charges were probably a cover for deeper bureaucratic intrigues, perhaps Himmler's ploy to break von Braun's loyalty to the Army and force him to transfer allegiance to the SS. The attempt failed when the influential Speer and Dornberger were able to free him.

By its first wartime use in 1944, the V-2 missile became one of the most sophisticated mass-produced weapons of the Second World War. It stood 47 feet tall, 5 feet in diameter, and weighed 14 tons fully loaded. The V-2 could launch 1 ton of explosives as far as 190 miles in five minutes: at an altitude of about 50 miles, and at a speed of 3,500 miles per hour, passing the speed of sound only 25 seconds into its firing. It was the highest and fastest human-made object to date.

The one great advantage of this formidable weapon was that it was invincible. No airplane or missile could defend against it. It was a weapon of total surprise. Since it flew faster than the speed of sound, the V-2 actually hit the ground and exploded before anyone heard the sound of its approach. The German military reinforced this stealth factor by often deploying it from camouflaged mobile launchers, well hidden from aerial or artillery attack. These launchers comprised dozens of vehicles, holding the command center and propellant storage tanks, and crews. The Germans most often used it against cities. Just over 3,000 V-2 missiles hit chiefly England and Belgium, with some 5,400 persons killed, many in London and Antwerp. In its most deadly attack, altogether random, it destroyed a movie theater in Antwerp, killing 561 people.

The V-2 was mostly a weapon of strategic surprise and terror. As such, it was a measure of both Germany's scientific-technological prowess and its glaring strategic weakness. It paralleled German surprise "lightning warfare" (blitzkrieg), as in its invasions of Poland, France, and the USSR; or of the German aerial bombings, the "lightning" aerial war of the "blitz" during the Battle of Britain. As at the start of the war, so now at the end, the Nazi regime depended on surprise attack and terror bombings to paralyze the enemy in the short term, fearing for Germany's own survival in a long and drawn-out war, especially given the industrial reserves of the US and USSR.

We see this same reliance on surprise and terror in the Japanese case: the war in the Pacific began with a surprise attack on Pearl Harbor by Japan's industrial-strength battleships, cruisers and aircraft carriers, outfitted with some of the world's most advanced fighters and dive-bombers. Yet that war ended with Japanese pilots and their best planes in desperate kamikaze suicide-terror attacks on American ships.

The paradoxes get even deeper. The V-2 was a weapon of indiscriminate destruction and mass terror because it was faulty and error-prone. Dornberger promised early on that it would be precise in its targeting: that 50% of the missiles would fall within a 0.5-kilometer radius. Yet in the end, the 0.5 kilometers turned into a 20-kilometer radius, and only for those missiles that made it close enough in the first place. The Peenemünde team never perfected the guidance systems for precise targeting. Even worse, roughly one-quarter of the firings failed, and others veered off course entirely and exploded in the sea or in forests and fields.

Yet another mark of paradox was in the ultimate name given to this missile in 1944, the "Vengeance Weapon" (*Vergeltungswaffe*) or V-2, by the Minister of

Propaganda, Joseph Goebbels. By this time, Nazi Germany was losing the war on the eastern front after the defeat at the Battle of Stalingrad in 1943. It was also losing the war on the western front, what with the Italian campaign, the victories of the Royal Air Force in destroying the Luftwaffe, the looming Normandy invasion, and the Allied strategic aerial bombing campaigns, unleashed with an ever-deadly ferocity between 1941 and 1945. The Nazi regime deployed the V-2 terror attacks in special vengeance against the Allied aerial bombings, which targeted Germany's largest cities, their residents huddled in underground bomb shelters trying to survive, though the resulting firestorms incinerated or suffocated them even there. Some 300,000 people perished. Hitler and his staff, too, were reduced to a subterranean existence and their demise in a Berlin bunker.

To discuss the V-2 missile at war is also to return to a theme we have raised here from time to time, our asides on "vertical horizons," above worlds and below worlds, although in this instance not as an aside but as a core issue. Note the ultimate paradox of all: the V-2 as emblem of both human expectation and destruction. On 3 October 1942, the date of its first successful test launch, Walter Dornberger proclaimed that "today we have created the spaceship." Yet the V-2 proved its specific and earthly purpose on 7–8 September 1944, when Nazi Germany used it as a ballistic missile to kill civilians in London, fulfilling Adolph Hitler's demand, made on 7 July 1943, in front of von Braun: "What I want is annihilation … annihilating effect."

We have even more depths to plumb. In August of 1943, an Allied bombing raid on Peenemünde killed some 700 people (men, women, and children), including several top rocket engineers, a raid ordered by Winston Churchill and the British Defense Committee. To protect the V-2 production sites from further Allied attacks, Hitler ordered their removal to a secret underground site in the mountains of central Germany, at the "Middle Works" (*Mittelwerk*), a network of tunnels about 1 mile long and 600 feet wide. SS officer and engineer, Hans Kammler, infamous for the razing of the Warsaw ghetto and building the gas chamber at Auschwitz, completed the Mittelwerk using Germany's supply of slave labor, prisoners of war and civilians from its vast network of concentration camps.

The SS also supplied the slave labor (between one-third and one-half of the workers there) to actually assemble and build the V-2, workers from concentration camp "Dora," a satellite camp serviced by nearby Nordhausen and Buchenwald. The operative principle here, as at related camps, was "annihilation through work," annihilation being the crucial omega point for Nazi enemies. An estimated 10,000 persons died in the assembling of the V-2: from overwork in the dark tunnels; from typhoid, dysentery, and cholera (with little potable water for drinking and hygiene); and from beatings and executions. The SS adapted the very crane at the entrance of the tunnel, otherwise used to carry the V-2 missile, for summary hangings of prisoners, sometimes several at a time, those accused of insubordination or sabotage.

The Mittelwerk represented a very modern marriage of sophisticated technology, slave labor, and mass murder. By early 1945, production of the V-2 missile reached between 600 and 900 a month, barely more than the number of people then being killed making it or being killed by it. The V-2 was complicated enough to build and assemble in normal conditions, not to speak of the desperate and inhuman wartime conditions, which contributed to the missile's many faults. What began as a significant economy of scale and efficiency, a model of German engineering, also became a significant drain on ever-meager state resources, this thanks to the war generally, and to the Allied strategic bombings. The V-2 was an ineffective weapon, a poor substitute for the jet fighters or atomic bombs that might have helped turn the war Nazi Germany's way.

All of these issues bring us to the legal and moral complicity of Wernher von Braun and his associates, a topic that historian Michael Neufeld has studied in detail. The Peenemünde rocketeers became willing collaborators of a brutal dictatorship. They would certainly have preferred not to, but they did. They built a weapon of terror and mass destruction. They were casual observers of the slave labor and death camps that bordered the Mittelwerk. Von Braun visited the main site at least 12 times after 1943, seeing the bodies of emaciated workers, and the open latrines, taking in the sights and sounds and smells of human agony. He once admitted that he took great pleasure in the first V-2 attacks on London, as vengeance for the Allied strategic bombings, especially the Peenemünde raid. He and his teams even drank champagne toasts after the first V-2 missiles began to kill civilians.

In the rubble at Peenemünde, Red Army investigators found a tattered manuscript, a German translation of Konstantin Tsiolkovsky's 1903 essay, apparently one of von Braun's cherished possessions. Here was one of the minor trophies of war, yet a perfect emblem for some of the contradictions and problems that lie at the distant foundations of the Space Age. Von Braun eventually came west to the US, redeemed of his past, and became one of the faces of the coming Apollo program. His book went back east, from whence it first came, to the K.E. Tsiolkovsky State Museum of Cosmonautics in Kaluga, a remembrance of darker times.

Vertical Horizons: Above and Below Worlds

Between 1931 and 1935, human beings conquered two new vertical horizons: the heights of altitude in the stratosphere; and the depths of the ocean seas. The first involved an international race for the stratosphere that began on 27 May 1931 with the Belgian, Auguste Piccard, who launched a "stratostat," a stratospheric hydrogen balloon. He called it FNRS, initials for the Belgian National Foundation for Scientific Research, carrying himself and an assistant in an airtight, spherical gondola. They reached dramatic new heights: 51,775 feet. This was the first of its three flights, soon vaunted in "Ballooning in the Stratosphere" (*National Geographic*, 1933) and Piccard's memoir, *Above the Clouds* (1933).

The FNRS was also the catalyst for a brief race between the US and USSR into the stratosphere, though initially they were racing to outdo Piccard rather than each other. The USSR launched three stratostats between 1934 and 1935, most dramatically the "Osoaviakhim 1" in January of 1934, to 72,178 feet (13.6 miles), though the balloon and craft crashed back to Earth, killing the three-person crew, the result of shoddy construction and too rapid a descent. The US answered with three stratostats of its own, breaking all world records with the ascent of Explorer II in November of 1935, to 72,395 feet (13.71 miles). The enterprising Captain Albert Stevens of the US Army Air Corps became the voice of these efforts, captured on live radio and in the pages of *National Geographic* magazine, which even offered its readers one of Explorer II's other dramatic firsts: a foldout photograph of part of the curvature of the Earth.

Before this race to the stratosphere, the media covered another record-breaking feat, only downwards into the ocean, off the coast of Bermuda. The engineer Otis Barton and the naturalist William Beebe conducted a series of deep-water dives between 1930 and 1934, in their submersible steel vessel, the "bathysphere" ("deep sphere"), which plumbed new depths by reaching to 2,200 feet, breaking the earlier record of 365 feet. The National Broadcasting Company (NBC) radio also carried the 22 September 1932 descent on live radio, with Beebe's fascinating descriptions of the black and blue waters, with flashes of light and color from weird glow-in-the-dark creatures. On 15 August 1934, the bathysphere reached a record 3,028 feet, all of this covered in *National Geographic* (1934) and in Beebe's bestselling book, *Half Mile Down* (1934).

In an interesting coincidence, Beebe coined new scientific terms for a variety of fish never before seen by the human eye, and never seen since. Critics accused him of falling prey to optical illusions, just like Percival Lowell, created by the deep-water conditions and the fogged porthole. In another coincidence, the Chicago Century of Progress Exposition (1933) featured exhibits for both Piccard's stratospheric flight and Barton's bathysphere. Between 1946 and 1948, always the glutton for media attention, the stratospheric Piccard built a series of "bathyscaphes" ("deep ships"), one of them named the FNRS-2 which, in 1960, reached to approximately 35,813 feet at the Mariana Trench, a new world record.

All of these feats helped to turn the 1930s, otherwise a most anxious decade of economic trauma and the approach of war, into one of intermittent confidence and accomplishment. Human beings were now testing the limits of their own vertical horizons. These various mechanical spheres, call them movable domes, reached record heights and depths, and return us to earlier vertical horizons. They remind us of the envelopes of balloons and wings that first ascended into the air. They also point to the future as well, to new kinds of spheres, like the space capsules and space stations that will soon orbit us around the Earth.

Select Bibliography

Brooks, Jeffrey. *Thank You, Comrade Stalin: Soviet Public Culture from Revolution to Cold War*. Princeton: Princeton University Press, 2001.

Bukatman, Scott. *Matters of Gravity: Special Effects and Supermen in the 20th Century*. Durham: Duke University Press, 2003.

Cantrill, Hadley. *The Invasion from Mars: A Study in the Psychology of Panic*. Princeton: Princeton University Press, 1940.

Cheng, John. *Astounding Wonder: Imagining Science and Science Fiction in Interwar America*. Philadelphia: University of Pennsylvania Press, 2012.

Clark, John D. *Ignition! An Informal History of Liquid Rocket Propellants*. New Brunswick: Rutgers University Press, 1972.

Clary, Robert. *Rocket Man: Robert H. Goddard and the Birth of the Space Age*. New York: Hyperion, 2003.

DeVorkin, David. *Race to the Stratosphere: Manned Scientific Ballooning in America*. New York: Springer, 1989.

Harrison, Mark, ed. "The Market for Inventions: Experimental Aircraft Engines," in *Guns and Rubles. The Defense Industry in the Stalinist State*, ed. Mark Harrison. New Haven: Yale University Press, 2008.

Holliday, George. *Technology Transfer to the USSR, 1928–1937 and 1966–1975: The Role of Western Technology in Soviet Economic Development*. Boulder: Westview Press, 1979.

Kilgore, De Witt. *Astrofuturism: Science, Race, and Visions of Utopia in Space*. Philadelphia: University of Pennsylvania Press, 2003.

Link, Stefan. *Forging Global Fordism: Nazi Germany, Soviet Russia, and the Contest over the Industrial Order*. Princeton: Princeton University Press, 2020.

MacDonald, Alexander. *The Long Space Age: The Economic Origins of Space Exploration from Colonial America to the Cold War*. New Haven: Yale University Press, 2017.

Matson, Brad. *Descent: The Heroic Discovery of the Abyss*. New York: Pantheon Books, 2006.

Melnikova-Raich, Sonia. "The Soviet Problem with Two 'Unknowns': How an American Architect and a Soviet Negotiator Jump-Started the Industrialization of Russia, Part I: Albert Kahn," *IA. The Journal of the Society for Industrial Archeology* 36/2 (2010): 57–80.

Naimark, Norman. *The Russians in Germany: A History of the Soviet Zone of Occupation, 1945–1949*. Cambridge: Belknap Press, 1995.

Neufeld, Michael. *The Rocket and the Reich: Peenemünde and the Coming of the Ballistic Missile Era*. Cambridge: Harvard University Press, 1996.

Neufeld, Michael. *Von Braun: Dreamer of Space, Engineer of War*. New York: Knopf, 2007.

Siddiqi, Asif. *Red Rocket's Glare: Spaceflight and the Soviet Imagination*. New York: Cambridge University Press, 2010.

Sutton, Anthony. *Western Technology and Soviet Economic Development, 1930 to 1945; and 1945–1965*. Stanford: Hoover Institution Press, 1968–1971.

The phrase rocket state (originally *Raketen-Stadt*) comes from Dale Carter, *The Final Frontier: The Rise and Dall of the American Rocket State*. London: Verso, 1988, which is a colorful, interpretive survey of the history, media, and literature, and who borrowed the term from Thomas Pynchon's novel about Nazi Germany and the V-2, *Gravity's Rainbow*. New York: Viking, 1973.

4 Satellite

Strategic Power and Geopolitics

This chapter brings us to a moment of convergence, when Allied armies approached central Germany from the west, and the Soviet Red Army approached it from the east, each side literally racing for the remains of Nazi Germany's ballistic missile program: the V-2. They were also in a race for a larger network of sites containing Germany's "wonder weapons" of the war, including the V-1 "flying bomb" and jet engine facilities at Peenemünde and the Mittelwerk. A defeated Germany was a treasure trove of new technologies.

Once the war was over, scientists and engineers understood that they were living at a dramatic break in history, the beginning of rapid and radical transformations in aeronautics and astronautics. Theodore von Karman and Wernher von Braun both wrote reports in late 1944 and early 1945 on the advent of supersonic jets and rockets, ballistic missiles, and intercontinental and even space-based weapons. This was futurism as rational military prognostication. Their conclusion: the US had to become a rocket state. The USSR soon afterward translated this very logic into a state decree for the promotion of ballistic missiles, "On the Questions of Reaction Weaponry," signed on 13 May 1946 by Joseph Stalin, earlier patron of the world's first rocket state in 1934. The sides of the new Cold War were drawn.

The fortunate turn of events in the Second World War was that the German aerospace "wonder weapons" were no match for US industrial capacity, battlefield rockets, and aviation technologies. The might of US industry supplied the naval and land forces for two global fronts, across the Atlantic and Pacific theaters. The US had the ability to produce the B-17 strategic bombers and P-51 fighters to break through German aerial defenses and destroy the country's cities and infrastructures. For the Pacific theater, the US developed and deployed the still more powerful B-29 bomber that wrecked much of Japan, with even more massive civilian casualties. US industry and R&D enterprises also created the vast scientific and industrial network of the Manhattan Project, culminating in the atomic bombs that leveled Hiroshima and Nagasaki.

Germany's military forces were also no match for the USSR, especially its Army of over 30 million strong, often mobilized in shock-wave assaults against

DOI: 10.4324/9781003494409-6

German troops, without regard for human life on either side of the battlefield, and which accounted for the significant losses of the Soviet military in the war. Soviet military-industrial capacity counted as well, in tanks and aviation, as well as in the Katyusha rocket, a mass-produced, short-range, solid-fuel rocket that forced German military forces into retreat. In economic terms, the Soviet eastern front was also an American third front, at least in terms of the decisive $11 billion dollars of "lend-lease" material that the US transported across the oceans and land bridges to support the USSR. This included food and clothing, metals, 30% of all Soviet trucks, and 30% of all Soviet fighters and bombers between 1941 and 1945. Once again, the US came to the rescue of Russia.

Although the Nazi V-2 failed to turn its country's fortunes, compared to how the American atomic bomb dramatically ended the war in the Pacific, it still set a standard of technological achievement. The V-2 ballistic missile program at Peenemünde was Nazi Germany's "Manhattan Project," or the closest it came to one. Both the Nazi ballistic missile program and the American atomic bomb project focused on building two of the world's most sophisticated military technologies, demanding radically new kinds of government and military, academic and industrial cooperation. Each created novel weapons dropped from the air for massive destructive effect. Each cost about the same investment: roughly $2 billion in wartime values.

The Cold War was a similar story of intense competition, but between the US and USSR, to seize and perfect these strategic systems, ballistic missiles and nuclear weapons. It was a story we so often describe in sports metaphors, of racing ahead and falling behind. It was a story of the US gaining and then losing its early monopolies, and the USSR gaining its own advantages at the expense of the US, all for geopolitical influence in a new bipolar world.

Cold War Strategies for Earth

The Cold War (1945–1991) was a conflict in three fundamental dimensions: territory, technology, and ideology.

In territory, by the terms Halford Mackinder had first defined in 1904 and revised in the intervening years, and supplemented by the interpretations of Nicolas Spykman, the Cold War was a new chapter in a mighty geopolitical contest. It pitted East against West, the Soviet "heartland" against the US as a "new world" continental and oceanic power. By 1945, both sides had divided the remains of the Nazi Third Reich and Japanese Empire, with continuing tensions over the fate of Germany and China. The US maintained strategic dominance with a ring of allies in the outer "rimlands" of Eurasia, in Europe, the Middle East, and South and East Asia. This was the realm of relatively peaceful American containment, what the USSR called "capitalist encirclement." The conflict persisted over the next 40 years, with the deadliest of consequences in Korea and Vietnam, where Stalin and his successors exploited and sacrificed their communist Asian allies to draw the US into distant and costly wars. This was the realm of violent Soviet engagement. The USSR took later risks and gambits in Cuba and Afghanistan.

Technology was at the center of the Cold War, especially the new atomic and hydrogen "superbombs." They were what made the war "cold," because nuclear war never broke out directly between the two new "superpowers." Mutual deterrence was the rule. A new battlefield map of the Cold War, in the azimuthal projection, centered on the North Pole, reduced the globe to a measure of the new aerial technologies of jet propulsion, in long-range bombers, and in supersonic jet fighters and ballistic missiles. Both the US and the USSR faced each other across the pole with these new weapons. Technology guaranteed mobility and speed over regions, continents, and oceans. Rockets joined the rimlands to form a powerful geopolitical bind between the superpowers.

Historians sometimes mistake these convergences and competitions, and the bipolar world that shaped the Cold War, as a set of parallel paths. We represent this parallelism in maps and charts, tables and narratives. Yet these parallels are not equivalents. Content matters. The US and USSR were at dramatic ideological cross-purposes in the Cold War. One represented civic rights and pluralist democracy, however imperfect. The other was a one-party dictatorship and police regime. One represented relatively free markets and media. The other is a command-administrative and closed propaganda state. There was no Cold War without these opposing values and ways of life.

None of these three dimensions determined the genesis and structure of the Cold War. People did. To recall our phenomenological approach in this textbook: territory, technology and ideology were fields of understanding within which people lived and thought, spoke and wrote, acted and performed. This approach is so important to understand the Cold War and its later space race because both replicated a cycle of perceptions and misperceptions. As each side raised military and technological capabilities to compete for global power, they calculated their own strengths and weaknesses against the enemy's intentions. Yet they often fell into a trap of misperception or misrepresentation, exaggerating their own or their enemy's position, raising capabilities and escalating the conflict even further.

Only one superpower emerged stronger from the destruction of the Second World War: the US. The USSR was a superpower partly by the size and ferocity of the Red Army, which had overrun Nazi military forces and brutalized Germany, and which imposed an oppressive rule on its conquered territories in Central Europe. It was a global superpower in promise, for the kinds of technological leverage it might impose on the US. We also ought to note that the end of the Second World War returned the USSR to the status quo ante bellum, to the state of affairs before the war. Germany was destroyed, of course. Yet Stalin was master of Central Europe, on terms not so dissimilar from those he negotiated with Hitler to begin the war in 1939, and that the two powers shared as their "bloodlands" for several years.

The US nuclear arsenal was decisive at the end of the Second World War. It remained decisive for the first four years of the brief US atomic monopoly. These were the years of President Harry Truman's "imperial" presidency, when American democracy experienced a dramatic accumulation of federal military and police powers through the National Security Act (1947). The Act was meant to fortify the US with the rule of experts and engineers. It created a whole series of new technocracies within the US executive branch: the Central Intelligence Agency; the Department of Defense; the Joint Chiefs of Staff; the United States Air Force; the Research and Development Board; the National Security Resources Board. An all-coordinating National Security Council was at the peak of this bureaucratic hierarchy.

One of the most crucial developments within the US executive branch and military was the decision to increase its number of atomic bombs (and in time hydrogen bombs), and also depend for their deployment primarily on the Strategic Air Command. It was a new office in the new United States Air Force to coordinate the fleets of new jet bombers, the B-47s and B-52s, that within a few hours could drop their weapons on the Soviet enemy. Stalin's regime answered this array after 1951 with long-range bombers of his own (the Myasishchev-4 "Bison" and Tupolev-95 "Bear") and at such a pace that the US military and media even raised the threat of a "bomber gap," that the US was falling behind.

The CIA played an essential role in all of this, especially in the consecutive "National Intelligence Estimates" that literally calculated and interpreted the USSR's political, economic and military capabilities over the next 40 years. The US, thanks to the USSR's closed society and secret government, often operated in ignorance or speculation. For the CIA, the safer and smarter course was to assume the worst threat: to overestimate, to exaggerate, and to overspend. This, at least, ensured an American advantage in the "race" that was the Cold War. It was, of course, also an imperative for the CIA's own bureaucratic survival, and a reward to the very politicians doing the overspending for industries in their districts.

American publicists, beginning with Walter Lippman's, *The Cold War* (1947), have criticized this new system as a "national security state," or as a "garrison" or "emergency" state. It violated, so they argued, the traditional separation and balance of powers in the US Constitution. It represented a shadow government of unelected experts and technocrats beyond the scope of everyday understanding and congressional oversight. It was prone to authoritarianism and extremism in a drive for Cold War advantages. In time, critics focused their condemnations upon the CIA, not only for its concentrated powers of analysis and advice, but also for its eventual covert capacities in funding and organizing coups, in backing pro-American dictatorships, in persecuting or assassinating America's enemies. In time, this critique even reached the US civilian space program, what Walter McDougall called little more than "a paramilitary operation of the Cold War."

1984

George Orwell, an admirer of Evgeny Zamyatin's *We*, was one of the first publicists to coin the term, "Cold War," in a newspaper article of 1945, just as he was drafting the settings and plotlines of his classic dystopian novel, *1984*. With the advent of air power and the atomic bomb, he was fearful of the immense technological powers now in the hands of US leaders, mistrustful of its managers and technocrats, who had amassed such extraordinary wartime governing power as well. Yet he was most distressed about the rising dominion of the USSR, this combination of ancient "slave empire," and modern "highly centralized police state," that had divided old Europe and was harassing the world. It was "at once unconquerable and in a permanent state of 'cold war' with its neighbors." Its confrontation with the free West was leading civilization to the paralysis of a "peace that is no peace," to a rivalry between "monstrous super-states." Orwell's *1984* predicted that this permanent state of war would defeat the democratic West, and enslave its people in a system of total deception and control.

The US monopoly lasted until the USSR tested its first atomic bomb in the deserts of Kazakhstan, at the Semipalatinsk "base" (*poligon*), on 29 August 1949. President Truman, at the Potsdam conference in July of 1945, which negotiated part of the postwar settlement, had hinted to Stalin about this new American weapon of massive destruction. Stalin feigned interest, already knowing about it thanks to his network of spies in the Manhattan Project, who were transferring enough information in order to enable his own atomic bomb. Once again Stalin was out to "catch and surpass" the best scientific and technological achievements of the West.

He now ordered his most trusted lieutenant and loyal executioner, as a member of the State Defense Committee, Lavrenti Beria, to manage three new priority "tasks": building an atomic bomb, building a long-range bomber, and building a medium-range, liquid-propellant ballistic missile. Stalin was taking no chances. He needed both the power to deploy and destroy. In each case, both he and Beria gave express orders that Soviet engineers reproduce an exact duplicate of the American atomic bomb (from its stolen schematics), an exact duplicate of its B-29 bomber (with one in their possession, which the Soviets precisely reverse-engineered), and an exact duplicate of its captured V-2 missiles.

Beria had quite a Stalinist resume. He headed the NKVD from 1938 and the Organs of State Security from 1946. He was the architect of the massacres of some 15,000 Polish officers at Katyn Forest in 1940, and of the violent 1944–1945 deportations of hundreds of thousands of nationalities from their homelands in the Black Sea regions (among them, the Volga Germans, Crimean

Tatars, Balkars, Karachais, Chechens, Ingush, Kalmyks, and Greeks). He was one of the major directors of wartime industrial and military production. Beria's atomic bomb task was a dramatic success, built from stolen US designs, by the mobilization of tens of thousands of scientists and technicians, and of hundreds of thousands of convict laborers and labor gangs in the uranium mines and processing plants.

The Race for the V-2

The American race for the V-2 ballistic missile actually had its own distant origins in the outbreak of the Second World War, as American military forces began to prepare and retool for the possibilities of jet propulsion in the "coming war." Already in 1938, the Army Air Corps approached the Pasadena California "GALCIT" group (the Guggenheim Aeronautics Laboratory at the California Institute of Technology) to develop solid-fuel rockets for "Jet Assisted Take Offs" (JATO), to assist traditional propeller-driven aircraft heavily laden with fuel, cargo, or bombs. Von Karman and his team flew the first JATO with success in 1941 and, after further improvements, formed the Aerojet Corporation in 1942, with new and expanding facilities, to serve as a contract agent for the US Army. GALCIT soon renamed itself as the Jet Propulsion Laboratory (JPL).

Similar developments took shape on the east coast. In 1942, several veterans of the American Rocket Society (John Shesta, James Wyld, and Hugh Pierce), having perfected one of their liquid-propellant rocket engines in static tests, took their invention to the US Navy's Bureau of Aeronautics, transforming themselves into "Reaction Motors" incorporated. Their initial contract was also to develop take-off boosters, but for the Navy's aircraft, which needed the extra power to fly from the short runways on aircraft carriers or on small strips of land on the Pacific islands. Robert Goddard actually ended his career sharing his knowledge with this Navy JATO program.

Once the Allies began to learn the secrets of the German "wonder weapons" in 1943 and 1944, the US military assigned the JPL, Aerojet and Reaction Motors the tasks of preparing America's first ballistic missiles. On contract, beginning in 1944, the JPL and Aerojet developed a series of Army research rockets, culminating in the Corporal ballistic missile, using innovative storable propellants, nitric acid and aniline. With access to shared US Navy technical reports, Reaction Motors built the liquid-propellant engine for one of America's "X" or "Experimental planes," the X-1, a joint project of the US Army Air Forces and Bell Aircraft. It was the first piloted craft to break the sound barrier on 14 October 1947, thanks to Chuck Yeager. Reaction Motors also eventually built the engines for the Navy's Viking sounding rocket and for the later X-15, the world's fastest and highest test plane.

For all of these foundations and achievements, none of the early products of the US aerospace industry equaled the proven capacities of the V-2. It became the target of all attention and goals. In March of 1945, as the Second World War wound down, the Pentagon ordered US Army investigators to look for von Braun. He and his team were also looking for them, desperate not to fall into the hands of the Red Army, no doubt because of their ideological and racial prejudices, but perhaps also because of the horrors Russian prisoners of war had suffered as slave laborers at Mittelwerk–Dora. The US race for the V-2 culminated when the US Army beat the Red Army and reached the Mittelwerk on 11 April 1945, hastily evacuating its treasures: 14 tons of documents, parts and wholes of the V-2 ballistic missile, and about 150 members of the Peenemünde–Mittelwerk team.

In an operation called "Project Paperclip," the Army delivered von Braun and "volunteers" to Fort Bliss, Texas by the fall of 1945, and soon afterward about 100 V-2 missiles for testing to White Sands Proving Grounds, New Mexico, only 120 miles from Goddard's former ranch, and only 50 miles from Alamogordo, site of the first atomic bomb test. Here at White Sands, the US achieved two important firsts of the Space Age: the first multistage rocket, an American-made modified Corporal second stage, mounted upon a captured German V-2 first stage. It was also the first rocket to reach well into space, to 244 miles on 24 February 1949.

These efforts were the foundations for the Army Ballistics Missile Agency (ABMA), within a few years headquartered in Huntsville, Alabama. Mostly isolated within the arsenal system of army R&D, von Braun's teams perfected their early masterpieces, the Army's Redstone ballistic missile; and by 1955 its variant, the Jupiter C intermediate-range ballistic missile. The Redstone was built by the Chrysler corporation, the maker of popular Plymouth cars and Dodge trucks. Its engines were the product of the new Rocketdyne division of North American Aviation.

At the end of the war, Stalin sent his own investigation teams west, in the wake of the Red Army, to gather what V-2 parts, documents, and personnel they could find. He enjoyed a small advantage here, as these teams included top engineers from the Reactive Scientific-Research Institute (RNII), the centerpiece of his early rocket state. They arrived at Peenemünde in May of 1945, later to the Mittelwerk, after the Americans took most of its prizes, and in time were able to amass their own fledgling German–Russian network of specialists to begin rebuilding and redesigning the V-2. Their spoils were nothing like America's, as they only gathered parts of the V-2, some documents, and mostly technical personnel.

Stalin's regime established the Soviet ballistic missile industry as a signal state imperative, including its scientific-research, manufacturing, and military dimensions. For these purposes, Stalin revived and reconfigured his earlier rocketry institutions, the survivors hardened by the terror and the war. He concentrated them in the central "Scientific-Research Institute 88" (NII-88). He tasked it to build intermediate-range ballistic missiles. A Council of Chief Designers, under Sergei P. Korolev's direction, beginning in 1947, coordinated a far-flung network of scientific offices and manufacturing industries to serve the task. The Council worked together to design and develop the missiles, engines, guidance systems, radio and automatic control systems, and the ground equipment and launch complexes. Supervising both NII-88 and the Council was Dmitrii Ustinov, the Commissar of the Ministry of Armaments. This veteran Stalinist, and future master of Soviet space and defense programs, oversaw the Soviet state's core imperative to develop ballistic missiles as an achievable, cheap and more efficient alternative to the expensive strategic bombers in both the US and USSR arsenals.

On 23 October 1946, in an effort to further centralize and coordinate efforts, Stalin's secret police rounded up what was left of the German rocketry teams, around 130 persons in all, including several top designers. The specialists were removed to the NII-88 offices in Moscow and to a secret location at Gorodomlya Island, 200 miles northwest of the capital, where they lived in secrecy and isolation, under some very trying physical conditions. In both locations, the Soviet state kept the Germans separated from any Russian specialists, assigning the Germans tasks, and then integrating their best designs and innovations into existing Soviet orders. The German's "G" rocket series, for example, impressed the Soviets with several important features: integral fuel tanks, a forward oxygen tank, a radio-controlled engine cut-off, and strap-on boosters.

Both the Germans and Russians of NII-88 worked on 16 different kinds of missiles between 1947 and 1956. They began with exact copies of the German V-2, now called the Soviet "Rocket 1" (R-1), first launched on 18 October 1947 from a desolate stretch of land in the Turkic steppe at Kapustin Yar. Engineers lived in railroad carriages and tents there, suffering from extreme heat and cold, from spiders and the windswept sands. Variations of the original prototype soon followed (the R-2 and R-5 missiles of intermediate range). They culminated in the version of the R-5M that, on 2 February 1956, successfully launched a model of an 80-kiloton nuclear bomb over a range of 750 miles, a forewarning of threats to come.

Projects and Objects

There was a dramatic difference in how the US and USSR operated their aerospace endeavors. The US depended on research and development (R&D) "projects." These were a function of the project system, reaching back to 1914, when the US government began to fund R&D project contracts with universities, agriculture, and industry. Projects ensured that government monies were being spent wisely. They were a uniquely American invention. The project meant the planning of an end product or process, by way of discrete teams and stepwise engineering, with clear-cut means of funding, techniques, schedules, and objectives in mind. A simple piece of paper, the project defined a reachable technological endpoint. It set terms and sealed a bond.

Projects became the R&D rule of the Second World War and early Cold War. They taught the military and civilian enterprises to work together. Projects also fostered multiple and dispersed nodes of free-market innovation and cooperation. As projects became more complicated, the military services depended on contract agencies (firms like TRW or the Aerospace Corporation) to better manage the subcontractors and ensure interfacing and reliability, as, for example, between the structure, propulsion systems, guidance and control, and payload of a ballistic missile. This became a new model of the vertical program management of discrete engineering projects, fortified by systems engineering and computerization.

If the US mastered sophisticated R&D projects, the USSR depended instead on scientific-technological "objects" (*ob'ekty*), Stalinist style. These focused on ends rather than means. They were *actual* products like the Soviet atomic bomb, or Myasishchev-4 bomber, or R-5 ballistic missile, which the command-administrative system tasked to Soviet design bureaus and their partner ministries and plants. Soviet object tasks had the advantage of speed and end results. "Objects" A through G (the first four letters in the Russian alphabet) were the first models of nuclear weapons placed on the Soviet R-7 missile. Object "D" was the Sputnik satellite. Object "K" was Yuri Gagarin (well, actually the Vostok craft that he sat in).

As scholars Andrew Aldrin and William Barry have argued, design bureau managers like Sergei Korolev enjoyed something like a "project" approach. They had autonomy and influence. They fostered horizontal cooperation and creativity between the bureaus, and within their teams. They also wielded vertical power, to influence the ministerial and party bosses above, and the plants and flight centers below. The priority financing from the party-state offered incentives and rewards. But object tasks were also beset by an oppressive culture of vertical "leaderism" (*vozhdizm*), where patron-client connections, obedience, and secrecy were the rule, all functions of the closed command-administrative system.

The Lure of Space Exploration

In terms of their promotion of rocketry and spaceflight, especially in the arenas of popular culture, the US and USSR were remarkably different. The US was a country of relative wealth and abundance, the homeland mostly untouched by war. The USSR, still recovering from the devastation of its own command-administrative policies and the wreckage of the Second World War, was mired in poverty and shortage. Both systems promoted rising expectations. In the US, this meant disposable wages, bungalow homes, automobiles, and home appliances.

In the USSR, this usually meant barely survivable wages, a room in a shared communal apartment or barracks, and only the most basic of domestic possessions (shoes, clothes, accessories).

The glamor of rockets penetrated into the expanse of American popular culture in ways it never did or could in the USSR. American car manufacturers patterned their new models of automobiles on the rocket style, as in the rare "concept" cars of Harley Earl, like the Firebird XP-21 or Cadillac Cyclone XP-74, and in mass-produced models like the Oldsmobile "Rocket 88." A Memphis, Tennessee set of bands, including Ike Turner and the Kings of Rhythm, even recorded a song about it, "Rocket 88" (1951), a top hit on the Billboard R&B chart, and one of the first recordings in the new genre of "rock and roll." The closest the USSR ever got to these artifacts, besides propaganda in word and image, were the metal playsets and gymnastic equipment, so often made in the shape of space rockets, and still found in communal playgrounds around Russia today. Both the US and USSR became superpowers with supersonics and super bombs, but only the US had Superman and superhighways.

In popular culture, both the USSR and US experienced the fear and anxiety of nuclear annihilation from above. The official Soviet media stoked these fears of an American nuclear strike over several generations, propaganda designed to instill a hatred of capitalism and militarism. In the US, this fear threaded the popular media culture. Thanks to John Hersey's bestselling expose, *Hiroshima* (1948), American audiences relived its horrors in everyday detail. The potential of long-range bombers and ballistic missiles to carry the bomb anywhere in the world, compounded by the speed of "push-button" war, only magnified the nuclear anxieties. Americans looked fearfully to the skies, as in the Unidentified Flying Object (UFO) craze of 1947, reminiscent of the alien spaceship sightings of 1897. It all began with Kenneth Arnold's 24 June 1947 UFO sightings in Washington state. He saw a series of bright disk-shaped objects moving at tremendous speeds. More and more Americans now reported visions of "flying saucers" all over the country, probably just the distortions of natural phenomena, or lighting, or aircraft in flight.

Both American and Soviet popular culture celebrated outer space as a site for human exploration and liberation. It was also a place where we would discover our true selves. The rocket was the way, according to Willy Ley's, *The Conquest of Space* (1946), and Arthur C. Clarke's, *The Exploration of Space* (1951). In film, the US finally caught up with the USSR and its *Cosmic Flight*, and with Germany's *Woman in the Moon*. It came in the postwar movie hit, *Destination Moon* (1950), based on Robert Heinlein's science fiction stories. The film depicted realistic scenes of "rocket engineering" and space exploration. It warned of a "race" with the Soviets for "control" of outer space, the Moon, and ultimately Earth. It screened physics professors at work with their blackboards and slide rules, pondering all the differential equations of "reaction mass" and "initial velocity." The popular film challenged America to achieve the "fantastic" and "incredible," to reach for the ultimate envelopes of "higher and faster."

Millions of youngsters watched spaceflight adventures on television: *Space Patrol* on ABC, featuring Commander Buzz Corry (sponsored by Ralston Purina); *Space Cadet* on ABC, with expert advice from Willy Ley and the astronomers at the Hayden Planetarium (sponsored by the Kellogg Company); and *Captain Video* and the "X-9" spaceship on the Dumont network (sponsored by General Foods). Accompanying them was a whole new way of speaking, an adolescent idiom of the "rocket age": phrases such as *blow your jets*, *spaceman's luck*, *smoking rockets*, *countdown*, and *blast off*.

For young adults and their parents, *Collier's* magazine became the go-to source for the coming Space Age. Its editor, Cornelius Ryan, helped introduce Willy Ley and Wernher von Braun to the American public. With 4 million subscribers, it was the perfect venue to dazzle readers with the possibilities of the US as a "spacefaring" nation. The rocket would take us there, as a "first step" toward infinity. In the form of a ballistic missile, it would also protect us against the totalitarian evils of Soviet communism. *Collier's* offered sketches and rationales for Ley's and von Braun's space planes and circular stations for Earth orbit. Walt Disney appreciated the attractions, hiring von Braun as consultant for a series of ABC television shows: "Man in Space," "Man on the Moon," and "Man to Mars" (1955–1957), with nearly 15 million television sets watching. The shows educated, entertained, and advertised the new Disneyland Park (1955) in California, with its "Tomorrowland" and "Moonliner" rocket rides, only 30 miles from Pasadena's JPL.

In the USSR, Stalin's regime gained new confidence in rocketry and spaceflight. Russian science was now its own gold standard, with its own inventors and innovators, its own "founding fathers" of world renown. Konstantin Tsiolkovsky joined this elite cadre of "firsts," the premier inventor of "cosmonautics" as a particularly national "Soviet Russian" version of astronautics. The Central Committee of the Communist Party stage-managed the revival and retooling of his personality cult in honor of the 19th anniversary of Tsiolkovsky's birth in 1947. It ordered biographies written, a portrait painted, scholarships and prizes funded, medals struck, his collected works published, films and radio and media presentations to be made.

Soviet popular science magazines from this time, in the long tradition of the genre reaching back to the late imperial and early Soviet periods, also celebrated Tsiolkovsky and rocketry in a cavalcade of images and texts. They were sometimes fictional, at other times factual, always patriotic. The magazine, *Knowledge is Power*, served these purposes, essentially becoming the Soviet version of the American *Collier's*. Speeches and portraits by Stalin punctuated the campaign around "Our Great Future," a Soviet society of streamlined and orderly cities, of helicopters and planes moving freely about, of rockets reaching into space. Stalinist popular science, such as Viktor Saparin's *New Planet* (1950), predicted that Soviet communism would soon fashion fully automated and computerized societies. They would have plentiful sources of energy and newly cultivated lands, diverting Siberia's rivers to flood Central Asia with water and life, making for a whole new planet. This coming USSR, already one-sixth of the globe, was also destined to expand on a "cosmic scale," to become the "first in the history of the Earth to launch a human being into interplanetary space." The first "cosmonaut" would then travel by rocket to the Moon, Mars, and Venus, and create new Earths there as well.

The Russian propaganda state turned the Tsiolkovsky rocket into the emblem of Soviet scientific-technical superiority, and its advance into space travel. Vasilii Zakharchenko's, *Voyage to Tomorrow* (1952), pictured the future USSR as a scene of colossal engineering sites, in skyscraper cities, aerial transport, wind and solar power, huge canals, the Tsiolkovsky rocket, and a wheel-style Soviet "space station" (*sputnik*). Boris Liapunov's *The Discovery of Worlds* (1954) also featured a colorful optics, the old fantasies of space travel and moon colonies as new opportunities. "The impossible has become possible," wrote Liapunov. "The might of Soviet science … will create miracles." Space meant infinite horizons. The rocket was "like the tail of a comet," only now human-made and outward-bound.

These Soviet venues were not that different from what American and European publicists were already doing in their rocketry and spaceflight works. They shared a vision of space travel and colonization. What Walt Disney did for the United States, Joseph Stalin was also doing for the USSR. Both became the patrons of competing "tomorrowlands:" one in the capitalist "Moonliner" rocket at Disneyland in Anaheim; the other in the communist imaginings of future Lenin monuments, cosmodromes, and space catapults in Moscow. Stalin took on the mantle of a Soviet Walt Disney.

Competing Paths to the First Satellite

The political approaches to Sputnik began innocently enough, with two reform initiatives, one in the US and one in the USSR. With a resounding electoral victory, Dwight Eisenhower assumed the presidency in 1953 and advanced his "New Look" policies of fiscal responsibility. Franklin Roosevelt's and Harry Truman's previous administrations, building a "New Deal" welfare state, had expanded the federal budget from $4 billion in 1932 to $85.5 billion in 1954, with just over half of that budget now devoted to military needs, most recently with a focus on a new "national security" state. Eisenhower did not want to dismantle these states but restrain them.

Eisenhower won office on a platform of bolder Cold War posturing and smarter rearmament. At times, he played the risky diplomatic gambit of nuclear brinksmanship, as in the case of the Korean War in 1953, and the Taiwan and Berlin crises in 1958, threatening the USSR and its communist ally, the People's Republic of China, with nuclear war in order to leverage American territorial advantage. His administration fortified the posture of nuclear "massive retaliation," approving fleets of new strategic missiles. At its core now was the nuclear "triad": the long-range bombers and fighters of the "Strategic Air Command," intermediate-range and intercontinental ballistic missiles (ICBMs), and the development of submarine-launched ballistic missiles. He meant to create a "redundancy of effect," ensuring that the US retained a second-strike capability, the power to inflict massive and unacceptable retaliation on the USSR, were it ever to strike with nuclear weapons first.

The USSR also experienced a dramatic change in leadership, what with the death of Joseph Stalin in March of 1953 and the rise of Nikita Khrushchev (1953–1964), whose regime corresponded roughly with Eisenhower's years and with John F. Kennedy's short term in office. Khrushchev seized power in a coup d'état against Stalin's immediate successor, the secret-police chief Beria, who with his leading cohorts was arrested and executed by the orders of a new "communal" leadership. It represented the communist party, the state apparatus, and the military, though the party-chief Khrushchev eventually took charge. He then proceeded with a series of political and economic reforms that amounted to "de-Stalinization." He dismantled the worst of the secret-police state and revived the leadership ranks of the party. He closed the most brutal of the labor camps and rehabilitated some of Stalin's victims. He instituted a social regime of material rewards: those who worked hard and contributed to his regime, while also maintaining political and public conformity, received better working and living conditions. He allowed a brief opening in a freer culture, the "Thaw," as writers and artists now had the freedom to be mildly and openly critical of the "Stalin" style of police dictatorship.

Yet on several fronts, Khrushchev was ever the Stalinist, a product of the very system that made him and which he now attempted to reform and control. As a political commissar under Stalin, he was himself guilty of many of the very crimes (mass arrests and summary executions) for which he now accused Stalin and Beria alone. In the Stalinist style, he advanced a repressive anti-religious campaign against the Russian Orthodox, Roman Catholic, and Muslim faiths. He continued Stalin-style investments in the command-administrative system of industry and collective farms. He maintained and revived Stalin's original investments in ballistic missiles. He adapted Stalin's personality cult, who was a "father" of aviators, into his own personality cult, as a patron of outer-space satellite and cosmonaut spectaculars.

Two developments accelerated both the Eisenhower presidency and the Khrushchev regime along the path toward the Sputnik crisis. Both had to do with the payloads of ballistic missiles, be they either nuclear warheads or Earth-orbiting satellites. By 1953, American scientists had discovered ways to make the new hydrogen nuclear bomb lighter in weight. These new kinds of thermonuclear devices carried much greater destructive capacities and wider radiuses of destruction, along with the development of more precise guidance systems and targeting. All this meant that a nuclear-tipped ICBM was now immediately possible. Scientists also discovered that blunt cones, which created a buffer of compressed air at high speeds, meant the warhead and missile could better survive the heat of re-entry from space.

With these developments, in 1954 the Eisenhower administration now committed the US to revive its earlier plans for the rapid development of an ICBM, Project Atlas, one more means to intensify the competition with the USSR. As part of America's "redundancy" of efforts, the Air Force built Atlas, and a second type of ICBM, Titan, along with an intermediate-range ballistic missile, Thor. Instead of such American redundancies and dispersions, the USSR preferred centralized, concentrated, and economical efforts. Shortly before his death, Stalin signed a decree in 1953 ordering the development and deployment of a Soviet ICBM, the "Rocket 7" (R-7), a powerful launcher, to carry the USSR's heavier nuclear bombs across continents and oceans, thus targeting America's cities much quicker and cheaper than a fleet of expensive ace strategic bombers.

In the Sputnik crisis that soon overwhelmed each of these governments, the US had one great disadvantage: its open society of free markets and a free media, which invited Soviet intelligence sweeps to gather open-access information, and espionage to steal industrial and military secrets. On the other hand, the USSR enjoyed its own great advantage: a closed society of sealed borders, media control, and secret-police surveillance against its own citizens. These prevented few viable American chances for actionable human intelligence from within the country.

In response to this dilemma, Eisenhower proposed Open Skies (1955), a bold initiative to mark the air and space above both the US and USSR as free and navigable. Such high-sounding words masked the president's ulterior motive: to leverage the US advantages in reconnaissance aircraft (and eventually spy satellites) so as to collect signals and technical intelligence on the USSR.

Such a crop of new intelligence materials would have compensated for the relative lack of human intelligence out of the USSR and provided the foundation for more rational US planning and policy, based on the enemy's real capabilities and probable intentions. No longer would the CIA and its analysts feel the pressure to speculate and exaggerate or fall into the trap of calculation-misperception-escalation. No longer would the US need to excessively fear the threat of a surprise attack from its totalitarian rival. Eisenhower wanted intelligence in the truest meaning of the term, to gather information on the USSR and to calibrate the best and safest responses for the American democracy and treasury.

Understanding his country's position of inferiority in science and technology, aviation and reconnaissance, Khrushchev rejected Open Skies. He saw its advantages for the US, realizing that the USSR was stronger in the world if the

US misperceived it as such. Manipulation and deception were essential elements of Soviet strategy, what the military called "reflexive control theory" and the party-state called "active measures," the capacity to force the enemy to make decisions against its own interests.

Yet Eisenhower had another card to play, if a secret one: the U-2 spy plane, which his advisers contracted to Lockheed Aviation's famous "Skunk Works," its special aerospace laboratory for cutting-edge technologies of overt and covert warfare. Famous for developing the first US operational jet fighter, the P-80, Clarence Kelly Johnson's engineers created the U-2 before its deadline and under budget: a turbo-jet aircraft able to reach 70,000 feet, at cruising speed of 460 miles per hour, for up to ten hours at a time, at a range of 3,000 miles. The U-2's sweeping photographs of Soviet territory, 300 miles in each direction, offered superb photographic surveillance of the Earth below. With its spy flights beginning on 4 July 1956, the U-2 soon discovered some truths behind presumed Soviet military capabilities, for example, that the USSR had far fewer strategic bombers than the US had earlier estimated.

Khrushchev had a strong card to play as well, one given to him by Eisenhower's own administration, whose scientific advisers, as part of the Open Skies proposal, supported an initiative of the International Council of Scientific Unions for the International Geophysical Year (IGY). Planned for the period between 1 July 1957 to 31 December 1958, the IGY was a worldwide cooperative effort to study the Earth sciences from the ground, atmosphere, stratosphere, and even orbit, by way of the communication and observational satellites already in design by the US and USSR since 1946.

Khrushchev rejected Open Skies but craftily joined the IGY. He now planned to launch a satellite and thereby send a perfect double signal: that the USSR had perfected an ICBM to launch it, and the sophisticated network of scientists and engineers to make and track it. Khrushchev had no qualms about linking a military launcher (the R-7) with a civilian satellite. On the contrary, it made perfect sense. From the shadows of his closed society, he could always claim it was a completely civilian effort, as Soviet news releases did, already a year before Sputnik actually flew, even announcing the satellite's probable orbital path and radio frequency.

The Eisenhower administration, fearing a media backlash in the US and the world, restricted the use of any military ballistic missile as a launch platform for the IGY, unless absolutely necessary. It had to be civilian. So, the president gave priority to the Naval Research Laboratory (NRL), a government office with a long tradition of civilian and scientific enterprises, and its Viking sounding rocket for high-altitude and upper-atmosphere research. First fired in 1949, Viking reached a world record for a single-stage rocket in 1954, reaching an altitude of 158 miles, and taking photographs of the Earth from space. It was outfitted with a powerful Reaction Motors engine, and innovative inertial guidance mechanisms created by the Massachusetts Institute of Technology. For

America's first satellite, the NRL created a three-staged rocket based on a Viking first stage, outfitted with a Vanguard satellite, containing power cells and radio transmitters for tracking.

Eisenhower's scientific advisers held a second US satellite program in a closeted reserve. It was the ABMA and von Braun modified Jupiter launcher, based on the operational Army ballistic missile, now readying for its own satellite program, Project Explorer. Von Braun even later claimed he was ready with a Redstone variant multiple-staged rocket to launch a satellite to orbit as early as September of 1956, a year before Sputnik.

The USSR beat both the Navy and the Army programs, on 4 October 1957, when it launched what it called its "simplest" Sputnik satellite, a metal sphere of 184 pounds holding little more than a radio transmitter to send out its "beeps" to amateur radio operators below. Khrushchev's regime reaped huge propaganda and ideological rewards. Headlines in the US and around the world took it as a sign of Soviet technological prowess, of American scientific-technological inferiority, and as a signal flare in the continuing competition of the Cold War. The USSR followed the first Sputnik with Sputnik 2 on 3 November 1957, carrying a scientific payload and the dog, Laika, to orbit; and with Sputnik 3 on 15 May 1958, carrying a larger scientific laboratory to space. From the perspective of the media, these were all patent Soviet victories.

Amid these impressive achievements, the US fell even farther behind. On 6 December 1957, before a live television audience, the "civilian" NRL Viking-Vanguard, carrying a three-pound satellite payload, exploded only seconds into launch. It was a bloodless drama. No one was injured. Yet it played out in the newspapers and on radio and television for all of America to see and hear. Pundits renamed Vanguard as Rearguard, Offguard, Puffnik, Kaputnik, Flopnik, Stayputnik, and Kerplopnik.

By the time the ABMA and von Braun's team launched the first successful US satellite to orbit on 31 January 1958, the previously rejected Project Explorer, Americans experienced less pride than exasperated relief, and even Explorer suffered insult. The Germans called it *Spätnik* (*spät* meaning "late").

This US close "second," Jupiter–Explorer, was never really able to overpower and replace the myth of Sputnik that had settled into American popular culture and commerce. Sputnik took on mythic proportions, largely because Americans received it in the context of their wealthy commercial republic. They were already expecting rockets cars and rocket planes, or rockets to orbital space stations. It shook their confidence. Yet Sputnik also fulfilled America's market expectations. Sputnik was a media event and commercial product, all its own. Newspapers and magazines realized significant profits in "selling" this space spectacular, if at the expense of American pride. Magazines featured it on their covers. Comics told jokes about it. Songwriters turned it into rock and roll music, like "Rock Old Sputnick," by Nelson Young (1958), and "Sputniks and Mutniks," by Ray Anderson (1959). The first wave of the "baby-boom" generation, millions of youngsters around the impressionable age of 11, were imprinted by the images and sounds of Sputnik.

The American public was not completely floored in fear about Sputnik. They offered a few ironic twists. When the pianist Van Cliburn wowed Russian audiences by winning the International Tchaikovsky Competition in Moscow (1958), the US media christened him the "American Sputnik." Audiences later nicknamed the brand-new 12 starlike chandeliers, at the Metropolitan Opera House in Lincoln Center (1966), as the "Sputniks." They certainly looked the part, made of "starburst" crystal, ascending and descending at the start and end of each performance. Americans had a soft spot for Sputnik.

President Eisenhower was surprised by all the drama, what he called "one small ball in the air." He was always more of an organization man, a classic insider, rather than a public political animal. He knew how to negotiate with Congress or move bureaucratic levers but had no real talent for media relations. He was confident in American national capabilities and prestige, reluctant to promote space spectaculars (firsts or bests). He also knew, from what his top-secret U-2 spy plane was discovering, that Sputnik was more symbolic than real, that the Soviets enjoyed no real lead in military or space capabilities. He also understood that Sputnik served American purposes more than the USSR's. It established the principle of freedom of space that Eisenhower had pursued with Open Skies. Maybe it was a propaganda victory for the USSR, but only in the short term. In the long term, Sputnik was a victory for the US, and soon opened an era of American domination in space surveillance (with the Corona spy satellites).

It is also worth remembering that, although Projects Viking-Vanguard failed on 6 December 1957, Vanguard 1 did reach space on 17 March 1958, with distinctions all its own: the first satellite launched by a civilian rocket, the first with solar power, and currently the oldest human-made artifact still in orbit.

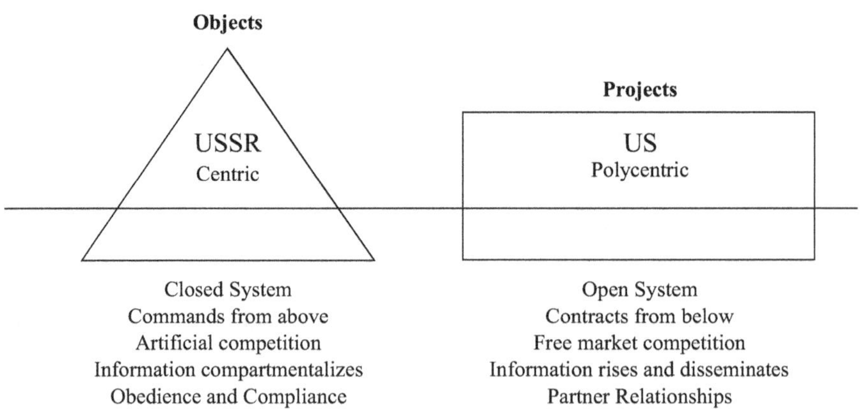

Figure 4.1 Objects and projects in spaceflight R&D. Author.

Vertical Horizons: Above and Below Worlds

The first Soviet and American satellites rose into orbit from a network of aerospace launch complexes and space-oriented cities on the planet. These places were staffed by a scrappy lot of Russian and American engineers and technicians, many of them veterans of the Second World War. They were characteristically tough talkers and hard drinkers and smokers. They were dedicated to their work and to their countries. These places were remarkably similar in terms of their isolated, peripheral locations and their aerospace functions. But they were dramatically different in how they operated. They also offer a cautionary tale. What happened in outer space ought never to be divorced from what happened here on Earth.

In the US, early testing and launch sites, both civilian and military, were located in remote places like the Navy's Point Mugu and China Lake (California), or NACA's Pilotless Aircraft Research Station, at Wallops Island (Virginia); or the Air Force's installations at the desolate White Sands Proving Ground (New Mexico) and at the Cape Canaveral Atlantic Test Range (Florida). The areas around the Cape enjoyed the most development after the 1950s. Originally a rather isolated and wild piece of land on a sparsely populated coastline, the Cape soon transformed into a vibrant tourist destination, filled with neighborhoods at Cocoa Beach, strip malls, and beachfront attractions like the Astrocraft, Polaris, and Starlight motels.

With the expansion of US project contracts for missiles and spaceflight technology, aerospace firms like Rocketdyne and Lockheed, Aerojet and Boeing created smart-looking research parks, often built with formidable glass and metal, in rural-like settings, near suburban communities and housing developments. Universities built such R&D parks too, beginning with Stanford's Silicon Valley, MIT's Technology Square, and Princeton's Route 1 (the Forrestal Center). With the formation of NASA in 1958, its research and development centers in Virginia, Maryland, Ohio, and California completed the process. The middle-class community of Clear Lake City formed near the Johnson Space Center in Texas. These R&D sites were under varying protective security and clearance regimes, from restricted to secret and various top-secret classifications. But they also operated as university-style campuses of a kind, for the relatively open exchange of ideas and information.

The USSR built two major military launch centers after 1945. At Kapustin Yar, in the Turkic steppe just north of the Caspian Sea, it launched early short-range and medium-range missiles. At Baikonur, in the Turkic steppe of Kazakhstan, near the lonely railroad station at Tyuratam, was the future cosmodrome for ICBMs and space launches. Both sites were colonial outposts, known by the Russian designation as a military "base" (*poligon*), a term that the Soviets also applied to the Stalin-era killing fields during the terror, and to Soviet nuclear test sites. At Baikonur, only 20 years before (1928–1933), Kazakh nomads experienced the nightmare of Soviet expropriation campaigns (for livestock, grain, wool and other goods). They meant forced settlement, collectivization, and an artificial famine that saw over 1 million deaths in the country. Kazakhstan was a Russian polygon for rockets and terror.

The USSR had research and development centers all its own. They were called closed cities: all of them closed to foreigners, some even closed to Soviet subjects, unless authorized to live and work there. The two premier ones for space were at Kaliningrad (now Korolev) and Star City near Moscow. Other closed cities were military bases and nuclear energy and weapons sites. One of them, at Akademgorodok, was meant to imitate the open American R&D model, though it never really took. Ukraine also played a significant role in aerospace, thanks to its industrial resources and intellectual talent, centered in cities like Dniepropetrovsk (present-day Dnipro), where Mikhail Yangel built his missile and spaceflight design bureau in 1954. Ukraine was the site of another artificial famine only 20 years before, the *Holodomor* ("Death by Starvation") aimed against Ukrainian peasants, in order to extract grain for export and purchase Western technology and expertise. Around 5 million people perished, accompanied by tens of thousands of executions, and several million deportations to labor camps and forced exile. Ukraine was a Russian polygon for rocketry and terror as well.

Select Bibliography

Afanasyev, Igor. "The Legacy of the V-2," *Journal of Slavic Military Studies* 11/4 (December 1998): 164–174.

Aldrin, Andrew. "Innovation, the Scientists and the State: Programmatic Innovation and the Creation of the Soviet Space Program." Ph.D. Dissertation, University of California, Los Angeles, 1996.

Barry, William. "The Missile Design Bureaux and Soviet Manned Space Policy, 1953–1970." PhD dissertation, University of Oxford, 1996.

Cameron, Sarah. *The Hungry Steppe: Famine, Violence, and the Making of Soviet Kazakhstan*. Ithaca: Cornell University Press, 2018.

Castells, Manuel, and Peter Hall. *Technopoles of the World: The Making of Twenty-first-century Industrial Complexes*. London: Routledge, 1994.

Danhof, Clarence. *Government Contracting and Technological Change*. Washington, DC: Brookings Institution, 1968.

Dickinson, Paul. *Sputnik: Shock of the Century*. New York: Walker, 2001.

Fox, William T. R. *The Super-Powers: The United States, Britain, and the Soviet Union.* New York: Harcourt, Brace and Company, 1944.

Gainor, Christopher. *The Bomb and America's Missile Age.* Baltimore: Johns Hopkins University Press, 2018.

Galison, Peter, and Bruce W. Hevly. *Big Science: The Growth of Large-scale Research.* Stanford: Stanford University Press, 1992.

Gordon, Michael. *Red Cloud at Dawn. Truman, Stalin and the End of the American Nuclear Monopoly.* New York: Farrar 2010.

Harrison, Mark. "A Soviet Quasi-Market for Inventions: Jet Propulsion, 1932–1946," *Research in Economic History* 23 (2005): 1–59.

Harrison, Mark. "New Postwar Branches (1): Rocketry," in *The Soviet Defence-Industry Complex from Stalin to Khrushchev*, J. Barber and M. Harrison, ed. Basingstoke: Macmillan, 2000: 118–149.

Hughes, Thomas P. and Agatha Hughes. *Systems, Experts, and Computers: The Systems Approach in Management and Engineering, World War II and After.* Cambridge: MIT Press, 2011.

Hunley, J. D. *The Development of Propulsion Technology for U.S. Space-Launch Vehicles, 1926–1991.* College Station: Texas A&M University Press, 2007.

Josephson, Paul. *Red Atom: Russia's Nuclear Power Program from Stalin to Today.* Pittsburgh: University of Pittsburgh Press, 1999.

Lowen, Rebecca S. *Creating the Cold War University: The Transformation of Stanford.* Berkeley: University of California Press, 1997.

Mackinder, Halford J. *Democratic Ideals and Reality: A Study in the Politics of Reconstruction.* New York: Henry Holt, 1919.

May, Elaine Tyler. *Homeward Bound: American Families in the Cold War Era.* New York: Basic, 1988.

McCurdy, Howard. E. *Space and the American Imagination.* Washington, DC: Smithsonian Press, 1997.

Mieczkowski, Yanek. *Eisenhower's Sputnik Moment: The Race for Space and World Prestige.* Ithaca: Cornell University Press, 2013.

Naimark, Norman. *Stalin's Genocides: The Chilling Story of Stalin's Crimes against Humanity.* Princeton: Princeton University Press, 2010.

Oreskes, Naomi, and John Krige, ed. *Science and Technology in the Global Cold War.* Cambridge: MIT Press, 2014.

Orwell, George. "You and the Atom Bomb" (19 October 1945), in *George Orwell: The Collected Essays, Journalism & Letters*, volume 4, Sonia Orwell and Ian Angus ed. Boston: D.R. Godine, 2000: 6–10.

Przybilski, Olaf. "The Germans and the Development of Rocket Engines in the USSR," *Journal of the British Interplanetary Society* 55 (2002): 404–427.

Rhodes, Richard. *The Making of the Atomic Bomb.* New York: Simon & Schuster, 1971.

Richelson, Jeffrey. *The Wizards of Langley: Inside the CIA's Directorate of Science and Technology.* New York: Basic, 2002.

Ryan, Cornelius. *Across the Space Frontier.* New York: Viking, 1952.

Snyder, Timothy. *Bloodlands: Europe between Hitler and Stalin.* New York: Basic, 2010.

Spykman, Nicholas. *The Geography of the Peace.* New York, Harcourt, Brace, 1944.

Taubman, Philip. *Secret Empire: Eisenhower, the CIA, and the Hidden Story of America's Space Espionage.* New York: Simon & Schuster, 2003.

Wang, Zuoyue. *In Sputnik's Shadow: The President's Science Advisory Committee and Cold War America*. Newark: Rutgers University Press, 2008.
Zhuk, Sergei. *Rock and Roll in the Rocket City: The West, Identity, and Ideology in Soviet Dniepropetrovsk, 1960–1985*. Washington, DC: Woodrow Wilson Center Press, 2010.
Zubok, Vladislav. *A Failed Empire: The Soviet Union in the Cold War*. Chapel Hill: University of North Carolina Press, 2009.

For studies on the Cold War, readers should begin with Walter Lippman, *The Cold War*. New York: Harper, 1947. Comprehensive studies include Aaron Friedberg, *In the Shadow of the Garrison State: America's Anti-Statism and its Cold War Grand Strategy*. Princeton: Princeton University Press, 2000; and Douglas Stuart, *Creating the National Security State: A History of the Law That Transformed America*. Princeton: Princeton University Press, 2012. More pointed and critical studies include Michael Hogan, *A Cross of Iron: Harry S. Truman and the Origins of the National Security State, 1945–1954*. New York: Cambridge University Press, 2000; and David Unger, *The Emergency State: America's Pursuit of Absolute Security at all Costs*. New York: Penguin, 2012.

Part II

The Space Race

5 Project

Vostok and Mercury in Orbit

The launch of Sputnik in 1957 multiplied the presumed "gaps" of the early Cold War. The bomb and bomber gaps now became science and space gaps, and missile and even Moon gaps. The USSR had laid down a gauntlet with its first spaceflight "first" and now the US found itself regularly falling behind, mired in a series of missteps, suddenly having to play by the Soviet rules of "catch and surpass." Space achievements became a means for either the US or the USSR to prove to its allies around the world that its political system was superior, the way of the future. Cold War strategy and geopolitics transformed into a strange realm of sports rivalry, media spectacle and psychological confrontation. The US was behind in another area: the media gap. Its advertising and publicity markets simply could not compete with the bombast of Soviet propaganda. The human dimension in spaceflight, beginning with Yuri Gagarin's flight on 12 April 1961, was about to dramatically change the stakes, opening the classic era of the "race to space" that lasted about a decade, to the last flight of Apollo in 1972.

Success or failure in the early years depended largely on the nature of each political system. Each state managed spaceflight science and technology as a new kind of politics, often through the prisms of media publicity and public opinion. Yet the USSR depended on surprise. This meant secrecy before and during most events, excitement and spectacle after the fact. Its communist dictatorship, closed political system, and propaganda state enjoyed signal advantages: in depending upon the surprise factor, in hiding faults and exaggerating strengths, in rapidly mobilizing resources and publicizing its best achievements. All of this made for what Walter McDougall has called a space program of "bluff and promise," one of "more scaffolding than concrete." The US depended on more transparent and gradual schedules. This meant excitement before, during, and after any launch. It openly publicized most of its tests and launches, buoyed by all the media publicity of a live event, and potential failure.

In this spaceflight contest, the era of Gagarin and the Vostok made for an impressive series of surprise Soviet firsts. These included the first intercontinental ballistic missile; satellite to orbit; dog in orbit; man in space; woman in space; spacewalk; unpiloted orbit of the Moon with safe return; spacecraft landings on

DOI: 10.4324/9781003494409-8

the Moon and Venus. Each of them enjoyed the power to turn the world's media away from the US. In these terms, the Soviets won the early race to space. Yet these were mostly advantages in the short-term, valuable in fostering the perception that the US had fallen behind. The firsts were more about quantity rather than quality. The Chief Designer, Sergei Korolev, recognized this truth in one of his favorite Russian proverbs: "the better is the enemy of the good." He knew that, given its scarce resources and economic disadvantages, the USSR needed to depend on cheap, efficient, and merely "good" technologies. They were the immediate "objects" of Soviet aerospace achievement, and they had the advantage of being original, sustainable, and renewable.

This chapter focuses on a concept that I introduced in Chapter 4 and that unites Part II of the book: the **project**. In defining objects and projects, the USSR and the US were both true to their different political and economic systems. Objects were fast. They served the goal of achieving spectacular firsts, building launchers and craft on the cheap, the Soviet way. They fulfilled Korolev's "the good." Projects took time. They served America's purposes to leap ahead and achieve sophisticated, superior technological feats, what Korolev condescendingly referred to as "the better." Both objects and projects were definite in time, if with an advantage to objects, which the Soviet system demanded to be built; rather than projects, which the American system dictated be bought.

Objects put the US at a distinct disadvantage. Through the course of a fast-paced seven months, as the Vostok successes mounted, the Kennedy administration committed itself to Mercury in order to match the USSR, announced Apollo to the Moon in order to surpass the USSR, and established Gemini as a kind of internal "stop-gap" to bridge the two. These were not merely reactions to the USSR's current challenges, but its challenges to come. With Projects Apollo and Gemini, President John Kennedy pretty much decided the wider space race, just as it was beginning, although no one knew this at the time. None of this speaks to an inevitable Soviet primitivism or American suprematism. The story does not end with this chapter, or with this book. These are not questions of inferiority and superiority. They are questions of difference, eloquent difference.

Missiles and Rockets

The Soviet "R-7" and American Atlas were two ICBMs that followed remarkably similar paths of development. Both were authorized in 1954, the R-7 first test-launched on 15 May 1957, Atlas on 6 June 1957. Both tests failed, though the R-7 successfully launched later in the year, and Atlas the next. Both were multiple-staged rockets (of the stage and a half variety), with core engines that fired in tandem with strap-on boosters, which then veered off in flight after emptying. Both burned kerosene as fuel rather than alcohol. Each was at cross-purposes in the Cold War: one meant to destroy Washington DC or New York, the other to obliterate Moscow. Each also found its most memorable achievement as a space launcher for peaceful payloads: the R-7 for Sputnik and Gagarin (and the Vostok series); Atlas for the flights of the Mercury program which first flew John Glenn into orbit. In each case, missiles designed for war and mass destruction became rockets for peaceful exploration.

The US and the Race to Space

From its very first propaganda successes after Sputnik, the Khrushchev regime was in a race to beat the US, exploiting every possible opportunity to be first again. It never called this a "race to space." It did not have to, in large part because it was winning from the start. The USSR instead had opened a "road to space" (*put' v kosmose*). Soviet propagandists accentuated that the US was the only one racing. From behind.

The single greatest casualty of the Sputnik crisis may have been President Dwight Eisenhower. The first months afterward saw him increasingly belittled by the media and political reaction. Against his own personal temperament, he was thrust into the media spotlight. On 7 November, just a few days after Sputnik

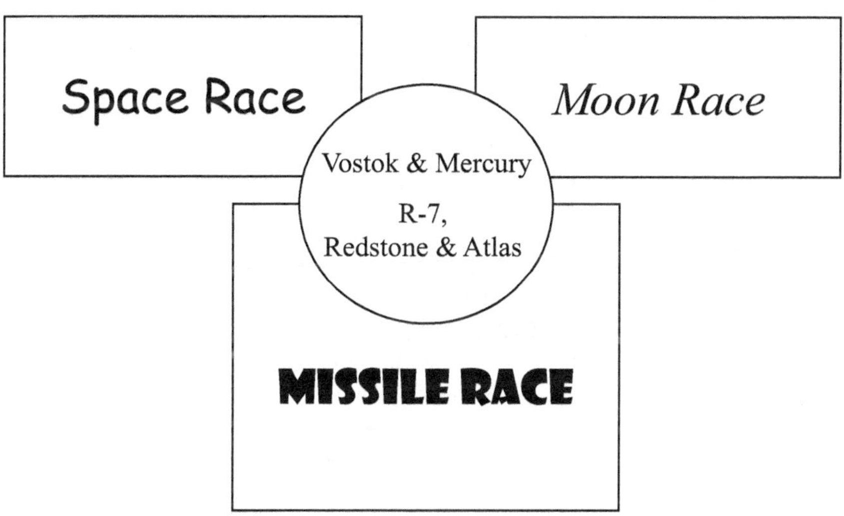

Figure 5.1 The conjunction of US–USSR spaceflight races. Author.

2, he offered a televised address to the nation, announcing several American initiatives, as well as trying to calm the media and other politicians to "avoid extremes." He suffered his third stroke several weeks later, no doubt partially a result of all the stress. Vanguard exploded on the launch pad several weeks after that. By November of 1958, through a media and political firestorm, about the USSR surging ahead and the US falling behind, the Democratic Party won an impressive number of new seats in the Senate, helping to set up John Kennedy's and Lyndon Johnson's run for the presidency the next year.

President Eisenhower may have underestimated the media and psychological effects of Sputnik, but he now answered the Soviets with a deliberate schedule of paced achievement. He created the President's Scientific Advisory Committee, headed by James Killian, president of the Massachusetts Institute of Technology, and a team of eminent American physicists and scientists. Their counsel helped to forge major legislative platforms enacted by the fall of the next year. The National Defense Education Act (NDEA) addressed what the media and politicians alarmingly called the "science gap," the US falling behind in science and technology education. It provided over 1 billion dollars over four years for government subsidies in undergraduate and graduation education in the sciences, mathematics, and foreign languages, especially Russian. Suddenly America had a lot more to learn, even from communist Russia.

Sputnik Moments

Like the events at Pearl Harbor at the start of the Second World War, the "Sputnik moment" of 1957 took on a powerful symbolic meaning: that of a surprise attack and of American vulnerability. Since then, publicists have called on the metaphor to mean a variety of weaknesses and threats in the American character. With every dip in student math or science test scores, educators and publicists sounded the Sputnik alarm. *Time* magazine (13 February 2006) warned of these low scores as a new Sputnik threat, America in decline. In his book, *Hot, Flat, and Crowded* (2008), journalist Thomas Friedman called for a new Sputnik moment to battle climate change and raise the US to greatness. Several iterations of the America Competes Act (2010–2022) have offered tens of billions of dollars to stem the tide of American decline, financing investments in solar energy, supercomputing, microchips, and hypersonics.

The promotion of education in Science, Technology, Engineering, and Mathematics (STEM) is an outgrowth of these Sputnik moments. It was an acronym invented by the National Science Foundation in 2001 to promote financing for these disciplines. The promotion of STEM meant making the US more competitive in the world, especially against China's rising power. The irony is that by 2017 the number of Chinese students studying in US universities peaked at just above 360,000, mostly in STEM majors.

The Russians have raised their own peculiar Sputnik metaphors. Sputnik became the brand name for cigarettes, cars, bicycles, transistor radios, and hunting rifles, to name a few. In 2014, the Putin regime launched a new state news service named Sputnik (spelled with Latin letters and broadcasting in English), though it delivers more propaganda and disinformation than news. Putin's regime also named the Russian COVID-19 pill Sputnik V, although the Russian public did not trust it, and US government agencies criticized its effectiveness. The recent Russian film, *Sputnik* (2020) turned the tables on Sputnik as a metaphor. It was actually a harsh portrayal of the Soviet-era space program, and achieved some box-office success, though it was also mostly a copycat version of Ridley Scott's *Alien* (1979). With all these twists and turns, Sputnik may be losing some of its symbolic punch. We need new metaphors.

Eisenhower's most enduring institutional legacy was the civilian National Aeronautics and Space Administration (NASA), already at work by 1 October 1958, established to coordinate US advances in flight technologies and space exploration. Its first director, T. Keith Glennan claimed not to be a "space cadet," but like Eisenhower a professional and administrator. The new body joined disparate institutions and offices from around the country, including the National Advisory Committee on Aeronautics (NACA) facilities; the Jet Propulsion Laboratory at Pasadena, California; several other military offices and installations; and eventually the Army Ballistic Missile Agency (ABMA) and the von Braun team at Huntsville, Alabama.

NASA became the country's most visible and vanguard R&D institution. In the established American project tradition of government-academic-business cooperation, NASA literally grew into a relatively free market of aerospace facilities and industries. It awarded competitive projects and contracts for its myriad spaceflight programs. Scientists and engineers at work making the country's aircraft and missiles, now turned to the challenges of spaceflight mission planning and astronaut preparations, spacecraft and launcher designs, launch and recovery operations, tracking stations, and control and communications centers.

Meanwhile, the US Army and Air Force attempted to seize the initiative and launch astronauts of their own. The Eisenhower administration brooked little tolerance for these grandiose proposals, though it finally approved something like the Army's "Man Very High," what eventually became NASA's Project Mercury, though on a slower schedule. As a rule, Eisenhower and his advisers were opposed to trying to "compete with the Russians on a shot-for-shot basis in attempts to achieve space spectaculars," as Glennan put it. They were against exploiting "technology as an instrument of propaganda and power politics." NASA had even drafted an Apollo Moon program, but the president and his circle tabled it in the fall of 1960, at the very moment of the US presidential election, pitting Eisenhower's vice-president, Republican Richard Nixon, against Democratic senator, John Kennedy.

Spaceflight issues were threaded through the election, as the Sputnik crisis and fears of space and missile gaps offered advantages for the Democrats, who exploited the Soviet firsts and Eisenhower's presumed failures at many opportunities. The Kennedy campaign turned missiles and rockets into political symbols, given added weight from running mate, Lyndon Johnson, who had served on the Senate Armed Services Preparedness Subcommittee and Special Committee on Science and Aeronautics. Kennedy's "New Frontier" slogan promised progressive political and economic reforms, centered on space technologies as a new measure of American might and prestige in the world.

These Sputnik-era challenges, and cinematic space metaphors, may very well have been decisive in the presidential election of 1960. It was no Kennedy landslide, instead one of the closest presidential elections in US history in terms of popular and electoral votes. Vice-President Richard Nixon held his own in the race, true to the Eisenhower standard of sober policy and stalwart confidence in America's place in the world. The loss was particularly annoying to the outgoing president, who warned the electorate about its flawed choice, in his "Farewell Radio and Television Address to the American People" (17 January 1961). Eisenhower cautioned against any space "spectacular and costly action" as a "miraculous solution to all current difficulties." He complained about the "conjunction of an immense military establishment and a large arms industry," filled with contract money and bureaucracy, calling it a "military-industrial complex." He criticized the "domination of the nation's scholars by Federal employment, project allocations," and even by their own "scientific-technological elite."

The media and academia, especially with the Vietnam War, soon applied the phrase "military-industrial complex" to their analyses and polemics. Eisenhower's biographers speculate that he applied it from his own experiences and promotion of the military-defense industry, especially the phenomenal growth in guided missiles. In this era of pop psychology, "complex" also had a clear psychoanalytical ring to it. Eisenhower was warning us about a failure of American values and attitudes. He may also have borrowed the phrase from his intelligence briefings about Soviet military power. From 1957, the USSR's "Military-Industrial Commission" (*Voennaia-Promyshlennaia Kommissiia*) was at the center of the Soviet system for weapons and space development. This was clearly not a model to emulate.

Thus, at the very end of his presidency, Eisenhower returned to its beginning, and the New Look platform of fiscal responsibility, limited government, and traditional American individualism. For all his very own grand initiatives in ballistic missiles, in defining massive nuclear retaliation, in U-2 spy planes and the NASA program, he still cautioned against the excesses of the national security state. By his rebuke of the "spectacular" and "miraculous," he also warned against what he should have called America's "media-commercial complex," its urge to match Soviet firsts with bigger, better firsts of its own. This was all beneath us, he warned, and undignified. It was a trap set by Sputnik. And it might be sprung again.

The Vostok Space Firsts

The USSR began preparing for a piloted Sputnik mission even before the first Sputnik. In the spring of 1957, Chief Designer Korolev and one of his lead engineers, Konstantin Feoktistov, planned what they called "Object K," an inhabited "Sputnik Ship" (*Korabl' Sputnik*). Its tasks, following the USSR's Cold War priorities, were to prove the possibility of human spaceflight and the potential for military reconnaissance.

The Vostok spacecraft, like its Mercury counterpart, sat atop a ballistic missile, in a compartment otherwise reserved for a nuclear warhead. Both crafts had limited functions. Neither could perform any change in orbit. The main engines in both were only ever fired to position them for re-entry and descent to Earth. Vostok weighed much more, at 5 tons, including the craft and instrument module, thanks to the carrying capacity of the R-7 launcher. This dramatically simplified Korolev's tasks, as his teams were able to fill them with both automated and failsafe redundant systems: 56 electric motors, 241 vacuum tubes, 800 relays and switches, 6,000 transistors, interconnected by 15 kilometers of cable. All of this protected the cosmonauts, enabling them to remain in orbit for up to a week's time.

In orbit, the re-entry "ball" (*sharik*), in which the cosmonaut sat, was attached to an instrument module that provided life support and power, as well the retrorockets that fired it into position for descent back to Earth. In a crucial decision that allowed for a quicker schedule, the ball was weighted on one side in order to position it for an automatically correct re-entry position. It was therefore "dynamically stable," without any need for a more challenging or complex aerodynamic design. The weight, with a special heat-resistant coating, provided protection.

The spacecraft had the simplest of avionics, only capable of roll, with jet thrusters that helped to keep it horizontal. By design, the cosmonaut was a passenger more than a pilot. The craft, oriented to the sun to maintain its proper bearings, was designed to fly automatically, with internal controls and with radio guidance, without any cosmonaut input. As Slava Gerovitch has surveyed, the Soviet spaceflight program trusted the machinery and ground control more than the man. The individual did not really count. Initially, there was also an inordinate fear of the physiological and psychological effects of spaceflight, in case the cosmonaut lost consciousness. In an emergency, the cosmonaut could only engage manual controls, for a visual retrorocket firing, using a special logic code.

For his cosmonaut corps, Korolev initially recruited about 3,000 candidates, out of whom 200 were screened, with 20 in the final group. They were all young fighter pilots in superb physical and mental health, between 25 to 35 years old. They had to be Russian by ethnicity with working-class backgrounds. They did not need any advanced college or engineering training, though mathematical skills were essential. What counted most was their reaction time and perseverance, their attitude and personality. On average, they were ten years younger than their American rivals. Given the physical constraints of the Vostok craft, they were 3 inches shorter and 20 pounds lighter. They were also much less experienced as pilots. Gagarin had only flown about 230 hours, compared to the very minimum of 1,500 hours required of the Mercury astronauts.

The USSR's Institute of Medical and Biological Problems, which doubled in size in these years (from 40 to over 100 staff members), set the standards to choose and train the first cosmonaut cadres. Its staff subjected them to extremes of heat and cold, to inner-ear tolerance tests, to isolation chambers. They were exposed to motion exercises on a "looping wheel"; to high acceleration forces on a rocket pole; to varying degrees of atmospheric pressure and oxygen; and to the "red and black chart" (a system of counting numbers amid confusing signals and irritations) in order to test mental acuity and stress tolerance. The cosmonauts also experienced brief moments of weightlessness, in an elevator at the high-rise building of the Moscow State University. From the 28th floor, it fell freely before coming to a cushioned stop, allowing the subjects to briefly float in the air.

In order to prove that human beings could tolerate the acceleration forces and weightlessness, Soviet aerospace physicians also tested a variety of animals. These experiments reached back to 1951, with mice, rats, rabbits, and guinea pigs; eventually with dogs launched to the edge of space and into sub-orbit. The Soviets placed Albina and Tsyganka in space suits, ejected them from a capsule at 52 miles altitude, and then recovered them safely. A number of failed dog flights followed into 1960, with final success as Belka and Strelka completed 17 orbits in August of 1960, the first living creatures to return to Earth from orbit. Between December and March of 1961, Vostok launched a final series of canine passengers, all of whom returned alive, setting up yet another coming Soviet first, a human being.

On the morning of 12 April 1961, at the Baikonur Cosmodrome in central Kazakhstan, Yuri Gagarin finished his meal of boiled eggs, bread and butter, mashed potatoes, and tea. This was the model of cosmonaut breakfasts to come. He then set off for his Vostok craft atop the R-7 rocket, into a seat made out of a cast of his own body, turning him into one more feature of the craft, in order to be efficiently ejected from it. He orbited the Earth for 76 minutes, taking off from Baikonur and landing near Saratov, the first human being to fly at

17,500 miles an hour. He did not control the craft at all, but flew on automatic pilot, as prescribed by the mission specifications, though Korolev gave him the logic code to open manual controls if necessary (it was not).

Gagarin's account of the flight was perfectly accurate, as scripted. His express role was to be a passenger and observer. He gave a precise narrative of the varying g-forces at launch, and of the separation of the strap-on boosters. Most of all, he described the blackness of space, the wondrous sights of a smallish Earth, and the thin blue line of atmospheric life that separated them. At some 50 minutes into the flight, heading home, he wrote, "I am in a good mood. I am continuing the flight. I am over America." He even scribbled some of these impressions down at the time, though he lost the pencil when it floated away into the cabin. As the retro rockets fired for his descent, he calmly described a near-death experience, what he generously called a swirling pirouette, when the instrument module failed to completely separate from the returning sphere, causing it to turn over and over upon itself, until several loosely connecting cables burned it off from the heat of descent.

From that moment, Gagarin's life became the stuff of legend, a personality cult. It was borne from his own achievement and quiet charisma, and from the genuine admiration people felt for him. It was also the result of the power of the propaganda state to magnify it all beyond normal proportions. Moscow radio called all stations to announce the flight, once it was ready for re-entry. This was a complete surprise to audiences within the USSR and around the world, introducing the "pilot-cosmonaut," Yuri Alekseevich Gagarin, a citizen of the Soviet Union, the "first human being in space" and his "Sputnik–Vostok" capsule. It followed with his exclamation upon launching, "we're outta here" (*poekhali*), a common enough and endearing phrase.

The newspaper headlines of the era, like the history books that came after, reveal just how handily the Soviets won the race to space in 1961. Yuri Gagarin's flight into Earth's orbit was the signal "global" event of the year, certainly for the USSR. Gagarin's flight seemed to propel it into the lead for scientific-technological superiority. He was the star of Soviet propaganda for decades to come. He was the star of the Warsaw Pact and the world alliance of communism, a significant factor in the Soviet drive for global allies near and afar. He was even the star of the American media, making the covers of its newspapers and magazines. Here was the double bind of the space race: that Americans not only had to watch Soviet spectacular successes from afar, but also suffer the celebration of their rivals right at home. Soviet firsts sold well in America.

True to its name, Vostok "the East" (as Sputnik had before) leveraged the power of rockets to move the rimlands, the Western European ones, to be exact. Public opinion polls from 1957 to 1961 revealed that significant numbers of French, Italians, and British really believed that the USSR had leaped ahead of the US in science and technology. The West and capitalism were fading. The heartland East and communism beckoned.

Soviet propaganda celebrated Gagarin's triumph in a strange marriage of past and future. Television and radio played Evgenii Dolmatovskii's song, "The Motherland Hears / *Rodina Slyshit*" (1950), a popular tune about the Second World War, which was now sung for him: "The Motherland hears, the Motherland knows, / Where her son flies into the clouds / Touched by her friendly caress and tender love." Yet media propaganda also featured Gagarin and his comrade cosmonauts with the musical sounds of the future, in the eerie Theremin, a Soviet invention. It was an electronic instrument, played by the musician's hands, which manipulated electric waves and touched nothing but the air. These were the haunting sounds of Soviet space victories, replayed on television shows and spaceflight documentaries. They may have reminded some listeners of less vaunted Soviet achievements, how the USSR launched Gagarin into space midway between two other military invasions: first into Hungary (1956) and second into Czechoslovakia (1968), depriving both nations of their independence, with several thousand dead.

A Tale of Two Movies

Both the US and the USSR remembered the early years of spaceflight as lived experience. From the US, *October Sky* (1999), based on a memoir, *Rocket Boys*, detailed the life and recollections of a rocket engineer, Homer Hickam. In this coming-of-age tale, Homer watches the menacing Soviet satellite, Sputnik, streak and beep through the October sky, alarming America that it has fallen behind. From his home in a bleak West Virginia coalmine town, Homer finds refuge in patriotism and in teachers and friends. They set him off on his journey of discovery, to model-rocket launches, science fairs, college, and ultimately NASA. The movie ends with a newsreel retrospective: a splice of documentary footage about America's (and Homer's) achievements with the Space Shuttle.

Russia had its own rite of passage story for the Space Age: Aleksei Uchitel's, *Dreaming of Space* (*Kosmos kak predchuvstvie*, 2005), a not-so-flattering imitation of *October Sky*. The "dreaming" refers to the broken promise of spaceflight just after Sputnik, and just before Yuri Gagarin's flight. The main character, Victor (nicknamed "Horsey"), lives in an Arctic Circle seaport hometown, a claustrophobic world of ice and fog, poverty and despair, secret-police and labor camps. He and his friends listen not to Sputnik's beeps but to radio transmissions and enticing music from the richer West, dreaming of escape. The movie ends ironically, with documentary footage of Gagarin's return to Earth, and his poses with the communist leadership at Red Square, its new political star.

Over the next two years, Korolev designed each of the remaining five Vostok missions for a maximum technical and propaganda effect. In Vostok 2 (6 August 1961), German Titov (Gagarin's backup) managed an impressive 17 orbits of the Earth in 25 hours and 18 minutes, even taking limited manual control of the craft for a time. Titov covered two times the distance to the Moon, a not-so-subtle challenge to the Americans. As the first human being to experience the full effect of weightlessness, he remembered some nausea and dizziness during his long flight, including some inner-ear pain. He also slept eight hours, surprised at not feeling any desire to turn on his side. By this time, no American had yet even flown in orbit.

The remaining Vostok flights were answers to the American announcements of its Projects Gemini, for two astronauts to rendezvous and dock in Earth orbit; and Project Apollo, for long-duration flights to the Moon and back. Korolev answered these with longer and longer stays in orbit, and with a paired sequence of pairs. Vostok 3 and 4 flew between 11 and 15 August of 1962, with Adrian Nikolayev reaching 64 orbits to Pavel Popovich's 48. They also parachuted back to Earth only six minutes and 193 kilometers apart. Vostok 5 and 6 flew between 14 and 19 June of 1963, Valerii Bykovskii reaching 81 orbits to Valentina Tereshkova's 48. Both pairs of craft flew only about 5 kilometers apart. In each case, the rendezvous was more of an illusion than real. Yes, they were technical feats, but more in the timing of the launches and in the mapping of the trajectories to achieve proximity. Like the other firsts, they made their mark in the world media.

Vostok 6 had the added attraction of Tereshkova, the first woman in space. An accomplished parachutist, she was expressly trained with a small cadre of five other women for this one purpose. None of them were a part of the official cosmonaut training corps. Valentina Ponomareva, a pilot, was originally slated to fly Vostok 5 to meet with Tereshkova in space, though this feat was over-ruled by party leaders, in part due to Ponomareva's independence and feminist views, even though she was an engineer and scored far better on tests than Tereshkova. Thus, the moniker of "first woman" was a propaganda tool more than anything else. In Soviet practice and its media, Tereshkova fulfilled her role as a traditional woman, subservient to her male cosmonaut brethren, the perfect domestic. Besides the flight, she completed one last ritual pairing when she joined cosmonaut Nikolaev in marriage on 3 November 1963. Nikita Khrushchev and party leaders, along with the directors of the space program, were in attendance, the rest of the country watching this wedding of heroes on television.

For all the scientific and technological achievements of the Vostok program, its propaganda functions were paramount. To hide any imperfections and inconsistencies, secrecy and deception threaded the official narrative. For example, by the terms of his flight plan, and all other Vostok flights to come, Gagarin was ejected from the Vostok craft at 23,000 feet, whereupon both he and the craft landed separately back to ground by parachute. Yet, according to the rules of the

International Aeronautical Federation, pilots needed to land inside their vehicles to gain a historical record. Since Gagarin ejected and landed by parachute, the Soviets lied about this over the next decade in order to clinch the feat. The Soviets also lied about various cosmonaut accidents, and even airbrushed their images out of media photographs, as James Oberg's work has studied.

For many years, the Soviets suppressed these kinds of details in the media, all to magnify their own successes and America's disappointments. The Soviet program did not announce a space launch until it had already achieved some measure of success. Yet this cloak of secrecy led to a whole series of rumors and conspiracy theories about cosmonaut deaths. The Judica-Cordiglia brothers (Achille and Gian Battista) even claimed to have captured the sounds of a dying cosmonaut, with "rapid heartbeat and heavy breathing" (February of 1961); or the sounds of a female cosmonaut perishing in a re-entry accident (November 1963), though researchers later discredited their claims.

Mercury to Sub-Orbit and Orbit

Less than a week into his tenure as director, NASA's T. Keith Glennan was already proposing a "Piloted Satellite Program" for human spaceflight. It was the foundational program for the new agency, what NASA renamed "Project Mercury" a month later, in honor of the god of commerce and travelers, the speedy messenger with winged sandals and helmet. Much like the Soviets, the Americans were thinking in Sputnik terms, to fit a human being where a satellite (or nuclear warhead) would otherwise be, and then send that person into outer space. The trouble was that Project Mercury was not as speedy as its symbolic mascot.

In comparison to the fast pace of the Vostok program, NASA suffered from a major disadvantage of slowness. Thanks to Eisenhower's imprimatur, its mission was not so much about getting to space first. Rather, the sober goal was to place a man in outer space and return him safely to Earth. For this purpose, and at a rather frugal project cost of $400 million, Chrysler built the Redstone, General Dynamics built the Atlas, and McDonnell Aircraft built the spacecraft (the contract was awarded in January 1959). There was little sense of a race, at least at first. "Go Fever" only seized the program later, after Gagarin's flight, when Americans became enthralled with the need to catch up to Soviet firsts and attain some breakthroughs of their own.

Nor did the Mercury program enjoy the reliability of Vostok's R-7 rocket. The Atlas was not yet ready to serve as a space launcher. Several Atlas failures highlighted the problem: the first, an explosion at launch on 19 May 1959, witnessed by the Mercury astronauts and an official government delegation; the second, a 25 April 1961 Atlas failure just weeks before Alan Shepard flew to space. NASA decided to limit the first flights to sub-orbit, using the Army's more reliable (if weaker) missile, the Redstone. Because neither the Redstone nor the Atlas could match the carrying capacity of the R-7, Mercury had to be much lighter than Vostok, at only one and a half tons. Hence, Mercury had to tolerate fewer redundancies, and was only able to remain in space for roughly 24 hours, supplied with a pure oxygen environment. It used no ejection seat but employed an innovative escape tower in case of a rocket malfunction. It soft-landed by parachute on water with the astronaut aboard, in both the Atlantic and Pacific oceans.

These weight restrictions, along with several other factors, convinced engineers to design Mercury less as a sphere, more as a conical shape. This meant that it was not dynamically stable like Vostok. Rather, it was more like an aircraft, needing a much more sophisticated avionics system to remain stable in orbit, and hold attitude and orientation. It also needed a more innovative, "ablative" heat shield to protect the craft from the heat of re-entry into the Earth's atmosphere. The astronaut was more a pilot than a passenger, at the helm with 120 controls, 55 electrical switches, 30 fuses and 35 mechanical levers, some of which operated three control jets for pitch, yaw, and roll. It had no window or escape hatch at first, but it did have several small portholes.

Korolev's Vostok program was cloaked in secrecy. Gagarin and his comrades were unknown to the public until his flight. America advertised. NASA presented its Mercury Seven to the world at a captivating press conference in Washington, DC, on 9 April 1959. The astronauts were overwhelmed by the attention, and a standing ovation from the reporters. They were, on average, between 32 and 37 years old, 5 feet and 11 inches tall (or less), and between 150 and 180 pounds. Like U-2 pilots, they were required to have bachelor's degrees in science or engineering, and to have served as test pilots for the military, with at least 1,500 hours of flight time. They had been chosen, so America learned, after a grueling week of tests at the Dr. W. Randolph Lovelace Foundation in Albuquerque, New Mexico, and then at Wright-Patterson Air Force Base in Dayton, Ohio. From 110 men, they were tested and prodded down to 69, then 32, then 7, enduring cardiograms and treadmill tests, electro-encephalograms and blood sugar tests, physical and mental stress tests, probes and samples of all kinds.

NASA and the media wrapped the Mercury Seven in all the glamour of America's new "space voyagers," men "daring and courageous." *Time* magazine wrote, "From a nation of 175 million they stepped forward last week, seven men cut from the same stone as Columbus, Magellan, Daniel Boone, Orville and Wilbur Wright." *Life* magazine offered a bonanza of $24,000 a year to each of them, over the course of three years (they otherwise made between $8,000 and $12,000 a year from their military salaries.) Life at Cape Canaveral and Cocoa Beach, Florida, had its own glamour and temptations. Most famous: the Starlight Motel with the space-themed Starlight Lounge, bedazzled with black lights, moonscapes, and flashing rockets. Here the astronauts were given free use of sports cars by sketchy car dealers, free rooms and meals from motels and restaurants, and all kinds of personal attention from the "groupies," who followed them around town at all hours of the day and night.

A Gender Gap

The US Air Force planned a gender match-up, with the "Mercury Thirteen" women, as an answer to the Mercury Seven men. These were female pilots and parachutists known as the "girl astronauts." Between 1960 and 1961, aerospace physicians chose and tested them at medical and flight facilities at the Lovelace Foundation, and at the School of Aviation Medicine in Pensacola, Florida. Geraldyn "Jerrie" Cobb, one of the most accomplished of the group, successfully completed several phases of the tests. She had been a flight instructor in the Second World War and flew military jets afterward, though always with a male pilot escort. Now she and her fellow pilots became the focus of widespread media attention and speculation as the coming "astronettes." The Mercury Thirteen never flew, rejected by Air Force leaders, several of the male astronauts, and Vice-President Lyndon Johnson. "Let's stop this now," he wrote in 1962. Johnson's "now" lasted for the next 20 years, until Sally Ride became America's first woman in space, aboard one of the first Space Shuttles. Besides a number of talented engineers (women like Yvonne Brill, Dorothy Simon, and Virginia E. Morrell), most women involved in the early space program were secretaries and administrators. Others were called "computers," since it was their job to compute complex mathematical equations, as, for example, a space capsule's re-entry point. Their story was told in the feature film, *Hidden Figures* (2016), about a group of African-American women who made important but otherwise silent contributions.

To prepare for spaceflight, the Mercury Seven endured another round of ordeals. Aerospace physicians starved them of oxygen to test their endurance; poured ice-cold water into their ears to test balance; sat them in saunas of up to 135 degrees for two hours at a time, to test stamina. They subjected their bodies to a jackhammer seat and centrifuge, and spun them around in circular metal cages, all to test their agility and reaction time. In isolation chambers, the astronauts faced the "busy box," filled with dials and switches, all of which challenged their mental acumen and patience with confusing and complex tasks. For analog physiological tests, NASA launched pigs, spider monkeys, and chimpanzees into space, "working our way up the evolutionary ladder," as Gene Kranz put it. Between 1959 and 1961, NASA and McDonnell Aircraft performed 16 equipment tests on the rockets, crafts and escape systems; and four animal tests (the chimps Sam, Ham, and Enos), all to prepare safety for the two sub-orbital missions of Alan Shepard and Gus Grissom.

Shepard finally flew on 5 May 1961, nearly a month after Gagarin. After a meal of orange juice and coffee, scrambled eggs with filet mignon wrapped in bacon, toast, and jelly on the side (a model of astronaut breakfasts to come), Shepard squeezed into the Mercury capsule, atop an Army Redstone missile, and then waited during four hours of delays. Yes, toward the end, after the official delay, for fear he might "short-circuit," he did relieve himself in his space suit, one last affront before his famed flight. Shepard spent only 15 minutes in a sub-orbital flight, rather modest compared to Gagarin's orbit. In some ways it was much more eventful and meaningful. At an apogee of 116 miles, he took over the controls and practiced roll, pitch, and yaw maneuvers, basically piloting the craft, and then spent about five minutes in weightlessness. He returned at a very steep, sub-orbital ballistic trajectory, splashing into the Atlantic some 300 miles down range from Cape Canaveral. It was a flight somewhat reminiscent of the shell of the Paris Gun that we met in Chapter 2.

Americans celebrated his return as a voyage into space. Millions watched television or heard the events live on radio. He made the headlines and magazine covers. Upon his successful touchdown, Shepard became a national hero, welcomed at the White House by President Kennedy, and honored with parades in Los Angeles, New York, and Washington, DC. *Time* magazine featured him on its cover, clad in his space suit, as if floating in space. A relaxed Shepard, his lap filled with fan mail, made the cover of *Life* magazine. The next month, over a half-million people came to see his Mercury craft at the Paris International Air Show and the Rome Science Fair, each time. The irony was that Shepard could have flown on 24 March 1961, weeks before Gagarin, to be the first human into space. Yet both von Braun and the NASA leadership delayed out of a sense of caution, turning it into one more test flight instead, which did result in continuing improvements to some potentially troublesome vibrations in the flight control system.

Gus Grissom's "Liberty Bell 7" (21 July 1961) experienced the first Mercury accident. This was only the second flight but the first to fly an adapted craft, with a control stick, window, and escape hatch, all installed at the astronaut's own demands. Grissom reached an altitude of 118 miles altitude and a range of 303 miles. Shortly after splashdown, the hatch blew open and a military helicopter saved Grissom from his sinking ship. Some news outlets and critics accused him of blowing the hatch open out of fear or clumsiness, neither of which was in his character. The recovery of the craft in 1999 did not completely resolve the issue, but NASA honored him best as one of only two astronauts (with Wally Schirra) to join the Mercury, Gemini, and Apollo programs. Grissom also gave one of world history's shortest and mightiest speeches, to 18,000 workers, at the Convair Division plant, part of the General Dynamics Corporation, in San Diego, California, in 1959. They were building the Atlas ICBM to take the Mercury astronauts to orbit. After a few nervous moments at the microphone, unsure of what to say, he proclaimed, to great applause: "Do good work."

John Glenn's "Friendship 7" (20 February 1962) was, after Shepard's first flight, the most celebrated of the Mercury missions. Atop America's first inhabited version of an Atlas rocket, he spent four hours and 43 minutes in orbit. The mission was flawless except for a loose heat shield on re-entry that could have imperiled the mission and him. Glenn was given an even greater hero's welcome, which included a ticker-tape parade down Madison Avenue in New York

City. President Kennedy awarded him with the NASA Distinguished Service Medal. *Time* and *Life* featured Glenn on their covers.

The last three Mercury flights achieved longer and longer durations in space, although only one of them broke a Soviet record. Each flight was remarkable on its own terms. Scott Carpenter (24 May 1962) flew three orbits of Earth over nearly five hours. Distracted by the sights, he did not follow the proper procedures, firing the retrorockets late, and landing over 100 miles beyond his pickup zone. Wally Schirra (3 October 1963) flew six orbits in nine hours and 13 minutes, in one of Mercury's most perfect flights. Gordon Cooper (15–16 May 1963) was the first American to fly in space for more than a day, reaching 22 orbits in 34 hours and 20 minutes. He very nearly did not make it back, experiencing a massive breakdown in many of the spacecraft systems, forcing him to fire the retrorockets manually for a quick descent.

One last Mercury mission was planned as a world record six-day flight, but it was sidelined by Cooper's troubles. NASA was too fearful of the scenario of an astronaut stranded and slowly dying in Earth orbit. That would have made bad press. But it also made a winning plot, in Martin Caidin's novel, *Marooned* (1964), just such a story of a failed retrorocket firing and a marooned Mercury spacecraft, the astronaut patiently waiting for death by suffocation. In this fictional tale, both the American and Soviet space programs improvised the necessary rendezvous, docking and spacewalk techniques to save him. It took a pulp paperback novel to get us to work together in space.

The most memorable account of the Mercury Seven was Tom Wolfe's playful, *The Right Stuff* (1979). It chronicled their lives within the high-speed culture of aviation test pilots in the 1950s. At the heart of the story was the quest for personal best: who had the right stuff to be America's first man into space? Wolfe painted it as a maniacal drive for superiority. He also suggested that the Mercury Seven shared a common glamour with President Kennedy. Both were at the top of their "ziggurat" pyramids, one of aviation, the other of politics.

Many Americans took these initiatives and challenges in stride. They had plenty of distractions, like keeping good jobs, raising families, making fine middle-class homes, and enjoying themselves. During the busiest years of the space race between 1960 and 1963, for example, there were no fewer than eight "Twists" on the Billboard Hot 100, including Chubby Checker's the "Twist," Sam Cooke's "Twistin' the Night Away," and the Isley Brothers's "Twist and Shout." The Soviets had these Twists as well, thanks to Voice of America radio, which the US surreptitiously beamed into the country. The Soviets may have been first in space, but they definitely remained second when it came to the stuff of consumer culture. The average Soviet yearned for the best of the American market. In response, the USSR provided its public with knockoff versions of many American favorites: Twists, bikinis, jeans, ball gowns, and even a copy of the Beatles, known as "The Jolly Fellows" (*Veselye Rebiata*) group.

The Soviets had their own traditions in popular culture. One of the most popular phrases of the Sputnik and Gagarin eras was "physics and lyrics" (*fizika i lirika*).

Young people liked to quote it to mean a genuine sense of national accomplishment, one of scientific achievement and relative cultural freedom. The phrase also expressed their new sense of community: the "physics" of their hard study and work; and the "lyrics" of rest and relaxation, as when they gathered to read poetry, sing folk songs, and play guitar. Andrei Voznesensky's poem, "Parabolic Ballad" (1959) captured this enthusiasm, comparing himself and his fellow youth to rockets, speeding into the heights with willful optimism. "Fate, like a rocket, flies in a parabola," he opined. For all this, American culture still remained very tempting. Many young people were not so fascinated with the R-7, but with another seven, *The Magnificent Seven* (1960), the popular Hollywood film starring Yul Brynner and Steve McQueen, screened in Soviet movie houses to great acclaim.

The Origins of the Apollo Program

Set within the full breadth of historical comparison, the timeline of the Vostok and Mercury programs highlights Soviet firsts and American seconds, winners and losers in the race to space. But the depth of historical coverage matters too. Time, we know, does not always move at the same pace or intensity, depending on the choices people make, and the events that happen, and the human emotions and investments in each. The Mercury years were not simply a sequence of responses to Soviet challenges. They were multivalent. They were a dynamic moment of nearly simultaneous decision-making as the Kennedy administration engineered not one but three spaceflight programs: confirming Mercury, reviving Apollo, and improvising Gemini. As lived experience, it was a dizzying array of maneuvers and proposals, hopes and promises. But these decisions also set the course for a decade of remarkable American achievements in space.

We tend to romanticize all of this, following President Kennedy's own stirring oratory about the Moon. Yet his motivations were not always so innocent. When the president first formulated his Apollo decision, he was not only embedded in the competition of the space race, but even more entangled in a gritty and deep Cold War geopolitics over Cuba. Kennedy first raised the issue of committing the US to some space spectacular only a few days after Gagarin's flight, at the moment of Khrushchev's victory parade in Red Square (14 April). Meeting with his newly appointed director of NASA, James Webb, and his advisers, he asked: "Is there any way we can catch up? What can we do?" It was an interesting use of the phrase, "catch up," given what we already know about the famous Leninist and Stalinist dictum of "catch and surpass."

An even worse blow came just a few days later with the failure of the Bay of Pigs invasion on 19 April. This was a covert US military operation, inherited from the Eisenhower administration, directed by the Central Intelligence Agency and elements of the military, to lead about a thousand Cuban paramilitary forces to overthrow Fidel Castro's new pro-Soviet regime. Part of the reason for the failure, as Irving Janis has argued, was overconfidence, or what he called "groupthink," the tendency for the president's elite advisors to please him and impress each other with unanimity and resolve.

The failed invasion was a monumental political and personal defeat for Kennedy. His aide, Kenneth O'Donnell, saw him "as close to crying" on the morning of 19 April, just as he took a long and lonely walk in the Rose Garden at 4 am. Another advisor, Clark Clifford, heard him say, "I have had two full days of hell – I haven't slept – this has been the most excruciating period of my life. I doubt my Presidency could survive another catastrophe like this." Jerome Wiesner, his science adviser, realized that "the president felt some pressure to get something else in the foreground."

Kennedy found himself, only three months into his presidency, with signal losses both in space and on Earth. The famous "New Frontier" was failing. He needed to revive his electoral campaign from the year before; that is, already begin the campaign for his second presidential term. The turning point came with the president's 20 April memorandum to his vice-president, Lyndon Johnson. It was a telling indication of his frustrations and fears, a rattle of questions about "beating the Soviets." Kennedy had lost his bearings. But this also shows the president finding his way, challenging his administration and the American public with a new sense of national purpose.

These events position Kennedy squarely as a man of his times who also tried to redefine them. Shepard's sub-orbital flight on 5 May offered a short respite for the administration, an event to celebrate, if also once again spotlighting how the US was still behind. The next weekend, President Kennedy met with his inner circle (Webb, Johnson, and Secretary of Defense Robert McNamara) to finalize and approve the Apollo initiative. Their advice was consistent: that "dramatic achievements in space … symbolize the technological power and organizing capacity of a nation"; and that "lunar and planetary exploration are … part of the battle along the fluid front of the Cold War." Their counsel was to transform American technology into a weapon of psychological combat.

All of this was a preface to one of President Kennedy's stellar moments of political theater, his speech to a joint session of Congress on 21 May, dedicated to "urgent national needs." This was a rather strange appeal. There was no imminent threat of war. Most of the speech was a rather restrained accounting of the Cold War rivalry. The president noted that the Soviets

> possess a powerful intercontinental striking force, large forces for conventional war, a well-trained underground in nearly every country, the power to conscript talent and manpower for any purpose, the capacity for quick decisions, a closed society without dissent or free information, and long experience in the techniques of violence and subversion.

He offered the basics of his new "flexible response" doctrine, transforming the US military into a more dynamic and adaptable force to answer Soviet territorial and threats around the world, as for example in Cuba and Vietnam. After all, the Cold War was "a battle for minds and souls as well as lives and territory."

Only at the end did he announce that in the contest "between freedom and tyranny," between the US and the USSR, "dramatic achievements in space" mattered. The world needed to know that the US would "one day be first." Therefore,

> I believe that this nation should commit itself to achieving the goal, before this decade is out, of landing a man on the Moon and returning him safely to the Earth. No single space project in this period will be more impressive to mankind, or more important for the long-range exploration of space; and none will be so difficult or expensive to accomplish.

The speech was a model of exaggerated threat and promise. Most Americans disapproved of this new, expensive, and seemingly impossible endeavor. A Gallup Poll in May of 1961 found 58% opposed to the Moon proposal.

Why did President Kennedy choose the Moon? Because it revived his "New Frontier," set a visible goal, and was also far off enough into the future to give the US a good chance to succeed against the communist foe. On the global market, he had turned the profitable Sputnik and Gagarin brands into one great new American brand: Apollo. Like his beloved Mercury astronauts, Kennedy was also "pushing an envelope," though his was a political and temporal one. In short, the Moon program gave him and America a new competitive edge. It also lent even more glamour to the Kennedy brand, this in an era of masculine prowess, of Hugh Hefner, Frank Sinatra, and James Bond.

The president also chose the Moon because it was achievable. The "Moonshot," as the US media rechristened Project Apollo, was not self-starting or self-made. Kennedy had already inherited a critical mass of technologies, networks, and processes from the Eisenhower administration. These included the initial NASA Project Apollo designs, the planning Space Task Group, the F-1 engines for the Saturn launcher, the plans for its J-2 engines, along with the propulsion systems for the other Saturn platforms. Kennedy had also inherited a vast R&D network of experienced and tested military, industrial, and academic teams, along with their techniques of project and program management and systems engineering. They had proven themselves with ballistic missiles, space launchers, and satellites. Now they were to be geared for the Moon. The Bay of Pigs invasion, another inheritance from Eisenhower, had not worked out well for Kennedy. Now Eisenhower's NASA and its Project Apollo would save him.

What followed over the next few months and years were a series of decisions and plans that compressed Mercury and Gemini both into one integral Apollo "moment." By October of 1961, before John Glenn's orbital flight, NASA had already awarded the project contract for the lunar spacecraft to North American Aviation. By December, almost as an afterthought, it announced Project Gemini to test the essential Apollo techniques of space rendezvous, docking and long-duration flights. By July of 1962, NASA had resolved to get to the Moon using the innovative and timely approach of lunar-orbit rendezvous. In September, it introduced the Gemini "Nine" astronauts. By the fall of 1963, NASA had approved the tested F-1 engines as the centerpieces for the first stage of the Apollo launcher, Saturn; and was about to endorse the "all-up" testing of its combined stages (rather than as separate stages), this as an efficient means of keeping pace with the president's schedule.

The Mercury years were also busy with President Kennedy's political campaigning, often taking him to the solidly Democrat but segregationist south. He needed its votes, though some southerners were increasingly restless with his overtures at home for African-American civil rights, and with his initiatives abroad for international agreements and cooperation. The campaign took him to Houston, Texas, in the fall of 1962, where he gave a speech at Rice University football stadium. It was the perfect sports backdrop for his declaration of a "space race" that he was intent on leading, this against the "hostile flag of conquest" and for the "banner of freedom and peace." As he continued, "we set sail on this new ocean because there is new knowledge to be gained and new rights to be won." The campaign took him a year later, on that fateful day of 22 November 1963, to Dallas, Texas. At an initial stopover the day before, he gave a speech on conquering and winning the "new frontier of outer space" at the new School for Aerospace Medicine at Brooks Air Force Base in San Antonio. The US was destined to be "second to none."

For American science, getting and staying first also meant thinking and investing in the future. Cybernetics was one approach. Norbert Wiener coined the term in 1948 to mean the shared capacity of machines and animals, including human beings, to process information and control themselves and their surroundings through feedback mechanisms. For two research scientists, Dr. Nathan Kline (a psychiatrist), and Manfred Clynes (a bioengineer), that meant eventually turning the astronaut into a cyborg, a half-human, part mechanical being outfitted with drugs, "biocybernetics," and radically engineered metabolisms, all in order to live and labor in the vacuum of space. The pair's expertise came from their laboratory experiments at the Rockland State Psychiatric Hospital in New York, one of the largest and most notorious in the country, where their patients had a hard enough time adapting to human life on Earth.

The philosopher, Hannah Arendt, contributed to the debates, questioning whether humanity, as she knew it, was ready to sacrifice our planetary home for the lifeless expanse of outer space. As a scholar of modern totalitarianism, Arendt understood how communist and fascist regimes depended upon alienated individuals and atomized societies to maintain power. She saw the same kinds of threats from outer space, where human beings would be wholly dependent on science and technology, on planning regimes, detached from the gravity of Earth, from the variety and communion of social interaction that makes us fully human. Embedded in the small and suffocating spaces of interplanetary crafts, Arendt warned, we would lose our Archimedean center, a place of our own.

Vertical Horizons: Above and Below Worlds

One of the developments in aerospace medicine, just before and during the Mercury program, were experiments with gravitational or "g-forces," the measurement of an object's horizontal or vertical acceleration, or "acceleration felt as weight." Hence 1 "g-force" is equivalent to Earth's gravity and our normal weight; 2 "g-forces" to two times our weight on Earth. Aerospace physicians concentrated on this gravitational factor as pilots began to reach ever-higher altitudes and ever-faster speeds in the 1940s and 1950s, especially with the jet fighters, X-planes, and supersonic aircraft. They studied all the strains that these craft produced on the human body and mind: temporary immobility, loss of vision or consciousness, or an increase in blood pressure and anxiety. Dr. John Paul Stapp, a colonel in the US Air Force, was a leading innovator, testing the punishing g-forces on his own body in rocket sled experiments at Edwards Air Force Base, California, between 1946 and 1958. He endured 25 g-forces, topping off at 40 g-forces in deceleration experiments, as when he stopped from a velocity of 632 miles per hour in only 1.4 seconds.

These studies were important for human spaceflight given the intense g-forces of launch and return. The astronauts of the orbital Mercury and Gemini programs experienced about 8 g-forces in their flights. Thanks to improved technologies, the Space Shuttle astronauts felt 3 g-forces on launch and re-entry. In one failed launch, the Soviet Soyuz 18a crew (1975) experienced 26 g-forces on their accelerated plunge back to Earth. They survived.

The rest of us earthlings sometimes enjoy the thrill of g-forces back on the planet. Formula One race cars can achieve 5 to 6 g-forces in their turns and brakes. The normal carnival amusement ride, like the "Gravitron," achieves up to 3 g-forces. Amusement parks boast the highest or the fastest or the steepest roller coasters, what we now call hypercoasters and gigacoasters. The "Top Thrill Dragster" at Cedar Point, Ohio, took its riders from zero to 120 miles per hour in less than four seconds, to about 4 g-forces. Batman: the Ride, at Six Flags in San Antonio, Texas, gave a 4-g experience. Kingda Ka, at Six Flags in Jackson, New Jersey, one of the tallest and fastest roller coasters in the world, pulled a maximum of 5 g's. One of the world's most demanding rides was the "Tower of Terror" at Gold Reef City Amusement Park, in Johannesburg, South Africa. The ride reached its maximum g-load in a descent, plunging into an abandoned gold mine at a paralyzing 6.3 g-forces. As advertised: "Ride the heart-stopping Tower of Terror 50-meter drop at a speed of 100 kilometers per hour straight into an open mine shaft!"

Select Bibliography

Ackmann, Martha and Lynn Sherr. *The Mercury 13: The True Story of Thirteen Women and the Dream of Space Flight*. New York: Random House, 2004.

Arendt, Hannah. "Man's Conquest of Space," *The American Scholar* 32/4 (Autumn 1963): 527–540.

Bromberg, Lisa. *NASA and the Space Industry*. Baltimore: Johns Hopkins University Press, 1999.

Byrnes, Mark. *Politics and Space: Image Making by NASA*. Westport: Praeger, 1994.

Clynes, Manfred and Nathan Kline, "Cyborgs and Space," *Astronautics* (September 1960): 26–27.

Cornish, Gabrielle. "Music and the Making of the Cosmonaut Everyman," *Journal of Musicology* 36/4 (2019): 464–497.

Dallek, Robert. *An Unfinished Life: John F. Kennedy*. Boston: Little, Brown, 2003,

French, Francis and Colin Burgess. *Into That Silent Sea: Trailblazers of the Space Era, 1961–1965*. Lincoln: University of Nebraska Press, 2007.

Fries, Sylvia. *Project Apollo: NASA Engineers in the Age of Apollo*. Washington, DC: NASA, 1992.

Gerovitch, Slava. "'New Soviet Man' Inside Machine," *Osiris* 22/1 (2007): 135–157.

Gerovitch, Slava. *Soviet Space Mythologies: Public Images, Private Memories, and the Making of a Cultural Identity*. Pittsburgh: University of Pittsburgh Press, 2015.

Giglio, James. *The Presidency of John F. Kennedy*. Lawrence: University Press of Kansas, 2006.

Harford, James. *Korolev: How One Man Masterminded the Soviet Drive to Beat America to the Moon*. New York: Wiley, 1997.

Hersch, Matthew. *Inventing the American Astronaut*. New York: Palgrave Macmillan, 2012.

Janis, Irving L. *Victims of Groupthink: A Psychological Study of Foreign-Policy Decisions and Fiascoes*. Boston: Houghton, Mifflin, 1972.

Kraft, Chris. *Flight. My Life in Mission Control*. New York: Dutton, 2001.

Kranz, Gene. *Failure is not an Option*. New York: Simon & Schuster, 2009.

Leopold, George. *Calculated Risk: The Supersonic Life and Times of Gus Grissom*. West Lafayette: Purdue University Press, 2016.

Logsdon, John. *John F. Kennedy and the Race to the Moon*. Basingstoke: Palgrave, 2011.

"Man Remade to Live in Space," *Life* magazine 49/2 (11 July 1960): 77–78.

McCurdy, Howard. *Inside NASA: High Technology and Organizational Change in the US Space Program*. Baltimore: Johns Hopkins University Press, 1993.

Nolan, Stephanie. *Promised the Moon: The Untold Story of the First Women in the Space Race*. New York: Basic Books, 2004.

Oberg, James. *Uncovering Soviet Disasters: Exploring the Limits of Glasnost*. New York: Random House, 1988.

Raleigh, Donald. *Russia's Sputnik Generation: Soviet Baby Boomers Talk about their Lives*. Bloomington: Indiana University Press, 2006.

Ryan, Craig. *Sonic Wind: The Story of John Paul Stapp and How a Renegade Doctor Became the Fastest Man on Earth*. New York: Norton, 2015.

Swenson, Loyd, et al. *This New Ocean: A History of Project Mercury*. Washington, DC: NASA, 1989.

Thompson, Neal. *Light This Candle: The Life & Times of Alan Shepard.* New York: Crown, 2004.
Watts, Steven. *JFK and the Masculine Mystique: Sex and Power on the New Frontier.* New York: St. Martin's, 2016.
Wiener, Norbert. *Cybernetics: Or Control and Communication in the Animal and the Machine*. New York: Wiley, 1948.

6 Terror

The Intercontinental Ballistic Missile Race

The essential weapon of the Cold War was the nuclear-tipped intercontinental ballistic missile (ICBM). Its predecessors were the shell of the Paris Gun; the short-range German V-2; the first medium-range Soviet and American ballistic missiles (the R-5 and Redstone); and the first intermediate-range ballistic missiles (IRBMs). When married to nuclear bombs, these missiles had the potential for the total destruction of an enemy's armed forces, missile fleets, and cities. Each of them translated into higher and faster means of projecting greater and greater firepower, over farther and farther ranges, culminating in the all-important ICBM. In its first perfected form by the mid-1960s, based on the specifications of the Minuteman I, this missile could carry a 1.2 megaton nuclear weapon about 700 miles high, into the reaches of outer space, in order to find its ballistic trajectory back to Earth within 30 minutes, delivering its payload to an enemy target up to 6,000 miles away.

The ICBM was one means of warfare in the Cold War; outer space was but one battlefield in that war. Territory, technology, and ideology were the three broader arenas of conflict. The ultimate prize of each superpower was the planet itself, the peoples, and territories of the Earth, for whose allegiance the capitalist US and communist USSR vied in proxy wars, as well as in all manner of ideological and propaganda competitions. In these contexts, Soviet space firsts in Sputnik, Gagarin, and Vostok were both practical and symbolic threats to the US. They were very public ways for the USSR to advertise its strengths and American weaknesses, especially to its existing and potential allies around the world. Behind these spacecraft were the launch systems that raised them into Earth orbit: the massive R-7 Soviet ICBM. Sudden Soviet space firsts were a constant reminder to the US and to the world that the USSR might very well be able to launch a surprise nuclear attack on the US, the infamous "first strike," and come out the winner.

As with Soviet and American spaceflight competition, events in the missile race began to move very rapidly after 1957. Sputnik had a cascade effect in the building of new varieties of ballistic missiles, new spy satellites to map and track them, and even new orbital space planes to someday replace them both. In these

DOI: 10.4324/9781003494409-9

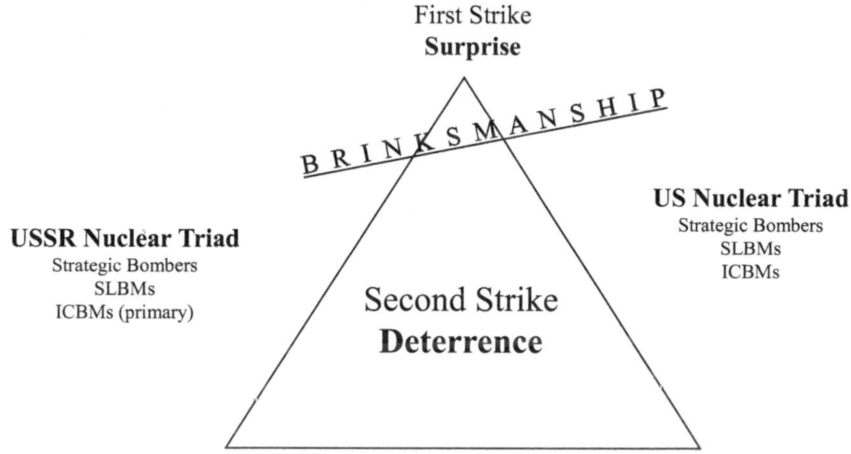

Figure 6.1 The balance of terror or mutual assured destruction. Author.

technological terms, the Cold War became a repeating contest of surveillance and vigilance, improvement and escalation, sometimes based on true capabilities, at other times based on false perceptions. Psychology tended to trump the technology. Fear governed both the estimates and the illusions. We can picture the contest in terms of the "balance of terror": the American maximal nuclear triad of strategic bombers, submarine-launched ballistic missiles, and ICBMs, matched against a Soviet minimal triad. It had the bombers and submarines, too, but made a premier investment in ICBMs. There was nothing stable about this balance. Each side in the conflict kept it in dynamic motion: focused on how to deter the enemy from making a surprise nuclear "first strike," mostly by maintaining reserves of enough weapons to destroy it in a retaliatory "second strike," though sometimes by leveraging national interest in power plays of "brinksmanship." All of this transpired within the ghastly new idiom of "Mutual Assured Destruction" (MAD), each side holding its own and the other's population hostage.

The Missile Gap

The R-7 shook US wartime planning to its core. It threatened deterrence, already institutionalized in 1950 as a safeguard policy. Intelligence analysts and government leaders now perceived that US strategic forces, its bombers and missiles, were vulnerable to a paralyzing surprise attack. They worried over the R-7's massive payload capacity, proven beyond US capabilities; and about its rapid flight time of roughly 30 minutes, beating the average early-warning times of American defenses. The fear was that only a few B-52 bombers would survive to retaliate, this when the US lacked a credible and operational ICBM fleet of its own. The R-7 introduced a new element of risk and uncertainty, a loss of control over when and where and how nuclear war might break out.

All of this represented a stunning reversal from the American nuclear monopoly of only a decade before, when Presidents Truman and Eisenhower played brinksmanship and threatened the USSR with massive nuclear retaliation for any real or perceived territorial threats. The R-7 threatened containment. It led to dramatic speculation that the USSR intended to break out of its established defensive postures and into leveraging territorial expansion around the world. Nikita Khrushchev played his own brinksmanship in his demands for a reunification of Germany under Soviet rule. Europe was at risk in this scenario. Did American allies have enough confidence in the US to remain loyal, or were Soviet territorial, technological, and ideological pressures breaking mutual expectation and trust?

Worse still, US intelligence analysts soon predicted a massive buildup of Soviet ICBMs over the next several years, one that would clinch its first-strike capability. The CIA's National Intelligence Estimate of June 1958 predicted that within one year the USSR would deploy 100 ICBMs; then 500 by 1960; followed by 1,000 by 1961; 1,500 by 1962; and 2,000 by 1963. The US would not be able to match these Soviet developments, either in the numbers of nuclear ICBMs or in their carrying capacity. The Soviet command-administrative economy would outstrip American industrial capitalism in numbers and sizes, the two measures of nuclear advantage that really mattered. This meant that the space gap and the science gap were also a "missile gap" that was bound to drive a new missile race. Like the first Soviet satellite, or dog, or human in space, now missiles also became a commodity of fear in the American marketplace.

Khrushchev magnified these American anxieties with a major military initiative in December of 1959: his very public announcement of the Strategic Missile Forces, a branch of the military equal in status to the Army, Navy, and Air Force. His priority was to invest in IRBMs and ICBMs to compete quickly and efficiently with what he saw as the strategic superiority of the US. The move represented a significant shift from emphasizing conventional land warfare to focusing on global nuclear war. By promoting missiles, so his logic went, he would also be able to reduce military spending across the board, centered on troop reductions, thereby leaving more state funds for consumer goods and services. The new Strategic Missile Forces also represented a significant advantage for the USSR: a centralized hierarchy in charge of both its military ballistic missile program and significant aspects of its civilian spaceflight program. Marshal Mitrofan Nedelin, a veteran of the Soviet artillery, headed the new Forces and worked closely with Sergei Korolev, among other lead designers, who were also guiding civilian programs. Both reported to their higher military and political authorities: Dmitrii Ustinov, Minister of the Defense Industry; and Leonid Brezhnev, the party chief of the military-industrial complex.

In his establishment of the Soviet Missile Forces, Khrushchev had several new designers and bureaus from which to draw, each competing for state funding and promotion. These new bureaus reached back to the very beginning of his regime, when he diversified Soviet missile programs into a variety of enterprises. Accompanying this diversification was a split between the missile designers themselves as to their preferred choices for missile propellants, a seemingly minor issue but one with implications for missile technology and application.

Representing tradition were Sergei Korolev and his loyal teams, who favored the standard cryogenic mixtures: liquid oxygen as the oxidizer and kerosene as the fuel. These had been favorite Soviet propellants reaching back to the 1930s, also partly inherited from the German V-2 program after the war. Korolev used them successfully in his R-7 ICBM and space launchers. He took them to his later development of a new series of ICBMs (the R-9 and GR-1), which he made more efficient with the designs of better pumps and with new means of handling the super-cooled liquid oxygen.

Representing innovation were the proponents of new hypergolic propellants, meaning the highly toxic and corrosive nitric acid as oxidizer with kerosene as fuel; or in later years nitrogen tetroxide as the oxidizer, and hydrazine (UDMH) as the fuel. Their significant advantage: they were more easily storable and ignitable and therefore allowed for quicker launch times. They were essential, for example, in Viktor Makeev's adaptation of the R-11 into the USSR's first submarine-launched ballistic missile. The masterful designer, Mikhail Yangel, perfected the use of hypergolic fuels (with Valentin Glushko's engines) for the USSR's first successful fleets of medium and intermediate-range ballistic missiles: the R-12 and R-14, with storable fuels and a new autonomous inertial guidance system (not needing any radio control) and with about a 1,000 or more mile range. By 1960, the USSR had positioned them at forward points in Belarus, Lithuania, and Kazakhstan, aimed at America's allies in Western Europe and Asia. Yangel was, by this time, also finalizing the development of his R-16 ICBM, a new missile based on the R-12 and R-14 as its two stages, and a crucial element in the country's competition to match and surpass US strategic reach.

These were an impressive series of advances in ballistic missile technologies, yet they also placed immense political pressures on the USSR's military-industrial leaders. Khrushchev punctuated all of this serious planning with his "rocket-rattling," boasts that the USSR was building a "nuclear rocket shield around the motherland." Or that "we can now send a missile to any point on the globe, carrying if necessary a hydrogen warhead." Or that "we are producing ICBMs like sausages." He really should have produced more sausages. The pressure peaked on 24 October 1960 in the single most-fatal accident of the Space Age: the R-16 or "Nedelin" disaster. It happened amid Khrushchev's warmongering, as one of the final tests before deployment of the R-16 against targets in the US, in preparation for a propaganda announcement about the missile for Bolshevik Revolution Day (7 November).

The missile had been on its launching pad at Baikonur's site 41 for three days (since 8 am on 21 October), being readied for its scheduled launch at 7:15 pm, already fully fueled for about a day, after many hours of intense preparation. At 6:45 pm, the teams encountered a problem with the launch sequence. Marshal Nedelin, in haste to fire the missile and prove both it and him worthy, ordered the teams to approach the pad to resolve the problem. This broke all safety protocols, which required non-essential personnel to leave the site while crews emptied the missile of its propellants. In fact, Nedelin himself left the safety of the control bunker and sat in a chair, not far from the missile, to oversee a rapid resolution. In the process, a backup command ignited the second stage of the missile. Someone had left a control in the wrong position after a series of tests, and when someone else switched it back to zero, this activated the fuel and oxidizer lines and launched the second stage right into the first stage. On the launch pad. With engineers and technicians servicing the missile. Nedelin overseeing it all.

The final investigation of the accident, in the coldest of terms, called this turning of a switch "the unscheduled operation of the Programming Current Distributor (PTR)," which "activated the electronically powered EPK VO-8 pneumatic valve." Yet the results were horrific. As the second stage fired into the first, the whole missile exploded, followed by 20 seconds of massive explosions and two hours of intense flames as the missile propellants and tarmac asphalt burned. The fire and poisonous fumes consumed those near the missile, many while trying to flee. Of the several hundred persons at the site, at least 90 eventually perished from the explosion or their injuries, including Nedelin, two of Yangel's deputies, one of Korolev's deputies (there simply to get a signature on an important document), and the deputy chief of the State Committee for Defense Technology. No one was reprimanded. As Leonid Brezhnev commented, "All the guilty have already been punished." In fact, the R-16 project resumed within weeks, and Nedelin's remains were set in a place of honor at the Kremlin wall. Most of the other remains were placed in a mass grave at the site, with the public announcement that they had died in an airplane crash. The Soviet government then kept the disaster top secret for nearly 30 years.

The Balance of Terror

At the moment of the Nedelin disaster, far from suffering the setbacks calculated by the CIA, the US strategic posture was steadily improving. Thanks to the Eisenhower administration's planning and investment, the American nuclear triad was posing an increasing threat to Soviet geopolitical power. The US was approaching a force of some 600 B-52 bombers, and 1,000 or more B-47 bombers, all stationed at Strategic Air Command bases in the US and around the world; a portion of which was always in the air, or at the ready, to attack the USSR and its allies within several hours of notice. The first US Polaris submarine was already in play, with 16 missiles by November of 1960, and 40 more submarines added by 1966. Its solid-fuel missiles had only a 1,000-mile range but enjoyed the immense advantage of underwater stealth, enough to ensure the Soviet fear of an American second strike. By September of 1960, the first Atlas ICBMs (with a 5,000-mile range) were operational at Vandenberg Air Force Base in California and at Warren Air Force Base in Wyoming. The Titan I followed in 1961 and 1962, from sites throughout the West, with improved range, payload, and reliability.

America's IRBMs helped to round out these far-ranging strategic capabilities. After 1958, the US stationed Jupiter missiles in Great Britain, Italy, and Turkey; and Thor missiles in Great Britain, under Royal Air Force command. They were all outdated within a few years, and retired by 1965, yet they offered a temporary means of defending Europe until US improved ICBMs were fully operational.

By the fall of 1962, because of these myriad developments, the Kennedy administration enjoyed what it called a "clear" advantage in strategic nuclear warheads and delivery systems as compared to the USSR. Given the nature of nuclear deterrence, the USSR's second-strike capability provided a measure of balance against these superior forces. Yet neither side was ever certain about its own or the other's advantages or disadvantages. Every new or superior capability created an imbalance in the system, along with the perception or misperception of strength or weakness. What appeared as a stable balance of terror was actually a constantly shifting, sometimes deceiving gauntlet of rivalries, as each side struggled for constant improvement and temporary advantage. The key to it all was the ICBM.

Beginning in the 1960s, both the USSR and the US competed to achieve some significant advantage in its numbers and kinds of ICBM fleets. "Throw weight" and range mattered: the imperative for more powerful missiles, to carry more destructive bombs, to farther distances. Quicker launch times were essential too, this to beat the opponent with a "surprise" factor in either first-strike or second-strike terms. Accuracy was crucial to precisely target the enemy's ICBM sites. Hardened silos became essential for both sides to protect those sites and ensure the all-important second-strike deterrence. In time, in a rather astounding abbreviation, the "Multiple Independently Targetable Reentry Vehicle" (MIRV) became the standard to beat, as these were missiles with multiple warheads, ensuring that one or more might break through an enemy's defenses, again ensuring the second strike. Space satellites were also essential through all of this, in finding enemy ICBM sites, targeting them, and coordinating attacks.

This pyramid of ICBM factors meant that as soon as each side began developing and deploying its first-generation ICBMs (the R-7 and R-16, or the Atlas and Titan I), the second and third generations could not be far behind. So, the USSR advanced its R-16, R-36, and UR-100 ICBMs; the US, its Titan II and Minuteman. As T.A. Heppenheimer has noted, the Soviet R-16 and American Titan II offer yet another fascinating study in parallel developments. Both were

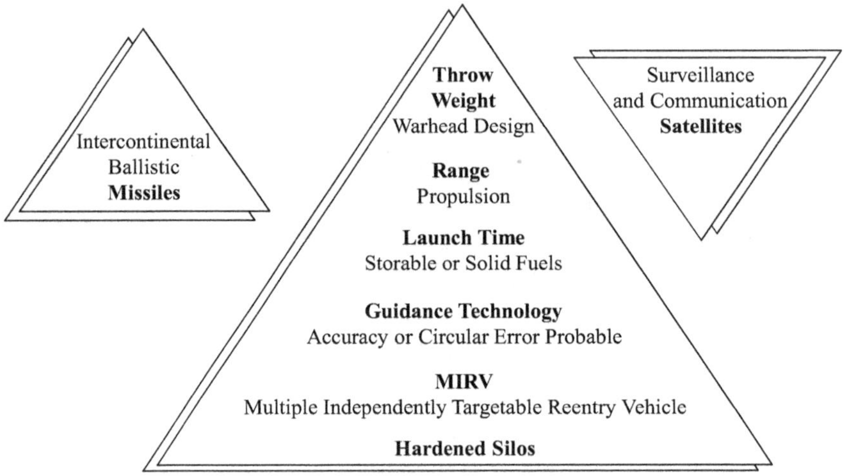

Figure 6.2 ICBM factors in the missile race. Author.

massive in size with massive nuclear bombs. Both used the new hypergolic, storable fuels for ready launching. Both were most effective in their deep, underground, hardened silos. With Titan II, for example, we find ourselves once again in the strange world of vertical horizons. It occupied the "above world" as a space weapon, reaching 200 miles into space for its ballistic trajectory, able to launch 8,850 pounds over a 6,000-mile range in about 30 minutes. It occupied the "below world" of its silo, 147-feet deep, covered by a steel and concrete door weighing 740 tons.

To complicate the scene even more, these years also saw the proliferation of nuclear weapons and delivery systems beyond the US and USSR. The United Kingdom tested its atomic and hydrogen bombs between 1952 and 1957, though relying on its own long-range bombers and US missiles for delivery. France followed with nuclear weapons of its own between 1960–1968, and the S-3 IRBM after 1980. With scientists educated in the US (Qian Xuesen and Deng Jiaxian), and with Russian scientific and technical assistance, the People's Republic of China developed its first nuclear weapons by the late 1960s, and ballistic missiles by the 1970s. These included the Dengfong-3 (Eastern Wind) IRBM and the Denfong-5 ICBM (1981). The PRC achieved these advances in the wake of the Great Leap Forward and Cultural Revolution, state-sponsored policies in collectivization, artificial famine, persecutions, and executions that counted tens of millions of human lives.

Given the nature of the USSR's closed society, the communist regime tolerated little to no dissent about the nature of this Cold War rivalry. The Soviet ICBM represented a necessary and inviolable defense against "capitalist encirclement" and for the "communist motherland." There was one notable exception. Mikhail Romm's, *Nine Days in One Year* (1962), a classic of Soviet cinema, though not directly about the threat of nuclear war, told the story of a nuclear scientist who, after a radiation accident, sacrificed his life for science and the state. The film carried the warning: "humanity has reached such 'perfection' that we can destroy everything alive on Earth within twenty minutes."

The West's open society of media and cultural freedoms was far more accommodating to criticism. As the strategic race for ICBMs accelerated after 1957, important works of literature and film questioned its logic and legitimacy. This was a new genre of cautionary tales, dramatizing just how simply and disastrously the US and USSR might collapse into a nuclear holocaust. Nevil Shute's 1957 novel, *On the Beach*, depicted the last days of a group of people in Melbourne, Australia, as they waited for their impending deaths from the radioactive fallout of a disastrous nuclear war. This bleak story, made into a 1959 film, even suggested that, under such circumstances, suicide was the best option.

By 1958, Peter George's *Red Alert*, also known as *Two Hours to Doom*, dramatized a mad US Air Force general launching a preemptive first strike against the USSR. The American government attempted to recall or destroy the bombers, but one managed to close in on its Russian target. Faced with the prospect of "Mutual Assured Destruction," the US president offered in compensation to destroy Atlantic City, New Jersey, to avoid an all-out nuclear war. A wide swath of southern New Jersey was an acceptable loss, though in the end it was saved. American readers found the same premise in the best-selling novel, *Fail-Safe* (1962), originally and coincidentally serialized in the *Saturday Evening Post* during the Cuban Missile Crisis, and later adapted to the screen. This time, though, as an American bomber made its way to destroy Moscow, the American president was forced to destroy his own New York City in return.

These frightful scenarios had their lighter moments, as in the case of Stanley Kubrick's black comedy, *Dr. Strangelove or: How I Learned to Stop Worrying and Love the Bomb* (1964), a refreshing and irreverent attack on Cold War mentalities. The film effectively subverted all the platitudes of those years: "the missile gap," deterrence, "Mutual Assured Destruction," and above all the US national-security leadership. Kubrick hilariously and unsparingly mocked the profound seriousness of the Cold War in this film, ending with an Air Force pilot riding a nuclear warhead cowboy-style to its intended target, followed by a montage of nuclear explosions, accompanied by the mellow tones of a nostalgic ballad from the Second World War.

Space Planes and Satellites

The US did leap ahead at times, with the world's first computerized defense networks, with the U-2 and SR-71 planes, with space planes like the X-15 and promised Dyna-Soar, and with spy satellites (the Corona series). These surveillance and test platforms were sometimes stabilizing, often not.

If Khrushchev fortified his R-7 and Sputnik successes with the creation of the Soviet Strategic Missile Forces, President Eisenhower raised the stakes against these moves with scientific and military initiatives all America's own. In a tour de force of university, industry, military, and government cooperation, the Eisenhower administration built a continental system of land, sea, and air-based radars, sensors, and trackers to defend against Soviet bombers. These were the Distant Early Warning (DEW), Mid-Canada, and Pinetree Lines. They reached from the US, through Canada, and into the Arctic Circle, as well as along the Atlantic and Pacific coasts. They were connected and partially controlled by the Semi-Automatic Ground Environment (SAGE) computers, which coordinated US defense systems, in cruise missiles, surface-to-air missiles, and jet interceptors. The "Ballistic Missile Early Warning System" (BMEWS) joined the network in 1961 to guard against a Soviet ICBM strike.

The mission of the Advanced Research Projects Agency (ARPA), created in February of 1958, was to bring still more satellite platforms to these arrays, for even better timing and coordination against a surprise attack: the "Satellite Military Observation System" (SAMOS) to further survey Soviet military sites; and the "Missile Defense Alarm System" (MIDAS), to scan the USSR using infrared sensors for Soviet ICBM launch signatures. With the Applied Physics Laboratory at Johns Hopkins University, ARPA also helped the Navy to create the Transit navigation satellite system to help locate and track its Polaris submarines. In yet another dimension of "vertical horizons," ARPA designed new weapons and approaches to fight the Cold War here on Earth, as, for example, with new kinds of rifle grenades and radio communications, or with the building of remote airstrips and the use of herbicides to reduce jungle cover. All were novel means of combating communist guerilla warfare on the ground.

By May of 1958, the North American Defense Command (NORAD) began the task of coordinating all these aerospace efforts, soon to be headquartered in the vast underground Cheyenne Mountain Complex at Colorado Springs, Colorado. The US had thereby constructed the first of several vast computer and data networks to unite the globe. Still pending were those of NASA and the National Oceanic and Atmospheric Administration, and of the Arpanet and internet, which we will explore in later chapters.

The missile gap also drove President Eisenhower's investments in new, high-performance, hypersonic planes. They were to test the best new aerodynamic designs and power plants for high-speed aircraft, but also as means to test the viability of space planes, aircraft that could break the envelopes of speed and altitude to serve as military space vehicles. In this, the US achieved record-shattering firsts with the experimental X-15, built by North American Aviation, with a superb liquid-propellant rocket engine by Reaction Motors. In a perfect "aerospace" moment, a converted NASA B-52 bomber raised the X-15 space plane to an altitude of some 8.5 miles, and a speed of roughly 500 miles per hour, before releasing it for its test flights. Between 1959 and 1968, the X-15 achieved several significant firsts, reaching into outer space to an altitude of 67 miles; and becoming the fastest piloted experimental aircraft on record, achieving a maximum speed of 4,520 miles per hour. The USSR had no analogs.

Aerospace

The term, *aerospace*, sometimes meant the natural envelopes of the atmosphere, stratosphere, and outer space. It also meant the actual new technologies that pierced these envelopes with altitude and speed records. Tsiolkovsky's essay of 1903 and Goddard's work of 1920, both written in the first years of aviation, were already reaching into aerospace with envelopes of new ideas and machines. Yet the term did not come into common use until just after Sputnik in 1957. Writers were hesitant at first, using it in hyphenated form as "aero-space," as if struggling to convince their readers that this vertical horizon, this territory above us, was conquerable. By the early 1960s, aerospace had three new essential meanings. It referred to a new physical realm of air and space, ready for human conquest by supersonic jet and rocket propulsion. It signified a new field of aerospace medicine to study how the human body and mind might tolerate its stresses. It also described a whole new industry to build jet craft, missiles, rockets, and spacecraft, all to take human beings through these realms, beyond the planet, ever faster and higher. Or as the US Air Force put it quite boldly in 1962, aerospace was the "operational, indivisible medium consisting of the total expanse beyond the Earth's surface." Its most practical and symbolic craft may have been the T-38 two-seater supersonic jet, which the original astronauts flew around the country in their preparations for spaceflight. Once over Cape Canaveral, they flew in "perfect formation" to herald their arrival for a coming space launch. The Space Shuttle pilots later flew and trained on the T-38 as well.

An experimental and civilian craft, the X-15 had a more practical and military successor: the X-20, or "Hypersonic Glide Rocket Weapon System," meant to become the US Air Force's first operational space plane. In design after 1957, and actually in construction long enough to give it a name, "Dyna-Soar" (for Dynamic Soaring), the craft was meant to serve as a spying and surveillance platform in space, and as a satellite destroyer. This was a one-pilot, reusable craft, to be boosted into space on an ICBM trajectory (by a Titan launch), then to orbit at 4 miles a second and at 96 miles altitude, eventually to glide back to Earth with its delta wings, gathering lift and bouncing on the way back. In terms of the fast-paced missile race, its most important function was as an "antipodal" bomber, the ultimate surprise factor to answer Soviet surprises, able to reach a target on Earth with only three minutes of warning to enemy sensors and radars. At $660 million, Dyna-Soar cost more than the Mercury program, but was perhaps worth the price in that, besides its other functions, one of its missions was space rescue: it might have saved a Mercury-type capsule in peril, like Gordon Cooper's perilous flight.

The Department of Defense canceled Dyna-Soar by 1963, given its high costs, also canceling the related and very expensive Air Force plans for an orbital Space Shuttle, space station, and possible lunar base. Instead, the US government gave precedence to the Gemini and Apollo programs, which in any case closely paralleled these plans. Moreover, the military and NASA found a common cause in the new "Gemini Planning Board," which joined them in partnership to promote the military uses and applications of Gemini technologies.

US civilian and military spaceflight programs were close-knit, if in secretive and hidden ways. We already know that NASA's space launches, those for human spaceflight and interplanetary probes, depended on the military's ballistic missiles for its launch platforms. This civilian-military alliance went further, though, as in the case of the National Aeronautics Space Council, comprised of the president, the secretaries of Defense and State, the head of the Atomic Energy Commission, and the head of NASA, whose central purpose was to promote space policy as a means of national security.

Eisenhower's more Earth-bound, military spy plane, the U-2, also enjoyed its continuing achievements into the 1960s. Yet it also had its limits, as in the infamous U-2 "incident" on 1 May 1960, when U-2 pilot Gary Powers fell from 60,000 feet back to Earth. Like the other U-2 pilots, he wore a space suit for protection. Powers was flying the main U-2 route from Peshawar, Pakistan over to Sverdlovsk, USSR when Soviet radar detected him, dispatching a volley of Soviet surface-to-air missiles to bring him down. One may have hit him, or several exploded near his plane in tandem, the blast effect forcing him into a fatal spin. Powers failed to activate the self-destruct mechanism of the plane and chose not to take the "voluntary" cyanide pill. He escaped from the cockpit and parachuted onto Soviet territory. He and his wreckage were recovered by Soviet authorities.

Eisenhower believed that the man and machine were destroyed, leaving no evidence. So, he lied. He lied to the American media and to the world, claiming that the U-2 was a weather-monitoring aircraft that had strayed from Turkish air space. When the facts came out, the president admitted the truth, proposing that the United Nations take over control of the program, an echo of his earlier Open Skies proposal. Khrushchev once again refused. As a result, U-2 flights over the USSR ended (they were about to at any rate). In their place, Lockheed prepared the SR-71 spy plane, which soon became the world's fastest and highest-flying operational jet-propulsion aircraft.

The US and USSR later exchanged Powers for the Soviet spy, Rudolf Abel. Powers had spied from altitude with some of the world's most sophisticated photographic equipment, piloting one of its highest-flying aircraft. Abel's instruments were more mundane: simple cameras and codebooks, and hollowed-out nickels, shaving brushes, and pencils in which he hid microphotographs of US military secrets. In their exchange, on 10 February 1962, Abel walked across the Glienicke Bridge to communist East Berlin, and Powers walked across it to the free West Berlin. Powers got out, but not with the pieces of his crashed U-2, which the Soviets put on permanent display, gathering dust at the Central Armed Forces Museum in Moscow.

The U-2 incident occurred just as the US was finalizing a replacement to the spy plane, in the new Corona spy satellite. As the Soviets sentenced Powers to prison, the Corona completed its first 17 orbits, releasing its film back to Earth for CIA analysis. This was a major American achievement: the world's first reconnaissance satellite and one of the country's top-secret or "black" programs for the next 30 years, well hidden behind the public spaceflight programs. The world only learned about it at the end of the Cold War: just as the Soviets revealed their secret disasters (like the R-16), the Americans revealed their secret successes.

Project Corona was a late product of President Eisenhower's Open Skies proposals, and became his premier covert intelligence operation, sometimes also known by the code names of its cameras, "Keyhole" (KH); or also by its diversionary public name, "Discoverer," falsely advertised as a biomedical research program to study mice and monkeys. It was a partnership between the US Air Force and the CIA, through the National Reconnaissance Office (NRO), which managed it and similar programs; joined by Lockheed, General Electric, Fairchild Industries, Hughes Aircraft, and Polaroid.

The USSR had a similar program in 1962, Korolev's "Zenit," although less sophisticated. At first, he simply placed a reconnaissance camera on a Sputnik satellite. Its spy satellites improved over the years, of course, as they were essential for mapping and targeting the US for a Soviet ICBM attack. As a rule, though, such photographic intelligence was somewhat less a priority for the USSR, which always depended more on human and technical intelligence, sometimes as simple as reading America's newspapers and magazines.

The methods of raising and retrieving Corona equipment offer us another classic aerospace moment, the meeting of high technology and low, the new and the old. The Air Force launched the system from Vandenberg Air Force Base, by way of a Thor IRBM with an upper-stage Agena rocket, which fired itself into orbit to photograph the Earth from its camera platforms. At the end of its mission, Agena fired into a spin at 600,000 feet, when the cone separated and began its descent into the atmosphere, with a parachute deployed at 65,000 feet. A unique retrieval system prevented Russian submarines from stealing the film canisters. Air Force JC-130, HC-130, and C-119 aircraft, out of Hickam Air Force Base, Hawaii, flew at very low speeds, 150 miles per hour and under 15,000 feet, using metal and wire winches extended below the aircraft, to catch the parachuting Corona film capsules in the air.

None of this was easy at the start, what with early failures of the launch and recovery system, and the several weeks needed to retrieve, transport, and analyze the data. The methods to stabilize and calibrate the cameras were a challenge, as the cameras needed to be lightweight but strong, able to withstand all the varying temperatures and conditions. In time, from its orbit at 100 miles altitude, several Keyhole (KH) rotating stereoscopic cameras successfully photographed swaths of an area 150 by 150 nautical miles. Soon after its first reconnaissance orbit on 18 August 1960, Corona had photographed roughly 1.5 million square miles of the USSR and its occupied states in Eastern Europe and Asia. CIA photographic analysts found 64 new airfields but no ICBMs. These and coming results were arguably far more important than the Sputnik achievements. By September of 1961, Corona had revealed the "missile gap" as an illusion, Khrushchev's rocket-rattling as a bluff. Photographs of Soviet ICBM bases at Baikonur and Plesetsk found not the estimated hundreds or thousands of ICBMs, but between 10 and 25 at most. Said one CIA analyst: "we were looking for a body and found a skeleton."

Over the years, the image resolution of the Corona cameras improved remarkably, from a 40-foot object visible using KH-1, to a 6-foot object visible from KH-4b. With 145 missions between 1960 and 1972, the system surveyed and mapped about 750 million square miles of the Earth's land surface, mostly of the USSR and Communist China, producing 39,000 canisters with 2.1 million feet of film, 800,000 images in all. Corona found almost all Soviet air and naval bases, major military-industrial facilities, atomic-weapons sites, and submarine production sites. It photographed and mapped its IRBM and ICBM bases for targeted destruction in case of nuclear war, and reassured US policymakers with fuller and more precise estimates of Soviet capabilities.

In reference to Corona during its heyday, President Lyndon Johnson confirmed its value to American policymakers. As he put it in 1967, the US investment of some $40 billion in overall space exploration was especially beneficial with regard to satellite surveillance.

> Because tonight we know how many missiles the enemy has and, it turned out, our guesses were way off. We were doing things we didn't need to do. We were building things we didn't need to build. We were harboring fears we didn't need to harbor.

In their reminiscences, Corona's founders called it a "constellation" of satellites, a "central nervous system" for the US intelligence community. Outer space became a "sanctuary" and "peaceful venue for intelligence gathering"; and Corona, said a later Secretary of Defense, William Perry, was our "jewels in the crown."

In time, Corona established a more stable balance of power between the US and USSR, but not before it also helped spur one of the Cold War's most dangerous moments, the Cuban Missile Crisis in the fall of 1962.

The Cuban Missile Crisis

Few episodes in the annals of Cold War history have captivated readers like the Cuban Missile Crisis of October 1962. Only 20 years from the launch of the first V-2 ballistic missile, we are already at a "missile" crisis threatening the global balance of power, millions of lives, and perhaps even the existence of the planet. In the 62 years since it happened, we have reduced it to the infamous crisis of "Thirteen Days." We have politicized it, dramatized it, and personalized it. We also need to properly historicize it, at least in the sense of positioning the crisis more squarely within the geopolitics of the missile race and space race.

We have seen that the US and USSR followed along similar trajectories, speeding along in an arms race for more, bigger, and better ballistic missiles. Yet they were at radically different points along those trajectories. The US was riding a wave of continuing military superiority; the USSR had fallen behind. This dilemma was compounded in the fall of 1961 when the Kennedy administration quietly but publicly released the news to the media that the infamous "missile gap" was really a hoax, that the Soviets had only produced an arsenal of no more than a few dozen ICBMs. By 1961, US intelligence concluded that the USSR had not and could not soon achieve strategic missile advantages over the US. Revealing the missile gap as a hoax escalated rather than pacified Cold War rivalries.

Khrushchev knew what Kennedy knew. Suddenly, Soviet technological and political achievements in space exploration (Sputnik, Gagarin, Vostok), along with their corollaries in strategic military reach, were in peril.

One of the lingering controversies about the crisis has to do with intentions. Why did the Soviets place IRBMs in Cuba? Was it for the defense of Cuba? Or was it to revitalize, in an instant, the Soviet strategic posture with the US? Khrushchev offered the best answer to the question when he wrote to Kennedy that the two countries had effectively tied themselves in a knot. Territorial and technological competition remained the two firmest woven strands of Cold War rivalry. They came together to form Khrushchev's knot. Several compelling accounts tell us that Khrushchev made his decision in April 1962, after a moment of reflection at the Black Sea, looking across the waters and thinking about the US Jupiter missiles stationed in Turkey, protecting this key US ally. His eureka moment, it seems, was in realizing he could position similar missiles in Cuba to protect that key new Soviet ally, as well as improve his leverage in places like Germany and Vietnam.

This version of events argues that his primary interest was in protecting Cuba. It makes sense. The Bay of Pigs had failed the year before, yet the Kennedy administration had also intensified covert operations against Cuba on the ground, codenamed "Operation Mongoose," which included acts of economic sabotage, military incursions, and even plans to assassinate Fidel Castro. At the time it happened, and ever since, the Soviets called the events a "Caribbean Crisis" rather than a Missile Crisis, this to highlight its support for communist Cuba, based on Soviet territorial and ideological strengths. Spaceflight history once again spotlights the geopolitics of rockets and rimlands, only this time a rimland of the US.

Yet Khrushchev was a global as well as local thinker. Those Jupiter missiles were also aimed at Soviet cities and missile sites, a threat to his country's power and existence, now even more vulnerable given the end of the missile gap. Khrushchev's placement of IRBMs in Cuba was a risky but quick way to offset the strategic imbalance, now that his bluff was over, and especially given the recent R-16 disaster. It was a way to gain some competitive parity with the US on the cheap: the USSR had a surplus of IRBMs and a shortage of ICBMs. It was also a brilliant tactical move, since US radar and defense networks, aimed at Soviet bomber and missile threats from the Arctic north, were weakest in the south.

Khrushchev implemented "Operation Anadyr" in July of 1962, a secret agreement with Fidel Castro's government to place some 50,000 troops, a naval base, various weapons, and ballistic missiles in defense of Cuba. The plan included the positioning of 24 launchers of the Soviet R-12 missile, and 16 launchers of the R-14, both within reach of the major cities of the US east coast, northward from Miami to New York City. A secret buildup of these forces followed over summer and early fall. By 10 October, ten missiles were ready for their warheads and fueling; and by 20 October, another ten more, though none were outfitted with the weapons. Khrushchev was thus readying one more "rocket" surprise for his American rivals.

The two weeks of crisis began on 14 October, when CIA analysts identified the telltale missile trailers and launch sites of Soviet IRBMs, based on the information supplied by the spy, Oleg Penkovsky, and by visual evidence from U-2 spy-plane photography. After a week of initial deliberations and negotiations, the rivals reached a high point of drama on 22 October, when President Kennedy delivered a televised address to the nation publicizing the Soviet moves to the American people and the world, announcing a quarantine of the island to prevent any further military and missile buildup. A whole series of crisis moments and tense negotiations followed. American and Soviet military forces went on high alert. The first Soviet naval vessel approached the quarantine line. A Soviet surface-to-air missile battery shot down an American U-2 spy plane over Cuba. The crisis only ended on 27 October when Khrushchev offered Kennedy a deal: the public recognition by both sides of the territorial and political "integrity" of both Turkey and Cuba; and a secret protocol on the removal of US missiles from Turkey (and Italy) and the Soviet missiles from Cuba.

Americans remembered these tense days in creative ways. Singer Bob Dylan began to write his classic, *A Hard Rain's A-Gonna Fall*, during the crisis, with stirring lyrics about the promise of doom, about poisoned forests and oceans and people after a nuclear war. Reverend McKendree Robbins Long, a Presbyterian minister and artist, painted several canvasses about the crisis, featuring the apocalypse and final judgment: Gagarin falling from space and Castro seized by one of the four horsemen, ICBMs arrayed and launched, the earth on fire. Lenny Bruce found some comic relief during the crisis, in one of his stand-up routines, warning his audience with the line, "We're all gonna die!"

Quarantine

By the codes of international law, President Kennedy's "quarantine" or blockade of Cuba during the missile crisis of October 1962 was an act of war. It was, in other words, a belligerent and preemptive action against the USSR and Cuba. Still, it restricted and eventually defused the confrontation, which was certainly preferable to attacking the missiles or invading the island. Thirty years later, under President George Bush Jr., Secretary of State Condoleezza Rice made a rare official justification of the move, just as the US government was preparing another preemptive strike (and war), this time against Saddam Hussein's Iraq, presumed to be holding weapons of mass destruction threatening the US and its allies. As Secretary Rice argued in an interview on the PBS *News Hour* (25 September 2002):

> The United States has always reserved the right to try and diminish or to try to eliminate a threat before it is attacked. It simply wouldn't make sense to sit and wait to be attacked if you thought that you could eliminate a threat. Let me give you an example from the 1960s. In 1962, we had the Cuban Missile Crisis. And the United States engaged in what is an act of war, that is, a quarantine of Cuba, a blockade of Cuba, because there were missiles that were about to become operational against the United States. The Kennedy administration didn't wait until there was an attack from Cuban soil against the United States. It simply doesn't make sense to say that you have to wait.

As a dramatic episode of Cold War rivalry, the Cuban Missile Crisis reveals key American strengths. Kennedy probably won the crisis. The US enjoyed the advantages in strategic nuclear weapons and their delivery systems, as well as in spy-plane and satellite surveillance systems, from the start to the finish. These allowed the president to leverage US power against the USSR. Khrushchev's gamble backfired. He was found out. He was the weaker party.

As a moment in the broader history of the Space Age, the Cuban "knot" reveals still more. Between 1960 and 1962, over the course of the "longer" Cuban crisis, both the USSR and the US were locked in an Earth-bound, territorial mode of thinking, each pondering their competitive strategies in either the missile race or space race. Khrushchev's Operation Anadyr was, in these terms, the Soviet equivalent of Kennedy's "Apollo" challenge to the Moon. Both were moments of weakness and fear, centered on both Cuba and outer space. As we saw in Chapter 5, Kennedy had failed with the Bay of Pigs and was humiliated by Gagarin's flight into orbit. Khrushchev was angered by Kennedy's continuing anti-Castro interventions, the missile gap revelations, and the announcement of the Apollo program. Both had long-lasting repercussions, for either spaceflight or for the arms race. Kennedy needed an American space spectacular to match the Soviet firsts. Khrushchev found his country wanting in the strategic nuclear balance, looking for a speedy advantage. His Cuban missile failure spotlighted Soviet weaknesses on the world stage.

There are several more threads to the Cuban knot. On 22 November 1963, the assassin Lee Harvey Oswald shot and killed President John Kennedy as he drove in a motorcade through downtown Dallas. Oswald was never tried and convicted of the crime. Jack Ruby murdered him two days later in a revenge killing. Yet the official Warren Commission report amassed credible and convincing evidence that Oswald was the lone assassin. The physical evidence of the gun and bullets, the eyewitness testimonies, the circumstantial evidence of his preparations and escape. All of this pointed to Oswald.

The commission report also offered a sweeping psychological study of the man, burdened with a long history of emotional problems and difficult family circumstances. By the time of the assassination, Oswald fancied himself a kind of Marxist guerilla warrior. A former US Marine who had defected to the USSR, he had also spoken publicly in defense of Castro's communist revolution, and just weeks before his crime had visited the Cuban Embassy (in Mexico City) in an attempt to flee to Cuba. Instead, he found an opportunity in Dallas to murder Castro's nemesis, President Kennedy, in what he likely considered as an act of revolutionary justice.

Here was one last thread in the Cuban knot that we have been following since the Bay of Pigs crisis of 1961 and the Cuban Missile Crisis of 1962. Far from pausing or delaying the Apollo Moon program, the president's death only strengthened it, as the political establishment, media, and public opinion galvanized, in a state of shock and mourning, to redouble their commitments. President Kennedy had tied his own personal and political fortunes to a program of great national purpose, Apollo. Now, with his death, the nation responded by renaming Cape Canaveral as Cape Kennedy, rededicating the Moon program to his personal memory.

Dramatic events soon happened in the USSR as well: in the statist coup that overthrew Nikita Khrushchev from power in October of 1964. A cabal of three leading communist party officials removed him from his posts, sending him to house arrest for the remainder of his life. It was an act bound up in political symbolism, Khrushchev was brought down and humiliated at the very moment of one of his public triumphs, another space first, in the Voskhod program that we cover in Chapter 7.

The new regime was now led by Leonid Brezhnev, the veteran party commissar in charge of the military-industrial complexes of the USSR. His key allies were Dmitrii Ustinov, the military chief of the country's defense and space industries, and Mikhail Suslov, a veteran Stalinist in charge of the country's ideological and propaganda interests. They had plenty of personal and political grievances against Khrushchev: his failed economic initiatives and cultural liberalizations, for instance. Yet also paramount were Khrushchev's foreign-policy failures, first among them the Cuban Missile Crisis, which left the country humiliated on the world stage. This dramatic spotlight on the USSR's strategic nuclear disadvantages, more than any other factor, convinced the new leaders to address the imbalance. Yes, they did continue to patronize Soviet space programs, even their own abbreviated "civilian" Moon program, but only so long as they did not deviate from the priority to build the fleets of ICBMs and other space military platforms to compete with the US in the Cold War arms race. It is quite a story, how the US soon headed to the Moon and the USSR invested in a fleet of ICBMs instead.

Vertical Horizons: Above and Below Worlds

One of the most celebrated personal and political confrontations of the Cold War, between Premier Nikita Khrushchev of the USSR and Vice President Richard Nixon of the US, happened at the American National Exhibition in Moscow (1959). These were the famous "Kitchen Debates." A moment when the "above world" of Soviet nuclear-tipped ICBMs fused with the "below world" of American everyday life. The exhibition was actually part of the warming of relations between the two countries, part of an exchange of showcase exhibits in each country, in order to introduce something of our national values and common humanity to each other, and maybe propagandize a bit.

The Soviets went to New York City's Coliseum in Manhattan, staging the Soviet National Exhibition that featured science and technology, centered on the Sputnik satellites. The US and USSR flags, stars and stripes alongside the hammer and sickle, adorned the outside of the coliseum. The American National Exhibition went to Sokolniki Park in suburban Moscow, featuring a wide array of consumer items: foods and drinks, literature, music, automobiles, home appliances, a model kitchen, and even a model home. The Soviets may have had Sputnik, but America boasted "Splitnik," the nickname for an average American split-level home on display in Moscow, though also cut in half to accommodate the 2 million Soviet visitors. They were used to standing in lines for shortage goods like fresh fruits and basic consumer needs. Now they stood in line to see and taste a part of America. The Pepsi tasting stand was a crowd favorite, with Russians enjoying the eye-popping, refreshing drink for the first time.

On 24 July 1959, Splitnik's kitchen was one of the scenes of the impromptu Khrushchev–Nixon debates: the Soviet leader continuing his superpower and rocket harangues; the vice president making a case for American consumerism and capitalism. Nixon conceded that the Russians had more powerful rockets, but defended American middle-class prosperity, symbolized by the color television set. Khrushchev was unrelenting, boasting of Soviet superiority in weapons and consumer goods. The truth was harder. At the time, the Soviet Gross National Product was less than half of the US.

Select Bibliography

Barber, John, ed. *The Soviet Defence-Industry Complex from Stalin to Khrushchev*. New York: St. Martin's, 2000.

Beschloss, Michael. *Mayday: Eisenhower, Khrushchev and the U-2 Affair*. New York: Harper and Row, 1988.

Bluth, Christopher. *Soviet Strategic Arms Policy before SALT*. Cambridge: Cambridge University Press, 1992.

Burrows, William E. *Deep Black: Space Espionage and National Security*. New York: Random House, 1986.

Collins, Martin. *Space Race: The U.S.–U.S.S.R. Competition to Reach the Moon*. San Francisco: Pomegranate Communications, 1999.

Day, Dwayne, John M. Logsdon and Brian Latell, *Eye in the Sky: The Story of the Corona Spy Satellites*. Washington, DC: Smithsonian, 1998.

Firth, Noel and James Noren. *Soviet Defense Spending. A History of CIA Estimates, 1950–1990*. College Station: Texas A&M University Press, 1998.

Gaddis, John Lewis. *We Now Know: Rethinking Cold War History*. Oxford: Clarendon Press, 1997.

Gavaghan, Helen. *Something New Under the Sun. Satellites and the Beginning of the Space Age*. New York: Copernicus, 1998.

Khrushchev, Sergei N. *Nikita Khrushchev and the Creation of a Superpower*. State College: Penn State Press, 2000.

Macrakis, Kristie. "Technophilic Hubris and Espionage Styles during the Cold War," *Isis* 101 (2010): 378–385.

Naftaly, Timothy, and Aleksandr Fursenko. *One Hell of a Gamble: Khrushchev, Castro, and Kennedy, 1958–1964*. New York: Norton, 1997.

Naftaly, Timothy, and Aleksandr Fursenko. *Khrushchev's Cold War: The Inside Story of an American Adversary*. New York: Norton, 2006.

Nash, Philip. *The Other Missiles of October: Eisenhower, Kennedy, and the Jupiters, 1957–1963*. Chapel Hill: University of North Carolina Press 1997.

Neufeld, Jacob. *The Development of Ballistic Missiles in the United States Air Force, 1945–1960*. Washington, DC: United States Air Force, 1990.
Oberg, James. *Red Star in Orbit: The Inside Story of Soviet Failures and Triumphs in Space*. New York: Random House, 1981.
Oberg, James. *Uncovering Soviet Disasters: Exploring the Limits of Glasnost*. New York: Random House, 1988.
Patterson, Thomas G., ed. *Kennedy's Quest for Victory: American Foreign Policy, 1961–1963*. New York: Oxford University Press, 1989.
Petroski, Henry, *To Engineer is Human: The Role of Failure in Successful Design*. New York: St. Martin's, 1989.
Pocock, Chris. *50 Years of the U-2: The Complete Illustrated History of Lockheed's Legendary Dragon Lady*. Atglen: Schiffer Publishing, 2004.
Podvig, Pavel, ed. *Russian Strategic Nuclear Forces*. Cambridge: MIT Press, 2001.
Pry, Peter Vincent. *War Scare: Russia and America on the Nuclear Brink*. New York: Praeger, 1999.
Quenoy, Paul du. "The Role of Foreign Affairs in the Fall of Nikita Khrushchev in October 1964," *International History Review* 25/2 (2003): 334–356.
Rich, Ben R., and Leo Janos. *Skunk Works. a Personal Memoir of my Years at Lockheed*. Boston: Little, Brown, 1994.
Richter, James. *Khrushchev's Double Bind: International Pressures and Domestic Coalition Politics*. Baltimore: Johns Hopkins University Press, 1994.
Richelson, Jeffrey. *America's Secret Eyes in Space: The U.S. Keyhole Spy Satellite Program*. New York: Harper Collins, 1990.
Roman, Peter J. *Eisenhower and the Missile Gap*. Ithaca: Cornell University Press, 1996.
Sambaluk, Nicholas. *The Other Space Race: Eisenhower and the Quest for Space Security*. Annapolis: Naval Institute Press, 2015.
Schwartz, Stephen. *Atomic Audit: The Costs and Consequences of U.S. Nuclear Weapons Since 1940*. Washington, DC: Brookings Institution, 1998.
Stares, Paul. *The Militarization of Space: 1945–1984*. Ithaca: Cornell University Press, 1985.
Stern, Sheldon, *Averting the Final Failure: John F. Kennedy and the Secret Cuban Missile Crisis Meetings*. Stanford: Stanford University Press, 2003.
Taubman, Philip. *Secret Empire: Eisenhower, the CIA, and the Hidden Story of America's Space Espionage*. New York: Simon and Schuster, 2003.
Wegner, Andreas. *Living with Peril. Eisenhower, Kennedy and Nuclear Weapons*. Lanham: Rowman & Littlefield, 1997.
Westad, Arne. *The Global Cold War: Third World Interventions and the Making of our Times*. Cambridge: Cambridge University Press, 2006.
Zaloga, Steven J. *The Kremlin's Nuclear Sword: The Rise and Fall of Russia's Strategic Nuclear Forces, 1945–2000*. Washington, DC: Smithsonian, 2002.
Zubok, Vladislav. *A Failed Empire: The Soviet Union in the Cold War from Stalin to Gorbachev*. Chapel Hill: University of North Carolina Press, 2009.

7 Rendezvous

Voskhod and Project Gemini

As with the Cold War intercontinental ballistic missile competition, so also in human spaceflight, the accent of the space race was on continuous improvement and competition. The Vostok and Mercury programs were foundations. Each taught important lessons to the managers and engineers who made them. Each created the networks and infrastructures that enabled the programs that followed. Yet the intensity of the space race also ensured that these dramatic programs were short-lived, as Vostok gave way to Voskhod, and Mercury to Gemini.

In this rivalry, the Soviets achieved an impressive and continuing series of object firsts. The advantage usually went to the USSR. Yet Soviet firsts were more about getting somewhere or doing something (tasks and objects) before the US. All of the Soviet firsts, from Sputnik to Gagarin and Vostok, were supremely political events, meant to impress the world and surprise and frustrate the US. They were limited technological feats with short-term political value. They were a sequence of exclamation points rather than turning points.

None of this is to deny Soviet priority but to better define it. We ought to step out of the moment and the rivalry, draw out the broader and deeper comparisons, weigh program against program, mission against mission, craft against craft. From these perspectives, the Voskhod program was merely one more set of surprise tactical firsts. Gemini far surpassed it in terms of the quality of the crafts that it flew, and in terms of the missions that it accomplished. The Gemini program was a truer historical turning point with long-range consequences. Building upon the Mercury capsule's superior piloting capabilities, and the coming Apollo program's interplanetary reach, Gemini had strategic significance, a rendezvous with the future. These three American programs amounted to a remarkable and unmatched series of accomplishments in just over ten years, from 1961 to 1972, President Kennedy's campaign era of purpose and promise.

The Voskhod Mission Firsts

The USSR Voskhod (the "Dawn") program had no other purpose than to answer the US Gemini. Soviet state authorities formally approved plans for Voskhod

DOI: 10.4324/9781003494409-10

in March 1964, well after Gemini's establishment in December of 1961. The task, of course, was to have the two Voskhod missions happen before Gemini, in late 1964 and early 1965. This was exactly what Voskhod was meant to do: a "stop-gap" effort to achieve a speedy series of Soviet firsts at the expense of America's slower and very public space projects. Its whole existence was to prove that the USSR was still ahead, if once again in quantity rather than quality, with firsts rather than bests. As Vostok was preoccupied with doubles, pairs of cosmonauts and of spacecraft, Voskhod was focused on a triple play, the world's first three-person crew; and on a dramatic new surprise, the world's first spacewalk (or in NASA's terms, Extra Vehicular Activity or EVA).

The Voskhod missions were planned and organized with haste. They meant that a series of later Vostok flights were shelved as irrelevant to the space race. These included plans for several week-long spaceflights, as well as several more for scientific experiments and military reconnaissance missions. What counted was the political value of beating the US. This also meant that the new Voskhod missions were fraught with danger, governed less by careful consideration than by a calculated recklessness, what the engineers and cosmonauts at the time both termed "risk planning." Soviet policymakers and designers both accepted a higher level of risk than ever before, increasing the pace of events and planning, always to keep up with Gemini, at the expense of safety and caution.

Voskhod 1 (October 1964) was the world's first three-person crew. It was built from the modular Vostok spacecraft model, with continued use of the automated controls as the primary system and with manual controls as the backup. Like Vostok, it used the same spherical re-entry module with the instrument module for propulsion and life support. Korolev appointed one of his best designers, Konstantin Feoktistov, who had contributed crucial innovations to both the Sputnik and Vostok programs, to modify the spacecraft. He also appointed him a member of the crew, a powerful incentive for Feoktistov to modify the craft well. He would have to survive his own Voskhod modifications. Several of the changes were safety-minded, including a new life-support system, which removed the carbon dioxide from the atmosphere to make it more breathable; and a new single-parachute landing by both the crew and craft together; and a backup retro-rocket.

In terms of risk planning, so as to accommodate the new weight of three cosmonauts, the craft was stripped of many of its earlier safety redundancies. Further, the cosmonauts wore no space suits, essential in the case of a breakdown of the craft's life-support system. Nor did they have ejection seats, essential for an accidental explosion at launch. To complicate matters even more, the new seats for the three members of the crew, given space constraints, were placed perpendicular to the spacecraft controls, meaning the cosmonauts had to twist their bodies and necks to reach them. For all these risks, the Voskhod 1 mission proceeded smoothly enough, hailed as another proof of Soviet space superiority: three men in space in one craft for the first time.

The very success of the mission challenged the Soviet space program to do still more, to keep up with its own pace. Risk planning continued in the preparations for the next mission. Korolev now arranged a test flight of an unpiloted version of Voskhod 2, some three weeks before its scheduled flight. It was destroyed when a bad signal triggered the self-destruct mechanism on the craft, meant to protect any populated areas in the case of a wayward craft during descent. This was the last test capsule, leaving mission control without important flight data. The program was at a crossroads: either launch the remaining Voskhod capsule as another test and fall well behind the expected US Gemini schedule; or launch Voskhod as a piloted mission and surpass Gemini, namely Ed White's spacewalk then scheduled for May of 1965. Korolev chose the risk.

Voskhod 2 (March 1965) was the Soviet space program's most challenging mission to date, a second redesign of the craft to accommodate the essential piece of new equipment: an ingenious airlock for Aleksei Leonov's spacewalk. Made of double-walled pressurized rubber, it opened accordion style as air was pumped through the small rubber tubes over the length of the canvas, expanding from 27 to 79 inches (nearly 7 feet). Several more internal spacecraft controls were moved in order to make room for the panel to work in this airlock chamber. Leonov and Pavel Belyaev also wore spacesuits (each with two pressurized membranes for extra protection), with the use of oxygen packs that lasted for 45 minutes, as Belyaev was supposed to rescue Leonov in case he lost consciousness out in space. And, lest we forget the priority purpose of this flight, Korolev made sure that the spacecraft had four cameras to capture what became the last great Soviet space first: Leonov's walk into outer space.

On that remarkable day, 18 March 1965, as cosmonaut Leonov emerged from Voskhod's airlock, he measured his body against the expanse of one-quarter of the planet: finding the Black Sea below him, Italy and Greece to his left, the Caucasus and Volga to his right, the Baltics above his head. He was the first human being outside a spacecraft to fly at 17,500 miles per hour. "What I saw as the hatch opened took my breath away," he later wrote, seeing the deep blue colors of the Earth set against the dark black of outer space, what no astronaut or cosmonaut had ever seen before, even from the restricted glass of a spacecraft porthole. Then, "with a small kick, as if pushing away from the side of a swimming pool, I stepped away from the rim of the airlock," assisted by a 16-foot cord. In this instant, Leonov became a modern version of da Vinci's Vitruvian Man (as illustrated on his space patch): imagining himself a "seagull with its wings outstretched, soaring high above the Earth." He "felt like a representative of the human race," at once "enormously powerful" yet also "insignificant," as if he was a "tiny ant" against the backdrop of Earth. The film footage of the event, shown on tape over Soviet television throughout the day, captured him in just this way: all-powerful yet fragile.

The trouble started when, after ten minutes, he was over the Pacific Ocean and preparing for re-entry to the airlock. On his first few attempts, he simply did not fit, exhausting himself in the process, his heartbeat racing to 193 beats a minute and his breathing 12 times greater than normal. His spacesuit filled with sweat. Then he remembered how, on exiting the craft, he had "felt like I shrank," as his "hands pulled away from his gloves" and his "feet pulled away from his shoes." In other words, his spacesuit had ballooned and stiffened from the pressure differential in space. Finally, after several minutes of desperate struggle, he released some of the internal pressure in his suit with a small valve, and made his way back in, though now headfirst, meaning that he had to curl his body around within the airlock in order to reach up and close the hatch. But he had made it back.

Still more troubles soon followed. With 22 hours left in the flight, after re-entering the cabin, the hatch did not fully seal, forcing an environmental control system to compensate by flooding the cabin with oxygen, the pressure rising to hazardous levels, which the crew finally reduced by regulating the temperature and humidity controls. Then, after the airlock was ejected by explosive bolts, the craft began to roll wildly, sunlight flashing in and out of the portholes and disorienting the crew. This briefly stopped as the automatic re-entry and landing program was activated, but then failed again, forcing the crew to make a delicate manual re-entry. Leonov had to lean over from his seat, and hold Belyaev by his legs, in order to adjust the optical device and position the craft for descent back to Earth.

The delays meant one extra orbit, followed by one last mishap. As with Gagarin's flight, the instrument module did not separate immediately from the re-entry module, causing the craft to spin out of control for a time. All of these troubles meant that the craft landed some 1,200 miles beyond its targeted site, coming down in the deep snow of an evergreen and birch forest near Perm, Siberia. In the first of several cinematic moments, still scalding from the heat of re-entry, the craft actually sank through the melting snow. Lodged against two trees, the cosmonauts also had to rock the craft back and forth to free the hatch. Rescue aircraft found their distress signal, though none could reach them. The crew spent the first night huddled in their craft, wolves howling around them, 22 degrees Fahrenheit below zero outside. By the next day, rescuers helped them ski some five miles to a hastily built log cabin, and after the forest was cleared for a helicopter retrieval, they were finally able to start their journey back to Baikonur.

These difficult orbital and re-entry issues put the plans for all of the upcoming Voskhod missions on permanent hold. Voskhod 3 was to have been a long-duration flight of 18 days, also serving as a military surveillance platform, and to be paired with Voskhod 4 (scheduled for 20 days) in an artificial gravity test. Both were to be outfitted with improved environmental systems for their longer durations. Voskhod 5 was planned as an all-female crew of two cosmonauts, with a spacewalk, hence another first, two women in outer space at the same time. Voskhod 6 was to have tested the world's first jet pack in a spacewalk. Korolev and his superiors planned, in one way or another, a whole series of new firsts to again beat the Americans in outer space. Yet all were either canceled or shelved over the next several years as undue risks. In trying to race ahead, the Soviets had instead fallen well behind. Two years passed until the next piloted missions, beginning with Soyuz 1 in April of 1967, though sadly it was fraught with a whole new variety of risks all of its own.

In these terms, the Voskhod program, along with Khrushchev's downfall, was a significant turning point in the history of the USSR and its spaceflight programs. The Soviets may have achieved the first three-person crew and first spacewalk, but these were temporary and superficial gains, pure publicity and without any long-term effect. In this sense, the Soviet space program reached its apogee with Voskhod, one last series of space firsts to clinch its lead before the coming of Gemini and Apollo.

Project Gemini Thresholds

Project Gemini, already over three years in the making, launched its first mission only four days after Voskhod 2 returned to Earth. What followed was a dizzying schedule and unmatched pace of flights: ten piloted missions in 20 months. These flights, though not without flaws or mistakes, also achieved almost all of their objectives with near-perfect results. They established an American preeminence in space technology.

First known as Mercury Mark II, NASA rechristened it as Project Gemini in January of 1962 to honor the twin spacecraft meant to rendezvous and dock in space, but also the twin astronauts who would fly together, and the twin program with Apollo. Its discrete goals were to establish the technologies and techniques in order to rendezvous and dock two spacecraft in Earth orbit; to place astronauts in outer space for EVA spacewalks; to simulate a Moon mission with long-duration flights of up to two weeks; and to make precise landings on Earth by way of guided re-entries. Each of the Gemini missions conducted a variety of scientific and biological experiments as well. For these purposes, three of the Mercury astronauts were eventually chosen to serve in Gemini (Grissom, Cooper, and Schirra), with the first new Gemini-9 astronauts chosen in September of 1962, and 14 more a year later.

Like Mercury and Apollo, Gemini grew out of NASA's newly minted institutional networks, which built upon its early and rather improvised Mercury successes with a growing workforce and a new set of hierarchies and procedures. The Office of Manned Space Flight, centered at NASA Headquarters in Washington, DC, now supervised the Manned Spacecraft Center in Houston, Texas. Their charge was to follow two parallel tracks: to watch and try to keep up with what the Soviets were attempting and achieving, but most importantly to follow their own timetables and imperatives, both in answer to President Kennedy's ten-year plan. Schooled in the successes and near failures of Mercury, NASA veterans Chris Kraft, at Houston's Mission Control Center, and Gene Kranz, director of its Flight Control Teams, led the effort to launch Gemini and prepare the way for Apollo.

NASA also grew into the existing aerospace industries. In terms of historical perspective and financial investments, Projects Mercury, Gemini, and Apollo were management and technical initiatives of remarkable complexity and challenge, the latest in a long line reaching back to the Manhattan Project in the Second World War and to the project system of the early Cold War, as covered in Chapter 4. NASA set the project timetables, negotiating the contract agreements with industry and academia, and ensured the reliability of the machinery. Here were all the advantages for the next round of American architectures and achievements in space.

Continuity and innovation took shape in the form of the Gemini spacecraft. McDonnell Aircraft built it, as it had the Mercury capsule, but with a new design for its new functions. Gemini was, quite literally, Mercury and more. It was a Mercury-type conical craft, with similar shape and ablative shield, but with myriad improvements. For example, Mercury's control and environmental systems had been fully integrated and interlocking within the capsule, the retro-rockets part of the capsule too. Gemini followed a new modular approach, on the order of the Vostok and Voskhod programs. Besides the control and re-entry cabin, Gemini included detachable equipment and propulsion modules, both jettisoned before final re-entry to Earth's atmosphere.

Building on examples from the aviation industry, NASA engineers (under the lead of Jim Chamberlin, formerly of Avro Aircraft) drafted these parts as "modular, accessible, and serviceable." They designed a craft that could be built easily and efficiently through a "routine and systematic" production cycle, reaching the launch site fully tested and ready to fly. They ensured that the control systems were easily accessible from interior or external panels, this for quick servicing or replacement. Gemini was, in these ways, a masterpiece of aerospace construction.

As to the control and re-entry cabin, with the close cooperation between Chamberlin's teams and the astronaut corps, especially Gus Grissom, NASA built Gemini like a jet fighter, to be piloted and controlled in ways both reminiscent and intuitive of the astronaut's own flight experiences. This was perhaps Gemini's most perfect aerospace moment: the symbiosis between Mercury's piloting controls and Gemini's advances, between the aerospace industry's techniques and the government's political priorities, between the astronaut's missions and their proficiencies as test pilots.

The central innovation of the Gemini craft was the Orbit Attitude and Maneuvering System (OAMS), which used 16 thrusters for attitude controls (pitch, yaw, and roll); for navigating the craft through rendezvous and docking with other craft (the Agena Target Vehicle); as well as for making its changes in orbit and altitude. To assist in these maneuvers, as well as in its precise re-entry schemes, Gemini used in-flight radar and a variety of other aeronautics instruments and methods. It also had the first onboard digital computer and software for these navigational procedures, especially for its precise re-entries; and in terms of continuous improvement within the program, the computer of the Gemini 3 had a 4,096-word memory, while Gemini 8 held enough for 500,000. Among Gemini's other innovations: ejection seats (rather than the Mercury escape tower), the superb Titan II ICBM as its launch platform, solid-state electronics, and fuel cells in the later flights to generate electricity for the control systems.

The early missions were devoted to assessing the machinery: Gemini 1 and 2 were unpiloted tests of the spacecraft and booster. In the first piloted mission, Gemini 3 (March 1965), Gus Grissom and John Young used the craft's OAMS to change their orbit, the first such maneuvers in history, and also achieved the first controlled re-entry, using the craft's thrusters to provide lift and also navigate to a precise landing site. James McDivitt and Ed White, aboard Gemini 4 (June 1965), matched a Soviet long-duration record with 62 orbits of their own. Gordon Cooper and Pete Conrad, aboard Gemini 5 (August 1965), far surpassed it with 120 orbits of Earth, nearly eight days in space.

Aleksei Leonov may have made the first human being to make a spacewalk, but White's Gemini-4 "American" first spacewalk was made on live television. He became a human rocket of sorts, too, maneuvering about with a jet gun fueled by two small tanks of compressed gas. Gemini 6 and 7 (December 1965) offered perhaps the most dramatic turning points of all: paving the way for the rendezvous and docking missions to come; and culminating with the greatest of long-duration flights to date. Frank Borman and Jim Lovell launched first to begin their marathon 14-day mission. Following a faulty launch sequence, later resolved, Walter Schirra and Tom Stafford launched eight days later to meet them in an orbital rendezvous, coming within 1 foot of each other. Gemini 7 was left to finish its two-week sentence. Borman remembered the boredom, most of all. "We waited for time to pass … checking off days … we just drifted and tumbled for days at a time," as he and Lovell sang a country and western song out of tune, to pass the time and regale the listening public.

The later Gemini missions were remarkable for challenges with docking procedures and spacewalks that, through trial and practice, the astronauts and NASA engineers resolved in time. Neil Armstrong and David Scott experienced a space emergency with Gemini 8 (March 1966) when, after docking with the Agena target vehicle, both craft began to spin out of control, which only worsened after they undocked, the spins reaching one revolution per second. "We have serious problems here," reported Scott. "We're tumbling end over end up here … and can't turn anything off." All attempts to steady the craft failed until Armstrong shut down the OAMS, discovering what turned out to be a stuck attitude control thruster. As for rendezvous and docking, Tom Stafford and Eugene Cernan made a partial success with Gemini 9 (June 1966).

Coming missions fared much better. John Young and Michael Collins, aboard Gemini 10 (July 1966) docked with the Agena and then changed their orbit to a record 475 miles altitude, where they then made another rendezvous with an older Agena. Pete Conrad and Richard Gordon in Gemini 11 (September 1966) made a rendezvous and docking with the Agena (which they reached on their first orbit, another first), then rose to a record 850-mile altitude. At the highest apogee of any orbit to date, they were the first spacefarers to see the full curvature of the Earth, the planet as a sphere. They also tethered two spacecraft together in an artificial gravity experiment that made them minor artificial planets all their own. In that summer of 1966, "Wild Thing," by the Troggs, which was "No. 1" on the Billboard "Hot 100" for two weeks in July, captured some of the excitement of these Gemini missions and their dynamic, shaking rockets in flight.

As for the five Gemini EVA spacewalks, after Ed White's successful first, Eugene Cernan's was burdened with too many complex tasks in too short a time window. When he finished, his spacesuit was soaked with perspiration, his visor fogged with moisture, and his heart racing at 180 beats a minute. "My spacewalk from hell," so he called it. Michael Collins spacewalked more successfully to the Agena target vehicle for retrieval tasks; as did Richard Gordon in his tethering tasks, though they wore him out. Besides making by-now routine rendezvous and docking procedures, the final Gemini 12 flight (November 1966) peaked with the most successful EVA of all. After months of practice on Earth, including underwater, Buzz Aldrin perfected the techniques of moving about and working in outer space with a whole series of Velcro fasteners, bungee cords, footplates, and attachments, celebrating what he called "suitable restraints and careful movement."

None of these were spectacular firsts. They were of a different order than the first three-person crew or first spacewalk. They were not as newsworthy. But they were firsts as "bests," besting what the Soviets had done in terms of the science and technology, and besting America's own timetable to achieve the Apollo missions soon to come.

Table 7.1 Project Gemini milestones in the space race. Author.

Project Gemini Milestones in the Space Race
Onboard guidance computer
Astronaut-controlled orbital maneuvering and changes
Orbital rendezvous of one craft with another
Docking of a spacecraft with a propulsion stage (six consecutive ones of these)
Use of a propulsion stage to move to a higher orbit
Long-duration flights of up to two weeks
Extra Vehicular Activity -- spacewalks (five in all)
Controlled re-entry and precision landings (seven in all)
Artificial gravity tethering tests

The irony in all of this was that Gemini might have been able to reach the Moon, either as a circumlunar craft or even as a combination lunar orbiter and lander, as Mark Wade has argued. In such a case, it would have served more as a smarter and more efficient ladder into space, building on Earth-orbit achievements to make way for lunar exploration. Gemini could have become a more lasting scaled program of foundations and achievements. But the US project system was built on planned obsolescence, on creating the new at the expense of the old, what Joseph Schumpeter had called "creative destruction." In the shadow of President Kennedy's Apollo program, Gemini was never meant to be permanent. It was only ever temporary. It was but one tentative step, not a bridge, to Earth's orbit. NASA abandoned the Gemini hardware on its fast track to Apollo, setting a trend for a peculiar American disposable culture of spaceflight, as happened later with the abandonment of Apollo and the Space Shuttle.

Carefully hidden within these Gemini ventures were a number of top-secret military tasks as well. In fact, behind its plans and objectives, Project Gemini was as much a military as a civilian program. The Department of Defense and Air Force came to depend upon it as a replacement for their canceled X-20 orbital space plane; as well as a complement to their plans for a Manned Orbital Laboratory (MOL). In planning between 1963 and 1969, MOL combined the Gemini spacecraft with a military reconnaissance platform for Earth surveillance, serviced by the Keyhole cameras of the Corona satellite program. Gemini provided the control systems and re-entry vehicle, with a hatchway to connect it to MOL. The laboratory was scheduled to orbit the Earth over the course of about a month at a time, its tasks to photograph and spy on the USSR, and alternately even serve as a satellite interceptor and destroyer. The US was primed to bridge the military and civilian use of outer space.

Toward the very end of the Gemini program, on 3 November 1966, the Air Force launched a prototype of the MOL, attached to the Gemini spacecraft, boosted into space by a Titan III ICBM. But this was the closest MOL ever got to orbit. The Pentagon ultimately canceled the program just a few years later, in part given the mounting costs of the Apollo program and the Vietnam War, in part given the proven achievements of Corona. The piloted missions, planned for 1972 to 1975, never flew. Yet MOL left two important legacies. The first came in the form of the space race competition: Vladimir Chelomei's "Almaz" space station (1973–1977), remarkably similar to MOL and an outgrowth of the Soviet Soyuz program. The second legacy was in the later Space Shuttle, which in many respects was more of a successor to Gemini than to Apollo. For example, Richard Truly, pilot and commander of two shuttle missions and a later NASA administrator, came from the first class of MOL astronauts; and six out of the nine later MOL astronauts went on the serve as shuttle pilots and commanders as well.

Paraglider

In a clever but ultimately unsuccessful bid to land the Gemini on solid ground, McDonnel and NASA engineers originally planned to replace the traditional Mercury parachute, which could only safely drop the spacecraft in the ocean, with a Gemini paraglider, to land it safely on land. By parachute, the astronauts were completely passive, unable to guide their capsules in any way. Splashdowns by sea were also hazardous and costly. One proposed landing system was known as the "paraglider" or "Rogallo wing," named after its inventor, Francis Rogallo, a NASA engineer who, along with his wife, Gertrude, had been working on various flexible wing designs since 1948. The glider was a type of self-inflating airfoil. If properly deployed, it would allow the crew to sit upright, controlling the forward motion of the craft. In addition, the Navy would not have to scramble so many ships (from 8 to 27) in order to rescue a wayward or sinking capsule. From the outset, the idea was controversial, the key question being whether the paraglider was safe and reliable. There was even a design for a winged Gemini proposed for a wheeled landing. When the development of the paraglider could not keep pace with the timetable for Project Gemini, NASA scrapped it in 1964 in favor of the conventional, life-saving parachutes. Korolev had his own ideas to avoid the pitfalls of parachutes: a helicopter-style set of rotor blades to give the capsule lift and control, though it was never developed.

America's Apollo Initiatives

By the very nature of President Kennedy's Moon decision in May of 1961, Project Apollo was concurrent with both Mercury and Gemini. In Chapter 5, we discussed how all of the essential decisions about Apollo and Gemini were made either during or just after Mercury. This meant that, at one and the same time, NASA proceeded with two or three separate but interrelated tracks in human spaceflight. The complete history ought to follow these polyvalent lines, narrating how people and hardware, procedures and techniques, adapted between one project and others. One human factor is the easiest to map, as in the genealogy of the astronaut corps. In the first four Moon flights, 10 of the 12 astronauts had served in Gemini with flight experience, two more were backups, and half of all Apollo astronauts had Gemini experience. Three Apollo astronauts (Grissom, Schirra, Shepard, and eventually Slayton) had served in Mercury.

In terms of Apollo's mission objectives and approaches, the "mode decision" and spacecraft architecture were the most important elements for success. They were intimately related. The "how" NASA planned to get to the Moon, its mode decision, was the key to deciding "what" it would need to get there, especially in terms of the weight specifications of the all-essential lunar module, necessary to land a man on the Moon and return him safely to Earth, within a decade, of course. It was also crucial in terms of the size of the launch system to boost the spacecraft into a lunar trajectory. The mode decision was the essential hinge upon which so much else depended.

Between 1958 and 1962, the Army, Air Force, and NASA debated several flight modes for a possible mission to the Moon. An early favorite was Direct Ascent, using Wernher von Braun's Army plans for a massive Nova booster, calling for a single launch of one combined platform directly to the lunar surface. A variation was to send two missions to the Moon, requiring less powerful rockets: one carrying propellant for the return trip, the other the crew. This Direct Ascent mode to the Moon was the most challenging and visionary of all the choices. It would have established the US as the preeminent spacefaring nation. Yet this option was also very costly and time-consuming to build.

Recognizing these enormous hurdles, NASA began to consider Earth Orbit Rendezvous in 1958: this was a plan to launch some 15 smaller rockets to raise the lunar modules and landers into Earth's orbit, using a space station to assemble them and then boost them into a trans-lunar trajectory. Von Braun and others in the scientific and engineering communities, especially the President's Scientific Advisory Committee, saw this as the single best approach. It was the safest option and would have established a more permanent US presence in space, fulfilling von Braun's vision of space shuttles and space stations as steppingstones to the planets. Yet it also was costly, time-consuming, and complicated.

By the summer of 1962, after heated debates, NASA administrators finally decided in favor of a third and winning proposal: Lunar Orbit Rendezvous (LOR). Among others, John Houbolt at the Langley center had been one of its most dedicated promoters. This was a plan to send a complex of craft to orbit the Moon, sending one of its small component parts down to the surface and back again using orbital rendezvous techniques. In one crucial respect, LOR matched perfectly with John F. Kennedy's Apollo decision, as a political charge to take a man to the Moon and return him back to Earth, all within ten years. Houbolt's brilliance was to speak in scientific terms what President Kennedy had already said in purely political terms. NASA soon came to advertise LOR by way of the elegant image of a figure eight, using the power of the Saturn booster and the spacecraft thrusters, as well as the gravitational pulls of the Moon, to fulfill Apollo's measure.

The risks of such an approach were significant. LOR had to keep a three-person crew alive over the course of a perilous journey covering two weeks. The greatest risks involved the entry to and exit from lunar orbit. These had to be perfect. On the journey there to avoid crashing into the Moon or bouncing away from it. On the way back to avoid burning up in the Earth's atmosphere at too steep an angle; or bouncing away from it into an orbit of the sun at too shallow an angle. Once in lunar orbit, a breakdown in the life-support systems might be disastrous, so far from the Earth and any hope of rescue. A breakdown in the spacecraft propulsion systems in lunar orbit meant possibly being stranded there or even crashing into the lunar surface.

Yet the advantages of LOR were remarkable, especially in saving money and time. As Houbolt put it, "lunar orbit rendezvous offered a chain reaction simplification on all back effects: development, testing, manufacturing, erection, countdown, flight operation, etc. All would be simplified." The cascade of efficiencies meant that the Saturn platform would likely be powerful enough to launch the system to the Moon. Its engines were already being tested and perfected by 1962–1963. LOR favored the modular and functional designs for the Apollo Command and Service Modules, having already been prepared under Max Faget's direction in 1960 for NASA's earlier Direct Ascent plans. All that was needed now was to adapt a lunar module to the original design. Plus, LOR meant that Apollo could apply a "free return trajectory," using the very gravitational force of the Moon to help slingshot the spacecraft back to Earth, if necessary. This was a key to the mission profiles of Apollos 8, 10, and 11, though each ultimately inserted their spacecraft into lunar orbit using their own propulsion.

In terms of space architecture, LOR organized the Apollo spacecraft into three parts.

First was a cone-like command module (CM), which held the flight control equipment and the crew living compartment, the essential piece to return them back to Earth. Given the imperatives of precise navigation to and from lunar orbit, as well as the essentials of rendezvous and docking techniques, the electronics and computer technologies of the CM were a constant and crucial challenge. It was also outfitted with a propulsion system to navigate its precise re-entry path down to Earth, with an ablative shield to protect it from the heat of re-entry, like Mercury and Gemini, and with a parachute system for an ocean landing.

Second was a cylinder-shaped service module (SM), which contained the propulsion and life-support systems for the CM, and which was attached to the CM in flight, as the CSM, until re-entry when they separated. North American Aviation built the CSM modules, for $77 million each, in 1965-dollar values.

Third was the lunar module (LM) to carry two astronauts to the lunar surface and back, the CSM awaiting them in lunar orbit. The LM was composed of two stages, each with an engine of its own. Both were to land on the Moon, but only the Ascent Stage would launch from it, using the discarded portion as a boost platform. Of all the pieces of the Apollo spacecraft, this was the most challenging and problematic, regularly over budget and behind schedule. The LM had to be light yet sturdy, able to be guided and maneuvered down from the CSM to a planetary body without an atmosphere, and back again to the CSM in orbit. Grumman Aircraft Engineering Corporation built the LM for the price of $50 million each.

Go Fever

From the Mercury to Apollo, NASA personnel referred to the intense pace of spaceflight preparations as "Go Fever." The phrase meant the thrilling and inspiring drive to beat the Soviets in space spectaculars, as well as to fulfill the Apollo directive to reach the Moon before the decade was over. It translated into a frenzied and sometimes reckless pace that blinded NASA and its contractors to obvious and avoidable faults in the machinery and procedures. At fault were the political pressures from outside and inside the organization, as well as a culture of bravado and competition in the astronaut corps, centered on the reluctance to criticize a spacecraft and lose the flight, as well as one's reputation. Astronaut Eugene Cernan remembered, "things were moving too swiftly" before his Gemini 9 spacewalk; and "faulty assumptions, overly ambitious goals, and the hurry-up attitude known as *Go Fever* were about to send me walking into an unknown and dangerous environment." Flight Director Gene Kranz cautioned after the Apollo 1 fire: "We were too 'gung-ho' about the schedule and we blocked out all of the problems we saw each day in our work." Astronaut Donn Eisele later applied the term to mean a kind of Orwellian "groupthink," peer pressure that turned people into conformists, not honest and constructive critics. "Go Fever," an advantage and spur to action, escalating the pace and the risk, became a disadvantage, blinding people to the perils of complex spaceflight.

With many delays and problems, NASA engineers worked with painstaking care on reducing the weight of the LM and ensuring its essential control and navigation functions. Meanwhile, the Apollo Spacecraft Program Office continued with intensive pre-flight tests of the CSM spacecraft at North American's facilities and at the Kennedy Space Center's altitude chamber. By January of 1967, NASA was ready for a "plugs out" test of the Apollo 1 CSM on launch pad 34 at Cape Kennedy. This was a simulation of a launch countdown, though the spacecraft and rocket were not fueled, calling for the transmission of power from "plugs in" with the launch complex power, to "plugs out," on internal power alone.

This CSM test was in preparation for a planned 21 February flight of the Apollo 1 crew (Gus Grissom, Ed White, and Roger Chaffee), to be launched by the Saturn 1B booster, to test the CSM in Earth orbit, though using a "Block I" design, unable to dock with the LM. Grissom was a perfect choice as Command Pilot of the mission. He had flown Mercury 2 and Gemini 3, and also enjoyed the reputation as NASA's "engineering" astronaut, who had made significant design improvements for the Gemini and Apollo spacecraft. His faith in the Apollo CSM was far from certain. He and the crew warned NASA and North American about the dangers of the flammable Velcro that filled the craft, as well as the risks of hundreds of modifications made at the Cape as the test approached.

Toward the end of a busy day of preparations on 27 January, and after four hours of tests, the Apollo 1 crew followed through a formal checklist and countdown, filled with minor problems for review and resolution. Then the instruments recorded a voltage spike at 6:30:54 pm. Within a short 15 seconds, a flash fire had consumed the interior of the CSM cabin, killing the three astronauts, who likely died of asphyxiation, though they suffered extensive burns as well. As the fire fed on the high-pressure oxygen atmosphere and its flammable materials, temperatures reached 2,500 degrees Fahrenheit, the flames burst forth through the seams of the cabin onto the launch-pad area outside. Launch-pad and rescue personnel were able to open the cabin hatches after some five minutes. They had to work their way through the bolts, made all the more difficult by the high pressures in the cabin, and by the absence of even basic or working fire-safety equipment. There were not enough fire extinguishers handy, or proper gas masks, or asbestos gloves to open the heated hatch. Some 30 men were injured. But they were too late.

The Apollo 1 fire sent the program into a 21-month hiatus before the next piloted launches. No one had anticipated the disaster: in a simple test countdown, on a launch pad, with an unfueled rocket. The Apollo Review Board, in a 3,000-page study, supplemented by a Congressional investigation, found that the fire probably began with a voltage spike in a bundle of wiring, whose coating was degraded from constant contact and small fluid leaks. The subsequent electrical arc, in a pure oxygen and high-pressure environment, surrounded by flammable materials in the cabin and on the astronauts' suits, caused the fire. The accident was perfectly avoidable and inevitable. Apollo 1 was "a fire waiting to happen," said Chris Kraft. "We had sort of stopped learning," remembered Gene Kranz. Basic physics and chemistry pointed to the causes. North American and a General Electric subcontractor had written to NASA about the fire hazards. General Samuel Phillips, head of the Office of Manned Spaceflight, had written to North American about its weak quality controls. However, NASA and the contractors did not heed these warnings.

As a result of the investigation, no one person or persons were held to blame or punished. Several managers were reprimanded. Harrison Storms was replaced as programs manager of the CSM and Joseph Shea as head of the Apollo Spacecraft Program Office. However, the management and engineering teams were otherwise too complex and intertwined to assign any one person the blame. The judgment simply pointed to a series of mistakes in the "Command Module design, workmanship and quality control." In a plain accounting of words, the review board applied these flat euphemisms to absolve the very bureaucratic mistakes and mismanagement responsible for the accident in the first place. This was the nature of the modern bureaucratic complex at the heart of NASA's spaceflight enterprise.

Table 7.2 The Apollo and Soyuz accidents (1967). Author.

Apollo US	*Soyuz USSR*
1967	1967
27 January	24 April
Apollo 1 (Saturn 1B)	Soyuz 1 (R-7 Soyuz)
Grissom, White, and Chaffee are killed in a spacecraft fire during a countdown test.	After completing 18 orbits, Komarov is killed by a failed reentry.

These very bureaucracies, though responsible for the disaster, found the means for self-renewal and progress, not only to improve the mechanics of the CSM and LM, but also to reconfigure how their management and engineering teams worked. Over the course of an intense six months, NASA transformed the spacecraft. It replaced the flammable materials in the cabin and on the spacesuits; placed switch guards on the wiring; and improved the welding of the fluid lines. The hatch was made easier to open from within and without. New fire-safety procedures were instituted for the pad area. The machines were overhauled, and management problems fixed, leading the way to the coming successes of the Apollo Moon launches.

The Origins of Soviet Soyuz

The Soviet space program answered its brief Voskhod program, as well as the dramatic series of American Gemini missions, with a long-term spaceflight investment of its own: Soyuz or "Union." Korolev had begun serious planning for it in 1962, but only received formal approval to fully develop the designs and build the craft in December of 1963. The first Soyuz (7K OK) was a Soviet answer to the Gemini spacecraft, though for a three-person crew and with automated guidance, meant to conduct orbital maneuvers in Earth orbit, to rendezvous and dock with other spacecraft, to perform precise and modified re-entries, and to serve as a platform for EVA spacewalks and scientific experiments.

In these early years, Soyuz was never just any one thing. Korolev and his teams were under severe demands from the political and military leadership to compete with the US on several fronts. Yet they were also constrained and often frustrated by mixed signals and meager funds, or by the change in leadership that overthrew Nikita Khrushchev, or by several rising competitors from within the space program. In these contexts, Korolev wisely adapted Soyuz for a variety of purposes, including as a military reconnaissance station, as an abbreviated circumlunar craft to orbit the Moon, and even as an expanded edition to follow the Apollo model of LOR and land a man on the Moon.

With all of these varied functions in mind, including his own success as the chief designer, Korolev's special design bureau developed a masterpiece piloted spacecraft: modular, serviceable, and adjustable. Mark Wade has called it "the longest-lived, most adaptable, and most successful manned spacecraft design" of the Space Age. Although Soyuz experienced redesigns and improvements over the years, its basic model has always been as a three-part craft, including: a spherical orbital module with the controls for docking and with room for experiments; a "headlight" shaped descent module, with heat shield and parachute system, the shape designed to provide some banking and lift for an easier re-entry; and a cylindrical instrument module with life-support systems, fuels, and communications equipment. By making these modules separable on re-entry, the Soyuz system saved precious space and mass for the crucial descent module; it had the absolute and most efficient minimums to bring back the three cosmonauts alive and well.

Much like the Gemini spacecraft, these modules were designed for mass production in the Soviet military-industrial complex, and for accessible service and replacement. Following the earlier experience with Vostok, the accent was on automated rendezvous and docking rather than on active piloting controls, though the pilots could always access manual controls, if necessary. The beauty of the whole system was that, with minimal redesign and effort, the modules could be replaced or adapted to perform any one of its possible functions. All it took was re-outfitting the orbital module with new equipment or the instrument module with new loads of propellants.

The Soyuz program, like the coming Moon program and all the other spaceflight ventures he directed, depended on the force of Korolov's personality and political authority. Yet his own biography offers a cautionary tale about building a whole spaceflight bureaucracy around one man. Korolev's unexpected and premature death (during minor surgery) in January of 1966, shortly after Gemini 6, threw his spaceflight programs into chaos for some months. His replacement, Vasilii Mishin, although a talented engineer and accomplished administrator, lacked the force of personality and managerial skills to lead such complex ventures. He was also an easy mark for the political leadership above him, and the several merciless competitors around him, as we shall see in more detail in Chapter 8.

By late 1966, Mishin and his teams were ready for unpiloted tests of the Soyuz system. The first three launches over that winter, conducted under the code name, Kosmos, were all failures of one kind or another. The first test on 28 November experienced failures in the control system and in fuel consumption while in orbit, where it probably self-destructed on command. The second test on 14 December resulted in a launch-pad explosion with at least one dead and several wounded. The third test on 7 February also experienced control failures in orbit. Mission control was able to guide it to a re-entry, though the life-support system and heat shield of the descent module failed and would have killed anyone on board. This Kosmos crashed and sank into the Aral Sea.

Regardless of this sequence of failures, and against the advice of some leading engineers, rather than conduct more unpiloted tests, the Soviet space program proceeded with the planned piloted flights of Soyuz 1 and 2. From hindsight, the decision seems almost inconceivable and suicidal. Yet the new regime of Leonid Brezhnev, through its lead overseer of the spaceflight programs, Dmitrii Ustinov, was campaigning for immediate results. They had their sights on a series of Earth-orbit successes in time for the May Day celebrations that spring; leading to a piloted circumlunar flight of Soyuz, just in time to beat the US to the Moon by the 50th anniversary of the Bolshevik Revolution on 7 November 1967. Here was a symbolic moment of the Communist Party's legitimacy to rule. Quite a motivation.

These political pressures were likely also influenced by the successes of the Gemini program, and by the long hiatus (two years) since the last Soviet space launch, with Voskhod 2. The Apollo 1 disaster probably had an effect too, inspiring Soviet politicians and managers with the lucky chance to catch up with the unlucky Americans. All of these factors drove the events leading to the USSR's first spaceflight disaster, and the world's first disaster in space: Soyuz 1, scheduled for 23 April 1967, only three months after the Apollo 1 disaster. Vladimir Komarov was the perfect choice for the mission. He had served as the backup for Vostok 2 and had flown with the two other cosmonauts in Voskhod 1. His mission for Soyuz 1 was bold and ambitious, to serve as a rendezvous craft for Soyuz 2, which was scheduled to be launched after him with three cosmonauts aboard, and two of whom were to spacewalk to his craft in a dramatic display of Soviet technical expertise and precise navigation.

Mission control and Komarov both understood that the spacecraft was in peril soon after it was in orbit, when a series of cascade failures began to paralyze the spacecraft and doom the pilot. They started when the solar array did not unfold, remaining wrapped up with the service module. As a result, the spacecraft received only about half of the solar power that it needed to function. In a short time, the cooling system failed, as eventually did the batteries that powered several control and habitat systems. Komarov was also unable to maneuver the craft properly, as the attitude control sensors failed, and several navigation systems interfered with each other (one manual and the other automatic). In essence, as he tried to control the spacecraft, he lost further control of it.

Soyuz 2 could have launched to save him, but it likely would have failed too, as each of the practice Soyuz missions had before it, so Mishin made the decision to cancel. Komarov was ordered to make a quick re-entry, but not before several poignant conversations with his friends, co-workers, and family below. No one, including Komarov, expected him to survive. When it came time to re-enter Earth's atmosphere, the retro-rockets failed, and he was forced to conduct a challenging manual re-entry orientation. The troubles sent him into a speeding ballistic trajectory back to Earth, which he might have survived, but the main parachute failed to open. Cosmonaut Komarov and Soyuz 1 crashed into the Earth approaching 150 miles per hour (or more). In lavish and heartfelt ceremonies of national mourning, his remains were buried in the Kremlin Wall.

As a result of the Soyuz 1 disaster, the Soviet spaceflight program came to a halt for 18 months. In a mirror image of the Apollo 1 disaster, the Soviets conducted a vigorous search for the causes. The government investigation commission made a thorough review of the design, production, mechanical, and mission-control systems for Soyuz. We see the same push for new and better standards of quality assurance and accountability as in the US. Yet the Soyuz 1 disaster did little to alleviate the intensifying Soviet–American rivalry in the coming race to the Moon. Just as the Soviet space program took Apollo 1 as an opportunity to leap ahead in the wake of an American setback, now CIA analysts took Soyuz 1 as a mark of Soviet determination to beat the US once again, seeing Komarov's sacrifice as the sign of an intensifying program. Each disaster had brought a period of recovery, accountability and improvement, but no real rest from the pace of competition.

Vertical Horizons: Above and Below Worlds

While NASA and the Gemini program logged a number of impressive space firsts in the 1960s, Americans enjoyed it all from the comfort of their own living room couches, or for kids, from the shag carpet in front of the TV. Launches and returns were covered live on television, although not in the vivid colors of nowadays, but in rather drab black and white. TV shows with spaceflight themes kept Americans entertained, fun-loving shows that sent us up into space or brought it down to Earth.

Starting in 1962, the animated comedy series, *The Jetsons*, took a clever look at family life in the future. Set in Orbit City, where all buildings are raised or float high above the ground, the Jetson family wizzes about in their flying saucer "aerocar." They have a robot maid and numerous futuristic gadgets and time-saving conveniences. *Lost in Space*, which aired from 1965 to 1968, offered another family in space, this time using real actors. It followed the adventures of the Robinson family, along with the spacecraft's cowardly doctor and robot, who could zap enemies with electric charges from its metal claws. *Lost in Space* portrayed many new and imagined Space-Age technologies, such as jetpacks and personal computers, email, and protein pills. The spacesuits were modeled on those used by the Mercury astronauts, complete with Velcro fasteners. More down to Earth was the sitcom, *I Dream of Jeannie* (1965–1970), about an astronaut and his magical genie at glamorous Cape Canaveral and Cocoa Beach. The show often parodied the astronauts and their command as bumbling and inept. They might be space pioneers, but they were also human, just like the rest of us.

Fantasy and science fiction on 1960s television was also a forum for our darkest fears about the unknown future and outer space. The *Twilight Zone* series aired from 1959 to 1964. Each episode told a different story that usually ended with an ironic twist of fate, and while many of the episodes had a science fiction context, they addressed very earthly problems such as racism or the threat of nuclear war. "The Monsters are Due on Maple Street" (1960), showed how easily neighbors could turn on neighbors, suspecting each other of being aliens in disguise, a comment on the suspicion-filled climate of the anti-communist "Red Scare." The 1962 episode, "To Serve Man," had aliens come to Earth and make it a veritable paradise, but they also took humans back to their own planet, where they ate them. *The Outer Limits* (1963–1965) placed an even greater emphasis on science fiction and horror. In "The Invisible Enemy," astronauts explore Mars in 2021 only to find it inhabited by an army of gigantic sharks that move and kill from beneath the planet's surface. In "The Inheritors," the narrator described poor Earth as a "tumbling grain of sand in the darkness of unending space."

These varied series were meant to amuse or awe TV viewers, allowing ordinary Americans the chance to participate in the strange new "above and below worlds" of the Space Age. With more than 73 million television households in the US by 1963, with an average viewing time of about six hours a day, the shows filled a market demand.

Although each country had a roughly equal population, hovering at just above and below 200 million people, TV in the USSR fell well behind the US at this time, with only 10.5 million televisions in 1963. This number doubled within a decade, and Moscow and Leningrad quickly reached US levels of television coverage. As Kristin Roth-Ey has surveyed, the television set became a symbol of Soviet technological progress and prosperity, as important as Sputnik in the national pantheon. As a rule, though, there were no equivocal or critical spaceflight programs, only quiz shows, children's cartoons, film re-runs, serious dramas about the Second World War, and propaganda newsreels and documentaries about the triumphant Soviet exploration of outer space. The communist party-state maintained strict party control over TV stations and programming.

Select Bibliography

Bergaust, Erik. *Murder on Pad 34*. New York: G. P. Putnam's Sons, 1968.

Bilstein, Roger E. *Stages to Saturn: A Technological History of the Apollo/Saturn Vehicles Launch Vehicles*. Washington, DC: NASA, 1980.

Burgess, Colin, and Kate Doolan and Bert Vis. *Fallen Astronauts: Heroes Who Died Reaching for the Moon*. Lincoln: University of Nebraska Press, 2003.

Cernan, Eugene, and Donald Davis. *The Last Man on the Moon: Astronaut Eugene Cernan and America's Race in Space*. New York: St. Martin's, 2000.

Causey, William. *John Houbolt: The Unsung Hero of the Apollo Moon Landings*. West Lafayette: Purdue University Press, 2020.

Collins, Michael. *Carrying the Fire: An Astronaut's Journey*. New York: Cooper Press, 2001.

Dizard, Wilson. "Television in the USSR," *Problems of Communism* 12/6 (November–December 1963): 38–45.

French, Francis, and Colin Burgess. *In the Shadow of the Moon: A Challenging Journey to Tranquility, 1965–1969*. Lincoln: University of Nebraska Press, 2007.

Hacker, Barton and James M. Grimwood. *On the Shoulders of Titans: A History of Project Gemini*. Washington, DC: NASA, 1987.

Hansen, James. *First Man: The Life of Neil A. Armstrong*. New York: Simon & Schuster, 2012.

Murray, Charles, and Catherine Bly Cox. *Apollo: The Race to the Moon*. New York: Simon & Schuster, 1990.

Roth-Ey, Kristin. *Moscow Prime Time: How the Soviet Union Built the Media Empire That Lost the Cultural Cold War*. Ithaca: Cornell University Press, 2010.

Schumpeter, Joseph. *Capitalism, Socialism, and Democracy*. New York: Harper, 1942.

Siddiqi, Asif. *Challenge to Apollo: The Soviet Union and the Space Race, 1945–1974*. Washington, DC: NASA, 2000.

Slayton, Donald K., and Michael Cassutt. *Deke! U.S. Manned Space: From Mercury to the Shuttle*. New York: St. Martin's Press, 1994.

8 Apollo

The Race for the Moon

As we approach the flight of Apollo 11, already knowing that it was one of the culminating moments in the history of the Space Age, it may be worth noting that the Soviets lost the race to the Moon because the Americans got there first. This states the obvious. Yet, in the space race, the spectacular, politics-driven firsts were the events that really only ever counted. This was the golden rule, however irrational, of the Cold War competition in space. The Soviets certainly appreciated this. They had really won the early space race contests with their firsts. The power of these moments was in their public surprise and self-evidence. First was first.

So many US firsts, as in the case of Gemini, were less palpable, more obscure. Apollo 11 stands out. Only Apollo really turned the Soviet tide. It was the great line of separation. Its Saturn V rocket served one purpose alone: a peaceful civilian mission to be the first to land on the Moon. The USSR had no choice but to answer, with its N-1 Moon rocket, gearing up for a race. Watching and listening from America's spy satellites and intelligence posts around the USSR, the CIA feared the N-1 enough to convince NASA to speed up its timetables several times, all to ensure that Apollo and America would be first. So many political and technological variables counted in the race to the Moon, but the launch system perhaps counted the most. The USSR offered no match. The Saturn V blasted off 13 times between 1967 and 1973, each time a mission success. The N-1 flew four times between 1969 and 1972, each time a failure or catastrophe.

Still, for all these dramatic contrasts, the story is not so simple. The US achieved a new threshold of achievement with Apollo, but mostly scrapped plans to build upon it with Apollo's "advanced projects" and "applications." And the USSR achieved some lunar concession victories all its own.

Apollo Technocracy and Public Opinion

Project Apollo, as we know from Chapter 5, had its origins in a manifestly political moment of the Cold War. President Kennedy conceived of it in the context of the space race with the USSR, following Yuri Gagarin's first launch into orbit. He was also motivated by the US geopolitical defeat at the Bay of Pigs, which

DOI: 10.4324/9781003494409-11

he had directed against Castro's communist regime in Cuba. Apollo was political leverage to overcome these deficits in national status and prestige. Once the decision was made, as the president announced to the nation on 25 May 1961, it had already entered a new realm altogether: that of his executive presidency, and all of its lines of power and influence with the Congress, the bureaucracy, the media, and the vast networks of the American project system and R&D. Project Apollo became a program, a system, an institution and infrastructure.

To build support within the US Congress to finance the Apollo venture, the Kennedy administration depended on James Webb, its new director of NASA. His job was not about science. He was not an academic. Webb's most essential task, and his single greatest achievement, was to keep the Apollo program funded. In this gritty politics, he was a practiced operator, a former aerospace executive, as director of McDonnell Aircraft, and a high-powered lobbyist, known for his combative personality and mastery of the art of threat and persuasion.

Webb also had powerful allies in the executive branch and Congress: Vice-President Lyndon Johnson, former Democratic senator from Texas, with a power base at Houston; Robert Kerr, Democrat from Oklahoma, chair of the Senate Committee on Aeronautical and Space Sciences, and owner of Kerr-McGee Industries, Oklahoma City, a conglomerate of oil and uranium interests; and Sam Rayburn, Speaker of the House of Representatives and Democrat from Texas. These powerful men helped to fund NASA in a truly historical buildup between 1961 and 1966, whose budget appropriations increased from $964 million (or 0.76% of US spending), to just over $5 billion (or 4.41% of US spending), its highest ever. They also helped to parcel out these monies to their own clients and business interests throughout the US south and west, oftentimes treasury rewards for the electoral and political base of the Democratic Party.

Webb offered Americans a sophisticated progressive ideology through the Moon program. With a number of other leading actors in the Kennedy administration, he promoted a new "technocracy" for the American republic, arguing that science and technology were the most important measures of its practical power and prestige status in the world. In the complex and sophisticated competition that was the Cold War, the US government ought to be directed by the best persons of talent and knowledge, especially scientists, engineers, and managers. Webb wrote this notion into the very recommendations for the Apollo program that he submitted to President Kennedy in May of 1961, seconded by Secretary of Defense, Robert McNamara. "All large-scale projects require the mobilization of resources on a national scale. They require the development and successful application of the most advanced technologies. Dramatic achievements in space therefore symbolize the technological power and organizing capacity of a nation."

Webb and McNamara cast the Moon program to the president as the perfect means of projecting American power in the world, and as the perfect expression of his New Frontier. Apollo became an experiment in American technocracy, what Walter McDougall has called a "technological revolution" in governance and statecraft. It became a metaphor for transformative social, cultural, and political change. It matched well with Kennedy's youth and vigor, as with his visions to perfect the progressive liberal-welfare state, and to turn the government's purposes to the common good. The Moon program would marshal the best of human knowledge and talent to focus on one great task, create new jobs for the working class, and rally the nation and the young for education and betterment. For McNamara, building the great Saturn boosters was the best way to overcome America's perennial inferiority to Soviet booster technology. The R-7 military and civilian payloads consistently outperformed America's. Building a Moon program, McNamara believed, also gave leverage and backing to the US in all varieties of Cold War rivalries, like the new guerilla wars in Central America, Africa, and South Asia (Vietnam).

The president invested in Apollo with a singularity of focus and purpose rare in American political life: a man, the Moon and back, ten years. Such grand, utopian ventures are rare in history, rarer still for the US. Most Americans were curious but skeptical. Committed, full support for the Apollo program never broke 50% of public opinion, but rather hovered just below, and in 1965 only reached 39%.

NASA cultivated this tenuous "39%" with great care over the years. One of its most important missions was in media relations, expertly promoted through its Public Affairs Office and History Office. Their task was not easy, in the context of all the media attention to Sputnik and Gagarin, those space artifacts and personalities that became new American commodities of news and exchange. They had to turn the Mercury, Gemini and Apollo programs into new brands of American ascent, rather than decline.

As a civilian agency, one of NASA's express tasks was to educate the public in the new science and technology of space exploration. These were challenging topics for most newspaper and television reporters. Each of the major news organizations (the Associated Press, United Press International, NBC, CBS, and ABC) had bureaus that covered the space program. But in a relationship that was rare in the US media, they depended on NASA personnel for expert commentary, information kits, reference guides, models of spacecraft, and access to the astronauts. In a trade of mutual benefit, NASA controlled the information and the spin, and the media received pre-packaged, ready-made news reports that were both accurate and patriotic. Exactly what the public wanted, and what sold news stories.

The Kennedy NASA, under Webb's rule, also established the Artist's Cooperation Program (1963), eventually gathering 52 artists to commemorate Projects Mercury, Gemini, and Apollo with public art, disseminated in magazines, books, and reproductions, displayed in several popular exhibits in 1965 and 1969 at the National Gallery of Art. Webb's Sustaining University Program infiltrated America's universities with hundreds of millions of dollars for science and engineering education and R&D, but also to embed government influence in them, especially for local economic development. Webb even had a plan to remake America's libraries, to turn the stationary shelves of sitting books into microfilmed and eventually computerized networks of scientific-engineering and technical sources, for ease and speed of remote access. The librarian would become a technician. Knowledge would become "information science."

NASA History

NASA sold the Moon to the US public in the early 1960s. It continues to sell American spaceflight history today. The USSR, as a closed society, packaged its spaceflight history as propaganda, weaving truth and deception. In contrast, as an open society, the US had its own much milder version of state sponsorship and official history. As an extension of its publicity and educational work, and to fulfill its technocratic ethos, NASA created a History Division in 1959 to write and promote its own history. This division has been a patron of over 281 books and studies, published mostly by its own press. In its massive seven-part series of interpretive historical essays and primary source documents, *Exploring the Unknown: Selected Documents in the History of the U.S. Civil Space Program*, John Logsdon et al., ed. (Washington, DC: NASA History Office, 1995–2008), NASA historians set the tone and content for nearly every facet of the space program: including presidential policies, international cooperation, military-civilian relations, government-industry relations, earth science, and space astrophysics. The NASA History Office has also partnered with the American Historical Association, Johns Hopkins University Press, and a variety of other presses, to underwrite historians and historical monographs.

From 1966, the National Air and Space Museum (NASM) and its Space History Department contributed to this vast library with works by its own curators, often published by their own Smithsonian Institution Press. It also has a fellowship program with the Guggenheim Foundation.

At both NASA and the NASM, general-schedule (GS) government employees, or historians under contract to them, have produced official histories of the US space program, usually peer-reviewed and of professional quality, often based on special access to government documents, and privileging either national narratives or museum collections. Their output has far outstripped books by independent scholars, or even the official histories of the military services, or books published by or about aerospace businesses. In rare moments of Cold War cooperation, NASA and NASM historians and Soviet historians, as guardians and gatekeepers, often met at the conferences of professional organizations, like the International Academy of Astronautics, the American Astronautical Society, and the American Institute of Aeronautics and Astronautics, to compare their state-sponsored spaceflight histories.

Apollo had its critics. Social dissent came from a variety of surprising places. One credible poll of 113 scientists, not associated with NASA, found that all but three of them "believed that the present lunar program is rushing the manned stage." Apollo was too much too soon. The leading sociologist, Amitai Etzioni summarized the most pointed critiques in his book, *The Moondoggle: Domestic and International Implications of the Space Race* (1964). Social and civil rights leaders criticized Apollo's massive funding upsurge in the midst of American poverty and racial prejudice. One of the more poignant pieces of dissent came from blues singer, Gil Scott-Heron, whose song-poem, "Whitey on the Moon" (1970) captured the plight of poverty and degradation in the inner city. The irony was that, thanks to President Kennedy's policies and President Johnson's Great Society initiatives, social welfare and nondefense funding rose at unprecedented levels beginning in the 1960s, eventually outstripping defense spending after the Vietnam War.

At both the start of the Apollo missions, and at the end, pop stars lamented the loss of human personhood and dignity in space. David Bowie's "Space Oddity" (1969) imprisoned his Major Tom in a metal machine, alone in the vacuum of the universe. Elton John's "Rocket Man" (1972) isolated his astronaut pilot in the cold, timeless expanse of space. American audiences were wise to the perils of spaceflight.

Program Management NASA Style

Mindful of all these partisan and public interests, NASA managed a system of competitive contracting to American private enterprise and academia. Its scientists and engineers found partners to design and develop Apollo machinery. Research contracts went to the Jet Propulsion Laboratory and Applied Physics Laboratory, to the Jet Propulsion Centers at Princeton and Purdue University. The proposals to build the Apollo spacecraft were already in place by October of 1961. Martin Marietta, North American Aviation, and three other consortia (spanning General Electric, Douglas, Grumman, Space Technology Labs, General Dynamics, Avco, McDonnell, Lockheed, and Vought), submitted proposals that averaged 2,000 pages long. The 192 members of the Space Evaluation Board picked Martin Marietta as its first choice. Yet North American won the NASA contract, in part to ensure the support of Senator Kerr, as that part of the deal was for the company to build its production plants in Oklahoma, in some cases on Kerr's land and using his banks and holding companies. This was a series of conflicts of interest that later came to haunt Webb's reputation, as he had once been employed by Kerr. This also turned out to be the largest single civilian contract in US history: $6 billion to build the Command and Service Modules, the Saturn engines, and the second stage of the Saturn V launcher.

Through all of this, NASA's power was in its vertical lines of funding and control. In time, its 36,000 employees ran the headquarters, the design and testing facilities, the spaceflight control centers and launch facilities, and oversaw a vast network of contractors. To name a few, besides North American: Boeing and Douglas helped to build the Saturn launch systems, Grumman Aircraft the Lunar Module, Bell Aerospace the Ascent Stage, Raytheon the onboard computers, Honeywell Corporation the stabilization and attitude control instruments, Bendix Corporation the Command and Service Module instruments, and MIT contracted for guidance and navigation. These businesses and offices represented 377,000 contract employees in at least 250 major enterprises. NASA and its contractors received the $25 billion in government funding that was Apollo.

If the US was to succeed in this Apollo venture, it needed to do more than just build the machines and train the astronauts. It needed to refashion NASA into an efficient organizational machine. The dilemma was how to manage a new, complex, and rapidly growing bureaucracy, as well as direct the contracting and supervision of sophisticated technologies. Its administrators, with special credit to George Mueller, chose a model straight out of the US military: program management. The country's horizontal aerospace projects now became vertical programs, managed as hierarchical, interactive systems, all in order to manage the people and paperwork and deliver dependable products on time. NASA borrowed the model from the US Navy Polaris and the US Air Force Minuteman ICBM program. The missile race was essential to the success of the Moon race.

NASA's Office of Manned Space Flight and its Apollo Program Office, responsible for the oversight of Apollo processes and hardware, integrated program management throughout NASA, especially the centers at Cape Canaveral and Houston. Von Braun's teams at Huntsville were less willing to adapt, as they had worked within the Army's self-contained "arsenal system," rather isolated from the wider and more dynamic project system. However, they eventually relented and adopted program management.

Samuel Phillips, director of the Apollo Program Office for the Manned Lunar Landing, was one of the masters of program management. A Brigadier General in the US Air Force, he was an electrical engineer by training and a veteran pilot of the Second World War. He had also served as a project officer on the B-52 strategic bomber program (at the Engineering Division of Wright-Patterson Air Force Base), had worked for the Strategic Air Command in negotiating the deployment of the Thor IRBM to the United Kingdom, and was director of the Minuteman ICBM program. In the relatively few years between 1963 and 1969, Phillips helped transform NASA practices in order to ensure that schedule and reliability were the two crucial standards of conduct. Apollo had to be on time and safe. Cost was third. His role was literally at the center of NASA, reporting to Associate Administrator for Manned Space Flight, George Mueller, above; and working closely with George Low, of the Apollo Spacecraft Program Office, below. He managed hundreds of lead engineers and contractors, often in daily contacts. He was responsible for overseeing a remarkable array of tasks: design and engineering, procurement and testing, construction and manufacturing; and logistics, training, and operations. Phillips's methods were precise and exacting, as for example his essential information "checkpoints" and "configurations" for the development of Apollo hardware. They reflected the new rationality and organizational logic that helped NASA and its contractors work like a well-oiled machine:

These standards allowed Phillips and his teams, including dozens of Air Force officers, to manage schedules, most of all. By 1965 they were constantly revising "delivery and launch timetables" for Apollo hardware, delaying action or speeding it up, depending on performance and reliability. Phillips warned contractors about serious problems; he planned the sequences of both unmanned and manned missions; he closely monitored changes throughout the cycle of production and delivery, this to ensure what he called "change control," the proper checks and balances in order to guarantee high-quality results. Program management, in sum, empowered NASA with results.

What about the Saturn launch platform? When NASA administrators finally decided on the Lunar Orbit Rendezvous (LOR) by the summer of 1962, and approved the Saturn V as the launch system for Apollo, they were banking on earlier achievements with space launching systems and rocket engines, both the legacies of the US ballistic missile program. In this sense, the early Apollo program takes us back to Sputnik too, and the shock of the R-7 rocket that launched it. In an effort to match its heavy-lifting power, the Army Ballistic Missile Agency (ABMA), US Air Force (USF) and Advanced Research Projects Agency (ARPA) coordinated their efforts to build a series of new American launchers and engines for the Cold War.

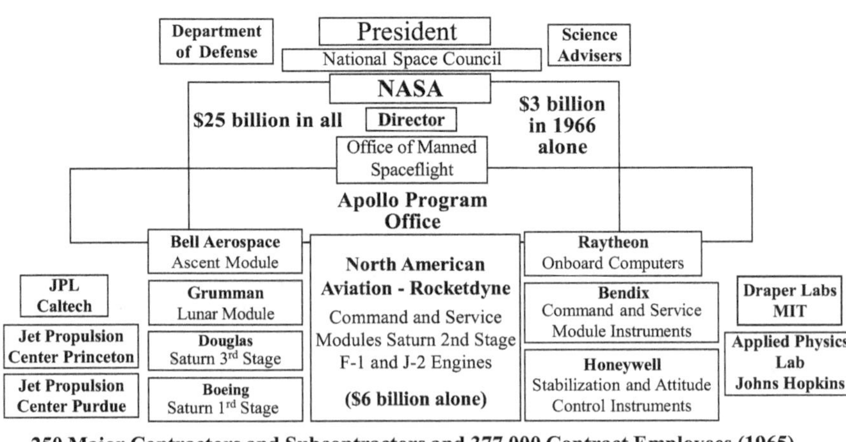

Figure 8.1 US Apollo program management. Author.

One result was the Saturn I, a "cluster" of eight Redstone boosters fixed around a central Jupiter fuel tank (with a second-stage Centaur rocket), an interesting "new" rocket platform based on a combination of the old. On 29 January 1964, the Saturn I launched with a liquid-hydrogen second stage, quietly sending "the largest satellite ever to go into orbit" up to that point (at 37,401 lbs.). This was a significant turning point. It finally beat all the Soviet heavy-launch records with the R-7. The Saturn I and the Saturn IB (already in development by 1960) both served as launch platforms for Apollo hardware between 1961 and 1968. The Saturn IB also launched Apollo 7 (11–22 October 1968) into Earth orbit for tests of the CSM, its only piloted flight in the Moon program.

Another result was the Air Force F-1 engine, one of the most powerful liquid-propellant engines ever built, contracted out to North American's Rocketdyne division and already test-fired in 1961. Five F-1s, burning liquid oxygen and kerosene, comprised Saturn's first stage. Five NASA J-2 engines, burning liquid oxygen and liquid hydrogen, comprised the second stage. One J-2 engine made up the third stage. Von Braun and his team were initially opposed to the novel liquid-hydrogen fuels in the J-2, but relented. By 1968, the Saturn V became the world's largest, most intricate, and most perfect space launch system ever built.

In their initial development of Saturn V, von Braun and his teams followed their normal, conservative approaches: incremental testing, verifying, and improving, then testing again. They static-fired the first stage, for the first time, on 16 April 1965 at Huntsville, Alabama. They then planned for their traditional course of measured preparations, with continued static tests to be followed by tests of each stage, one at a time. Yet the Cold War and space race intervened. George Mueller, head of the Office of Manned Space Flight, NASA's point man in charge of schedule, instead favored a speedier and riskier "all-up" testing. It meant combining the stages and test-launching them all together, all at once.

Von Braun, for all his futuristic promises of spaceflight, all his dazzle in the media, was rather closed-minded in his engineering and management visions. He and his teams at Huntsville were initially opposed to all-up testing, just as they had been opposed to Lunar Orbit Rendezvous, to the J-2 liquid-hydrogen engines, and to program management. Each of these was a momentum shifter in Apollo, a key to speeding the project along and ensuring success. In each case, von Braun was overruled. In each case, NASA succeeded.

With time of the essence, and the decade clock ticking, Mueller proceeded with the first all-up Saturn V launch as Apollo 4 (on 9 November 1967); and a second shortly after, as Apollo 6 (on 4 April 1968). Along the way, he and his teams resolved an issue with excessive vibrations in the rocket. Thus emboldened, Mueller approved the next all-up launch, the remarkable piloted mission of Apollo 8 to the Moon, the first such mission in what became a nearly perfect record of achievement.

Saturn was not just a rocket. It was an infrastructure as well, requiring the engineering and industrial contractors, as well as a trained workforce, to build it. To this extent, the program spanned a network of production, transportation, and testing facilities across the US. Boeing built the Saturn first stage at the Michoud Assembly Facility near New Orleans; then sent it by barge up the Mississippi River to Huntsville, Alabama, for testing; and back again to Cape Kennedy by way of the Gulf of Mexico and Florida coast. North American built the Saturn second stage mostly at Seal Beach, California and sent it east on a barge by way of the Panama Canal. Douglas Aircraft built the Saturn third stage at Huntington Beach, California and usually sent it east by way of the world's largest cargo plane, a converted Boeing 377 Stratocruiser called the "Pregnant Guppy," yet another interesting aerospace moment.

The Saturn V found its ultimate home at the new Launch Complex 39 (with two pads) built on Merritt Island just north and west of Cape Canaveral. This $800 million project began in November of 1962 and took all of four years to complete, under the expert design of the same team that built the Nazi launch complex at Peenemünde (Kurt Debus, Georg von Tiesenhausen and Theodor Poppel). It was a huge civil engineering project, to bury the pylons and pour the concrete on land that was half sand and half swamp, set between the Indian River and the Atlantic Ocean, in an area prone to hurricanes. At the center of the complex was the Vehicle Assembly Building (VAB), with enough room for four Saturn V rockets, and for many years a building with the largest indoor space (of cubic volume) in the world. The famous "Crawler" transported the vertical and fully assembled Saturn V (and its Mobile Launcher Platform), at 1 mile and hour, by a roadway to the launch pad some 3 miles away, just at the ocean's edge: one of the slowest "automobiles" in the world to move and position one of the fastest.

The capstone piece to all of these grand Saturn V developments was the smallest and lightest of the Apollo hardware, the lunar module (LM). All of the Saturn rockets and infrastructures were designed and built for this key component of the Apollo system.

The challenge was in building a very light but sturdy craft, ultimately without seats or heavy windows and hatch, all in order to save on the all-important weight factor. The LM had to fold inside a module of the Saturn third stage, with the CSM, and with which it was to rendezvous and dock in a series of maneuvers after the third stage had already begun its path to the Moon ("trans-lunar insertion"). All of this was to fulfill the imperative of the LOR mode decision: to carry two astronauts to and from the Moon by way of the CSM in lunar orbit.

Grumman Aircraft won the contract to build the LM. Rocketdyne built the descent engine. Bell Aerospace built the ascent stage. But the LM also needed sophisticated computer and radar equipment to allow it to descend to the lunar surface, where the LM commander was trained to take over manual controls, still guided by computer navigation. David Mindell has surveyed this new machine, the Apollo Guidance Computer and its cybernetics of human–machine interaction, in full scope. For these complex operating systems, computer scientists and engineers created a set of interfaces (hardware and software), linking landing and rendezvous radar, via a computer display and keyboard, to guide and land the LM, in conjunction with astronaut manual controls.

Given this sophistication, it seems appropriate to compare the "first" first of the space race, Sputnik, to this "last" first of the space race, the lunar module. Sputnik was an 184-pound metal ball, 23 inches in diameter, filled with two radio transmitters, three batteries, and a ventilation fan. The LM, in contrast, weighed 33,500 pounds of initial launch mass, was filled with advanced avionics, and fit a crew of two astronauts. It was only 12 years older, but a formidable rival, to be sure, as humanity's most complex and far-reaching spacecraft to date.

As the most challenging part of Apollo, the LM was also the last major piece of hardware to be built, finally ready for landing on the Moon in the very summer of Apollo 11. There were several significant delays in its planned schedule, namely in preparing the LM for its first unpiloted test with Apollo 5 (on 22 January 1968) and its first piloted test in Earth orbit with Apollo 9 (between 3 and 13 March 1969). The problems and time lags could have easily imperiled the Kennedy ten-year plan. Yet the extra time and attention to detail translated into reliability and a nearly flawless performance, as proven with the first test of the LM at the Moon during Apollo 10 (on 18 May 1969), and in each of the six lunar landings that followed, as well as in the one rescue of Apollo 13.

America's Missions to the Moon

As NASA practiced LM test flights in 1968–1969, it also prepared for one of the most dramatic flights of the Apollo program: Apollo 8 (between 21 and 27 December 1968). Originally scheduled as a test of the Apollo spacecraft in Earth's orbit, the mission changed dramatically in August 1968 when NASA redefined its flight profile as a circumlunar flight of the CSM.

Several factors entered the decision-making process. CIA intelligence analysts were predicting a possible impending Soviet mission around the Moon or even a Moon landing. They recognized the risks that the Soviet spaceflight program had taken with Soyuz 1. The Soviets had also launched the space probes, Kosmos 186 and 188, the first automated craft to rendezvous and partially dock. They had launched the Zond 4–6 series, lunar-distance and trans-lunar automated craft. Corona spy satellite imagery also showed the N-1 Moon rocket being readied in its launch tower at the Baikonur cosmodrome.

In light of these worrisome Soviet developments, and in the context of its own fast-paced program scheduling, NASA made a calculated and bold decision to risk the circumlunar flight after having tested the CSM only once, and after only two unpiloted launches of the Saturn V. Kennedy's decade was waiting.

Apollo 8 was a turning point. It was the first piloted flight of the Saturn V, the most powerful such launcher to date. It was also the first piloted spacecraft to reach parabolic escape velocity, at about 6.95 miles a second, faster than any human beings had ever moved, leaving the gravitational envelope of the Earth for its "trans-lunar injection." An American, not a Russian or Soviet rocket, fulfilled Konstantin Tsiolkovsky's dream of 1903 for "vertical ascent," humanity's true reach for outer space and the planets.

Apollo 8 was also the first piloted spacecraft to orbit another planetary body: the Moon. It was a journey of contrasts. What the crew saw up close at the Moon, a surface that was battered, gray and lifeless, was nothing like the Earth. On its fourth day in flight, as the craft came around from the far side, the crew delivered their momentous Christmas Eve message to one of the largest global television audiences ever. They read the first lines from the book of Genesis. "In the beginning God created the heavens and the Earth." They offered a prayer for the "good Earth." Bill Anders captured this moment in one of the most famous photographs of all times, "Earthrise," of a fragile blue planet seemingly alive and alone in the cosmos. "We came all this way to explore the Moon," Anders noted, "and the most important thing is that we discovered the Earth."

After Apollo 8, NASA proceeded with three remarkable spaceflights in a short period of time, reminiscent of the pace of the Gemini flights from just a few years before, if much more. These flights were a balance between caution and risk. With the LM finally ready for spaceflight, NASA tested the whole CSM and LM system in Earth orbit with Apollo 9 (between 3 and 13 March 1969). Yet the Moon race was on again with Apollo 10 (between 18 and 26 May 1969), a full rehearsal of the coming Moon landing, without the actual landing. Originally scheduled for July, CIA intelligence analyses of the Soviet Moon program, centered on the February tests of the N-1, even though it failed, convinced NASA to move Apollo 10 up to May, and Apollo 11 up to July. By now, with the USSR seemingly in gear to beat the US to the Moon, the schedule was everything. Apollo 10 was the first Apollo mission to rendezvous and dock the CSM with the LM after trans-lunar insertion. In lunar orbit, Tom Stafford and Eugene Cernan undocked the LM from the CSM and descended to within 9 miles of the lunar surface to test the LM.

Apollo 11 (between 16 and 24 July 1969) was the great turning point, the American "first" of the space race, though with one crucial difference from earlier Soviet firsts. This one was no real surprise. Everyone expected it. Sure, there was an inherent drama to the flight, filled with all the risks and dangers of any of the Apollo lunar landings. It attracted another global television and radio audience, with an estimated half a billion people tuned in, an ultimate simultaneous moment of the twentieth century. And yet Apollo 11, like many of the Apollo flights (excepting Apollo 13) was a nearly perfect mission, fulfilling most of its precise insertions, corrections, un-dockings, rendezvous, descents, spacewalks, ascents, returns, splashdowns. It was also the last of the great firsts. None would follow with the same weight and resonance. Apollo 11 was the culmination of the Moon race.

What began as a political charge to prove American scientific-technological superiority, a chapter in the Cold War, also ended as a series of scientific missions to explore the Moon, a first chapter in what would become the longer history of the Space Age. Apollo 11 was the first of six round-trip voyages to the Moon's surface. It began with a modest experiment on the Moon's seismic activity and collected nearly 50 pounds of rocks and surface samples. Later missions saw a rich variety of further scientific studies: studying the Moon's seismicity and radioactivity, or its physical and chemical makeup.

Several other Apollo discoveries are worthy of note. In November of 1969, Apollo 12 retrieved the camera of Surveyor 3, an early unpiloted probe that had soft-landed on the Moon on 20 April 1967. NASA scientists found a common and harmless bacterium, about 100 specimens of the streptococcus mitis bacteria, lodged in its polyurethane foam insulation. It had obviously been there from the original Surveyor flight, surviving for the 31 months between Surveyor and Apollo, without air or water, exposed to harmful radiation and extreme temperatures. As Pete Conrad said, "I always thought the most significant thing we ever found on the whole damn Moon was that little bacteria who came back and lived and nobody ever said s _ _ t about it." This discovery was especially interesting given the extreme precautions NASA took to protect humanity back on Earth from any possible viruses or germs brought back from the Moon, taking great care to disinfect the Apollo crafts and spacesuits with iodine and bleach solutions, quarantining the first Apollo crews until they were cleared of any possible contagions. The lesson of Apollo 12: we thought the Moon might contaminate us, but we ended up contaminating it.

Apollo 16 discovered a happy coincidence between the US and USSR space programs, otherwise engaged in such a fierce rivalry. From the surface of the Moon, a dead planetary body, Apollo 16 took an ultraviolet image of Earth that revealed the glow of the Earth's geo-corona, caused by the fluorescence of hydrogen atoms escaping the planet. Yet this phenomenon was already known from an experiment encased in the Soviet Luna probe of 12 September 1959, which included four "traps" set to measure and relay back to Earth the presence of charged particles in outer space, precisely like the ionized hydrogen atoms photographed by Apollo 16. Soviet astronomer, Iosif Shklovsky, analyzed Luna's findings to theorize about the geo-corona, a layer of these ionized gas particles escaping Earth's gravity and reaching out some 14,000 miles into space. Luna 2 and Apollo 16, a decade apart and from opposite sides of the space race, discovered this rare terrestrial effect, representing the slow loss of the Earth's oceans, on the order of 1 millimeter every 1 million years. Here was a true threshold moment for humanity.

The final three Apollo missions (15, 16, and 17) contributed to both lunar and terrestrial science thanks to the Lunar Rover Vehicle, an automobile that weighed 480 pounds and fit snugly into the LM. Though it only ever reached a maximum of 11 miles per hour, it enabled the astronauts to cover increasing distances around the Moon, gathering soil and rock samples, peaking with Apollo 17's impressive 18.6 miles of distance covered, under the supervision of geologist-astronaut, Harrison Schmitt. All these geological findings contributed to a radical new conception of the Moon's and the Earth's planetary origins and compositions. By 1984, the leading lunar scientists were already in agreement on the "Great Impact Theory," that a large planet-size object, maybe as big as Mars, collided with the Earth when it was around 100 million years old, and the debris of the collision formed into the Moon. The US went to the Moon to plant a flag but discovered the remains of an alien planet and part of our own.

One of the most dramatic of all the Apollo flights was the one that failed, at least in getting to the surface of the Moon. Apollo 13 experienced a near-catastrophic breakdown of its life support and control systems during its flight between 11 and 17 April 1970. Several days after launch, well on its way to the Moon at 200,000 miles from the Earth, an explosion and the rupture of an oxygen tank in the SM caused a sequence of failures in the power and electrical systems, in life support, and in the maneuvering and control of the CM, now crippled. NASA had never tested or simulated such a scenario. The initial crisis meant that the expelling gases of the explosion sent the CSM out of control, demanding a series of maneuvers and trajectory corrections. NASA decided that the crew should save themselves by finding refuge in the LM and save precious propellants by continuing on to the Moon and partially using its gravitational pull to swing back to Earth.

The Apollo figure-eight trajectory saved Apollo 13, as did NASA's improvisations, turning the entropy of accident back into the energy of solution and recovery. Flight Control and the Apollo 13 crew remade the LM life support system, meant to keep two astronauts alive for two days, in order to keep the three of them alive for three days. This required using plastic bags, cardboard, and duct tape, in order to fit the square Lithium Hydroxide canisters of the CM spacecraft into the circular slots of the LM system, which cleansed carbon dioxide from the air. Thus preserved for the flight home, the crew was also the first to suffer the chill of deep space, as they huddled together for warmth, their own body heat casting a strange human aura around them within their lifeboat spacecraft.

The drama of Apollo 13 revived the Moon program's television audiences, if only temporarily. The story eventually entered the realm of popular culture with the thrilling movie, *Apollo 13* (1995). At the time, as the remaining Apollo flights became ever more routine, neutralized as technology and science rather than dramatized as political competition, people simply lost interest. The space race as sport and entertainment lost its attraction. Another kind of vicarious thrill took its place. Worldwide television audiences, for example, did not watch the flights of Apollo 16 and 17, but what came in between: the Munich Summer Olympics of 1972, which saw Israeli athletes perish in a hostage-taking by Palestinian terrorists, political theater of a different but an increasingly common kind.

The Vietnam War

As spectacular as the Apollo moment was in 1969, the attention of American leaders and the public was focused on a tragic earthbound affair. The war in Southeast Asia absorbed much of Washington's energies and political capital. Presidents John Kennedy and especially Lyndon Johnson had increased the number of US troops in the region, providing ground and air support to the South Vietnamese in their efforts to contain the North Vietnamese communists and their guerilla insurgents, the Viet Cong. In 1968 alone, the US commitment to the war matched the whole cost of the Apollo program: $25 billion. That year also saw the death rate rise by 50%: to 15,000 American dead in 1968 alone. The American public and press also became increasingly skeptical about the progress of the war and about the hope of a speedy resolution. In January of 1968, the North Vietnamese Army and Viet Cong launched a massive surprise assault on the South. American and South Vietnamese forces soon repulsed this "Tet Offensive," but its sheer scale was shocking. Newspapers in the US soon began to characterize the war as a "quagmire" and the public began to speak of a "credibility gap" between what the government was saying and what was really happening. The progress of the war also captured the public's attention on the nightly news shows. Vietnam became America's first "living-room war." While the worst horrors of the war were rarely shown, as the networks worried about American sensibilities, there were stunning moments, such as when NBC News displayed a Vietnamese colonel summarily shooting an enemy prisoner in the head during the Tet Offensive. Television had captured both the triumph and beauty of the Moon walk and the savagery and ugliness of the war. The spaceflight breakthroughs of Apollo turned into America's geopolitical breakup in South Asia.

The Soviet Moon Race

If the US approached the Apollo program with one early and overriding goal, the USSR answered Apollo with a significant delay and many goals. President Kennedy's decision to mobilize American resources to the Moon reached back to May of 1961. The Soviet leadership did not make a final decision for their Moon program until August of 1964, a full three years later. Several factors contributed to this delay: the troubles with the Voskhod flights, bureaucratic competition and personal rivalries between Soviet spaceflight designers, and the Cuban Missile Crisis and Soviet military priorities. The overthrow of Nikita Khrushchev in October of 1964 pushed the Soviet Moon program still further behind.

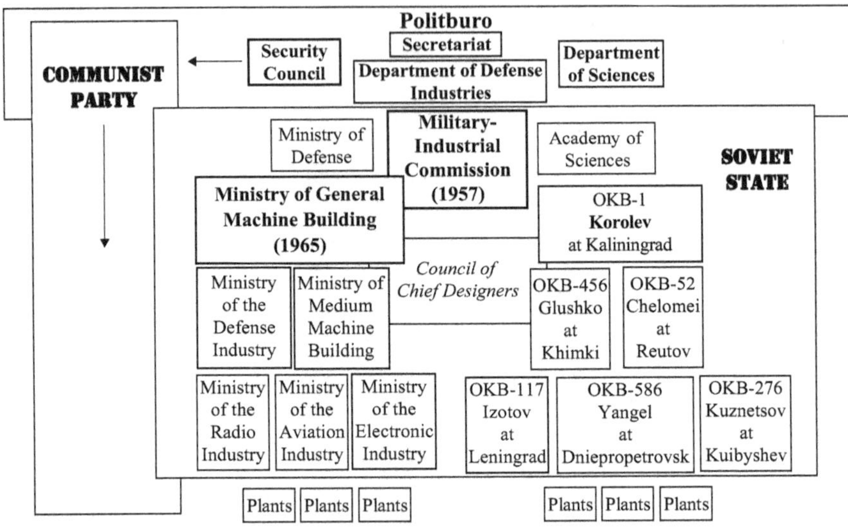

Figure 8.2 The USSR command system. Author.

A fundamental political-ideological prejudice also contributed to this delay. Several leading Soviet party and military leaders did not believe that the US could achieve such an ambitious goal as the Moon landing in such a short amount of time. They simply underestimated American spaceflight capabilities, especially the development of the Saturn V booster. Worse still, they overestimated their own capabilities. These added up to fundamental flaws in perception and calculation. In essence, some Soviet spaceflight managers believed in their own propaganda: that their firsts had made them best. Yet this was arrogance that made them weak, not strong.

The nature and organization of the Soviet spaceflight bureaucracy may have been the single most important brake on the Soviet Moon program. At its center was not the powerful and efficient Apollo Program Office and its principles of program management, but a relatively new government bureaucracy: the Ministry of General Machine Building, sounding like something straight out of an Evgeny Zamyatin novel. Founded in 1965, it was devoted primarily to military priorities, like ICBMs and spy satellites and military space stations, and only secondarily to civilian programs, like the Soyuz and the Moon missions. Its longtime director, Sergei Afanasiev (1965–1983), known as "the Fist," was famous for his fierce loyalty to the political and military chiefs above him, as well as his obstinacy and severity to those around and below.

Afanasiev also presided over a network of bureaucracies that were rent by fragmentation and conflict. By the middle of the 1960s, the USSR had lost the forceful and centralized Stalinist model from the early 1950s, replaced by a bulky, awkward, and often uncoordinated system of some 26 state departments and 500 state enterprises. This aging command-administrative system was underfunded and overtasked, weighed down by plans, decrees, and new plans; by programs denied, or revived, or reassigned. The system was also dominated by strong personalities, the bosses who ran their own enterprises like fiefdoms, doling out patronage and favoritism. They ordered tasks without explanation. They shuffled people around without offering cause.

Nor was this system well-suited to the modern imperative for open and free scientific exchange and innovation. Given the USSR's intense codes of secrecy, it was severely close-minded and compartmentalized. Major design bureaus and offices were not known by their actual names, but by "post-office box numbers." This was meant to hide their true identities from Western spies, but it also hid and isolated them from each other. Secrecy and isolation were the rules within bureaus and enterprises as well. Their departments were boxed apart, with separate rooms and wings, or laboratories and libraries, each assigned to different ranks and teams of specialists. Even they were often unaware of each other and their work. Communist party commissars watched all of these units for ideological obedience, with the plenipotentiary power to conduct loyalty checks, and to fire anyone within the enterprise, at least below the office of director.

The astronomer Shklovsky had his own explanation for the failures of this state system, and of Soviet spacefaring ventures: vodka. He meant Soviet alcoholism. Or to be precise, 5 billion vodka bottles, what he measured as the distance to the Moon in one-liter bottles. Soviet drinkers got the bottles, lost the Moon.

The strangest phenomenon of all, especially for what was a communist command-administrative system, was the internal competition within the Soviet spaceflight program. This was partly a legacy of the Stalin years, when the communist party promoted artificial market competition: rivalries between design bureaus to encourage incentive and innovation. There were actually four Soviet Moon-rocket programs which vied for funding through the early 1960s. Two were peripheral to the main goal, but still important alternatives. Sergei Korolev's improved R-7, for example, was not just the launch system for the Soyuz program, whose prototypes tested the circumlunar and even lunar-landing craft, all during Apollo. Early on, it was also considered for a possible Earth-orbit rendezvous approach to the Moon. From time to time, Mikhail Yangel proposed to build a massive R-56 launch system for a Moon-landing program, based on his successful work with the R-16 and R-36 ICBMs, though this proposal never went far.

The Soviet government eventually funded two serious Moon-rocket programs, both of which were in design and development since 1960–1961, and which Nikita Khrushchev initially approved in early 1962: Vladimir Chelomei's Proton rocket and Korolov's Nositel' 1 ("Carrier 1" or N-1). Both were given final approval, with major funding initiatives, by late 1964. Each was manifestly different: in design, engines, fuels, and guidance systems. Each vied against the other for priority status and funding. The two rocket programs were also separated by a personal and professional drama, centered on one man: the engine designer, Valentin Glushko. He had once partnered with Korolev on their masterpiece, the R-7 rocket. By the early 1960s, the two had fallen out, partly for personal animosities, but also in disagreement over the best propellants for a possible Moon rocket. Korolev wanted to continue to apply cryogenic propellants (liquid oxygen with kerosene or liquid hydrogen). Glushko was already converted to the advantages of hypergolics (nitrogen tetroxide and hydrazine or UDMH) and he became a successful partner in building engines for Chelomei's high-priority ICBMs. Korolev had little choice but to rely on a novice rocket-engine designer, Nikolai Kuznetsov, to build the new engines for the N-1, though in combination they were never up to the task.

Chelomei's Proton rocket, originally known as the "Universal Rocket" or UR-500, was the most successful of all the Soviet Moon rockets, thanks in large part to the Chelomei–Glushko partnership. Glushko's falling out with Korolev turned into a windfall for Chelomei. Originally designed as an ICBM (which, with SLBMs, were Chelomei's specialties), the Proton comprised four stages: the first stage of six Glushko RD-253 engines, burning hypergolic nitrogen tetroxide and hydrazine (UDMH); the second of four engines; and the third stage of one. The fourth stage, known as "Block D," was shared with the Korolev N-1, and burned liquid oxygen and kerosene.

During the Apollo competition, the Soviets earmarked Proton for a piloted circumlunar flight, with the L-1 spacecraft. This mission never flew. Yet Proton was a significant success through these years, launching the automated Zond (4 through 7) uncrewed circumlunar test probes in 1968 and 1969. The Proton launched the lunar-landing probes, a whole series from Luna 15 to Luna 24 (1969–1976), including the significant firsts of an automated soil return in Luna 16 and the automated lunar rovers, Lunakhod 1 and 2. The Proton also launched the Salyut 1 space station in 1971. Here were a series of distinct consolation achievements coincident with Apollo.

The Proton was also the foundation for what might have been the only hope for a successful Soviet-piloted Moon landing: Chelomei's 1965 proposal for a massive UR-700 launcher, outfitted with nine Glushko RD-270 engines in its first stage, and with the Proton (UR-500) as the second, third and fourth stages. Afanasiev at the Ministry of General Machine Building supported this project for a time, even seeking out contractors for its guidance system and launch-facility designs. But it did not last; the proposal was shelved. To build it, Chelomei would have needed ten years. It was a Soviet "Apollo program" five years too late.

The centerpiece for a Soviet Moon landing was Korolev's N-1 rocket, which was in design and development between 1960 and 1964. It was based on an American-style Lunar Orbit Rendezvous approach, by way of a spacecraft system known as the L-3 "Stack." It was similar to the setup of the Apollo

Command, Service and Lunar Modules, only smaller and lighter. The weight differential was due to the less powerful N-1, which could only launch 70% of the Saturn V's total payload. These weight restrictions on the L-3 Stack impacted its design. To save weight, the system required a cosmonaut to make a spacewalk from the lunar orbiter to the lunar lander. The lander was also only big enough to accommodate one cosmonaut, who was only allotted about 1.5 hours on the Moon's surface, and who was even outfitted with a pop-up metal device, in case he fell on the Moon and was unable to get up, without any crucial helping cosmonaut hand. This approach was filled with risks, even if the N-1 had ever launched successfully, which was the plan, for the first tests in 1966 and for lunar landings in either 1967 or 1968.

The challenges of the N-1 rocket itself were immense. One of them was personal: the death of Korolev in January 1966. The death of one man should not have affected a program of such caliber and scope. Yet when he died, the N-1 program lost its leader, its institutional memory. All the parts were still in place, but they no longer had a whole. Not surprisingly, after Korolev's death, Chelomei and Yangel re-entered the scene and engaged in a prolonged battle to replace the N-1 with their own boosters. Within a year, though, the communist party reaffirmed the N-1 program and confirmed Vasilii Mishin as director, but he was no Korolev.

The key challenges of the N-1 were technical. Like the Saturn V, it comprised three stages, but that is where most of the similarities ended. The Saturn had five superb F-1 engines in its first stage (hence the moniker "V"). Yet the USSR was simply not able to produce such engines at this time. So, the N-1 needed 30 engines altogether in its first stage, with eight in its second stage and four in its third stage. As a result, the N-1 was heavily overloaded with all these smaller, less powerful engines, together with all the pieces and joints needed to hold them together, and all the piping and valves necessary to power them. Numerous welding complications made the N-1 heavier still, and prone to leaks. Lastly, given the intensity of the Moon race with the US, and lacking in the time as well as in the logistics and funds, the Soviets never static-test fired the first stage of the N-1. This was a fatal error.

To make these matters worse, and the N-1 even heavier, the Soviets did not have the money or skill sets to create lighter metal alloys, or to manufacture miniaturized electronics and high-tech computerization. The US had its first guidance computer in the piloted Project Gemini (1965). The USSR had its first guidance computer in a piloted craft a full 15 years later, in a Soyuz T-2 (1985). Sophistication, the "better," mattered.

We have no better proof of the gravity of these technical challenges than the four failed launches of the N-1, a 100% failure rate due to mechanical failures and substandard quality control in production and assembly. It flew twice in 1969. On 21 February it exploded 70 seconds into its launch, due to a ruptured oxidizer line, resulting in a fire and crash. On 3 July 1969, the N-1 exploded at

Table 8.1 US–USSR factors in the moon race. Author.

	US	*USSR*
Priority:	The Space Race	The Missile Race
Decision:	May of 1961	August of 1964
Leverage:	Kennedy's Legacy	Khrushchev's Ouster
Investment:	$25 billion	$12 billion
Approach:	Program Management	Design Bureaus
Market:	Private Enterprises	State Industries
Launcher:	Mueller (Saturn V)	Korolev-Mishin (N-1)

launch, in a catastrophic failure with secondary explosions that destroyed the launch pad, and which probably resulted from "slag" in the first stage liquid oxygen tank, which was then ingested into a turbine pump. After this second failure, the Kuznetsov engines were sent back to his bureau for redesign. The N-1 then flew again twice, for 50 seconds in 1971 and 90 seconds in 1972, failing both times due to faulty propellant lines and excessive heat and vibrations. In sum, the N-1 never flew higher than 70,000 feet.

One other factor is notable. If the US spent roughly $25 billion on Apollo and related programs, the USSR probably spent only roughly $12 billion on its Moon program, which was only a fraction of the tens of billions of dollars it spent on its priority ICBMs at the time, according to the calculations of space analyst, the late Maxim Tarasenko. In this context, perhaps the true competitors to the Saturn V were not the Proton or the N-1, but the second-generation Soviet fleets of ICBMs, in the Chelomei UR-100 and Yangel R-36. The USSR deployed them in force by the early 1970s, to make up for the humiliation of the Cuban Missile Crisis, and concurrent with both the Apollo and N-1 programs, enough to bring the country to rough parity with the US.

The Moon race was not a simple case of victory and defeat. Project Apollo was transient, to the Moon and back, a kind of economic bubble. Projects by definition were only meant to last to a point in time, to be surpassed by new projects. Multi-variant Soyuz, in all its facets between 1966 and 1972, though it may have lost the race to the Moon, was made for the long term. Credit is due to some Soviet objects. They were meant to be first, and to last.

Vertical Horizons: Above and Below Worlds

In the US, the Apollo 11 achievement emboldened NASA, at the peak of its popularity and power, to extend its mission well beyond the Moon. Under the direction of Vice-President Spiro Agnew, the Space Task Group (part of the President's National Space Council) capped off its work in 1969 with a report outlining a $100 billion proposal for a Space Shuttle, a 12-person orbital station, a 100-person Moon base, and a long-term mission to Mars. Yet over the next few years, as the national excitement wore off and budget constraints took hold, President Richard Nixon canceled Apollo 18 in September of 1970, well before Apollo itself ended. Then in May of 1971 he ordered a freeze on the NASA budget for five years; and in January of 1972 finally rejected the Space Task Group's grand vision of a mission to Mars, or a Moon base, or even the orbital stations. He funded only the Space Shuttle, which he described as "a space vehicle that can shuttle repeatedly from Earth to orbit and back. It will revolutionize transportation into near space, by routinizing it." In the meantime, the remaining Saturn V rockets were either repurposed for the Skylab missions or eventually put on display: as museum pieces at the Marshall Space Center in Huntsville, Alabama; at the Kennedy Space Center in Florida; and at the Johnson Space Center in Houston. Over the years, the Saturn V on display at Houston fell into the most disrepair and decay, an embarrassment to the national memory, at least until a restoration project after 2004. Its paint and coatings flaked under the weather conditions and city pollution. Mold and plants grew on its edges. Birds and small animals nested in its nooks and crannies. The mighty Saturn, meant to fly into a lunar trajectory, sank into some earthly ruin.

In the USSR, the Brezhnev regime decided on a similar scaling back of its spaceflight visions, if with something of its usual delay. In 1974, a state commission removed Vasilii Mishin from his post as chief designer of Sergei Korolev's design bureau (OKB-1) and as director of the Soviet Moon program, replacing him with V.P. Glushko, Korolev's old rival. Glushko forthwith dismantled the edifice of Korolev's spaceflight bureaus and offices, integrating them under his new umbrella organization, "Energia," and rededicating government finances and focus to his preferred launch and spacecraft systems. He sponsored new histories of the Soviet spaceflight program, now to emphasize his contributions at the expense of Korolev's. He promptly decreed the cancelation of the next N-1 launch. By 1975, he ordered the destruction of all remaining N-1 rockets, scattering them to the four winds. Their parts were thrown into trash heaps or resold for scrap, though the hulls of the N-1 also made good barracks and garages, pig barns and sun shelters, even playground pieces in and around Baikonur. Here was Glushko's supreme act of physical destruction and historical revisionism, a dramatic gesture expressing the awkward truth of the Space Age: that only the "firsts," only the space spectaculars, really counted. Thus, just a few years after it had attempted to beat the US to the Moon, the N-1 became a ruin too, like some ancient imperial monolith reduced to pieces.

Select Bibliography

Beattie, Donald. *Taking Science to the Moon: Lunar Experiments and the Apollo Program*. Baltimore: John Hopkins University Press, 2001.

Bilstein, Roger. *Stages to Saturn: The Technological History of the Apollo/Saturn Launch Vehicles*. Washington, DC: NASA, 1996.

Bizony, Piers, and Mike Acs. *The Art of NASA: The Illustrations That Sold the Missions*. Beverly: Motorbooks, 2020.

Chaikin, Andrew. *A Man on the Moon*. Alexandria: Time-Life Books, 1994.

Collins, Michael. *Flying to the Moon: An Astronaut's Story*. New York: Farrar, 1994.

Compton, W. David. *Where No Man Has Gone Before: A History of Apollo Lunar Exploration Missions*. Washington, DC: NASA, 1989.

Cooper, Henry S. F. *XIII. The Apollo Flight that Failed*. Baltimore: Johns Hopkins University Press, 1972.

Fendley, Caitlin. "First Contact: Apollo 11 Astronauts as Field Workers and Test Subjects," *Quest: The History of Spaceflight Quarterly* 29/2 (2022): 23–38.

Fisher, Steven C. et al., ed. *Remembering the Giants: Apollo Rocket Propulsion Development*. Stennis Space Center: Monograph in Aerospace History, 2009.

Gisler, Monika, and Didier Sornette, "Exuberant Innovations: The Apollo Program," *Society* 46 (2009): 55–68.

Johnson, Stephen B. *The Secret of Apollo: Systems Management in American and European Space Programs*. Baltimore: Johns Hopkins University Press, 2002.

Kaufman, James Lee. *Selling Outer Space: Kennedy, the Media and Funding for Project Apollo, 1961–1963*. Tuscaloosa: University of Alabama Press, 1994.

Lambright, Henry W. *Powering Apollo: James E. Webb of NASA*. Baltimore: Johns Hopkins University Press, 1995.

Launius, Roger, and Stephen Garber, ed. *Research in NASA History: A Guide to the NASA History Program*. Washington, DC: NASA, 1997.

Lazier, Benjamin. B. "Earthrise: Or, the Globalization of the World Picture," *The American Historical Review* 116/3 (2011): 602–630.

Lipartito, Kenneth and Orville R. Butler. *A History of the Kennedy Space Center*. Gainesville: University Press of Florida, 2007.

Mailer, Norman. *Of a Fire on the Moon*. New York: Little, Brown, 1970.

Mindell, David. *Digital Apollo: Human and Machine in Spaceflight*. Cambridge: MIT Press, 2008.

Murray, Charles A. and Catherine Bly Cox. *Apollo. The Race to the Moon*. New York: Simon & Schuster, 1989.

Needell, Allan. "James E. Webb, Technocracy, and the New Deal Roots of Space Age Management," *Technology and Culture* 58/3 (2017): 790–814.

Orloff, Richard W. and David M. Harland. *Apollo: The Definitive Sourcebook*. Berlin: Springer, 2006.

Poole, Robert. *Earthrise: How Man First Saw the Earth*. New Haven: Yale University Press, 2010.

Scott, David Meerman, and Richard Jurek. *Marketing the Moon: The Selling of the Apollo Lunar Program*. Cambridge: MIT Press, 2014.

Shklovsky, Iosif. *Five Billion Vodka Bottles to the Moon: Tales of a Soviet Scientist*. New York: Norton, 1991.

Smith, Michael L. "Selling the Moon: The US Manned Space Program and the Triumph of Commodity Scientism," in *The Culture of Consumption: Critical Essays in American History*, Richard Wightman Fox, ed. New York: Pantheon, 1983.

Spencer, Brett, and Nancy H. Dewald. "NASA Reshapes Libraries: The Space Program's Impact on Academic Libraries, Special Libraries, and Archives," *Science and Technology Libraries*, 34/4 (2015): 299–314.

Spudis, Paul D. *The Once and Future Moon*. Washington, DC: Smithsonian, 1996.

Usdin, Steven. *Engineering Communism: How Two Americans Spied for Stalin and Founded the Soviet Silicon Valley*. New Haven: Yale University Press, 2005.

Webb, James E. *Space Age Management: The Large-Scale Approach*. New York: McGraw-Hill, 1969.

Wisnioski, Matthew. *Engineers for Change: Competing Visions of Technology in 1960s America*. Cambridge: MIT Press, 2012.

Zimmerman, Robert, *Genesis: The Story of Apollo 8*. New York: Dell Publishing, 1999.

Part III

The Fragile Species

9 Portraits

Our Expanding Universe

This chapter reminds us of a central premise of our text: that many of the historical ventures into outer space are really all about life here on Earth, about our home planet. We have seen this mirroring dynamic in the early pioneers, who calculated rocket trajectories as they competed with their rivals for national favor and international priority. We have seen it in the confrontations of the Second World War and Cold War, as the warring parties devised ballistic missiles to reach the edge of the atmosphere, or into outer space, in order to destroy or threaten their enemies on the ground. We definitely saw it in the classic era of the race to space, as the US and USSR competed for satellite and spacecraft firsts, all for status and prestige here on the planet.

This brings us to the theme of Part III, the planetary **portrait**. Not of other planets, but of our own, the fragile Earth and we the fragile species upon it. Apollo 8's "Earthrise" as emblem. Looking backward from outer space, we find the Earth and ourselves. NASA publicists often moved farther and faster than the portraits, as in Rick Guidice's, *View to the Future* (1970), a painting commissioned from the Ames Research Center. It illustrated not planet Earth, but the human being writ large, holding the planet in their arms, venturing together into space. The portrait of Earth is really one of humanity's expanding knowledge of the expanding universe.

In the 1960s and 1970s, so soon after the first human flights to orbit and to the Moon, the US and USSR mastered the science of sending out machines into space in search of life. Each of these new spaceflight ventures was a reach into space, but also a measure of return back to Earth. We sent out signals by radio telescopes, and mined the radio emissions received here, to communicate with possible extraterrestrial life. We launched interplanetary probes into deep space in order to surveil the far planets and travel beyond the solar system with messages from Earth. We landed spacecraft on Mars to detect possible forms of life there. Humanity set out to be heard and seen, perceived and understood, in the universe. We also found ourselves unable to break out of the world of human perceptions.

In each of these cases, scientists and engineers attempted to prove the age-old plurality of worlds thesis: that the Earth is but one living planet among many

DOI: 10.4324/9781003494409-13

others, that life is everywhere in the universe, and that biology and intelligence are common. Alien life may be more advanced than our own, a proposition that turns the plurality thesis into something of a mediocrity thesis, as if humanity and Earth are nothing special, perhaps not ordained by God or by evolution with any guiding place in the universe. Leading the charge were scientists in Russia and the USSR, where the revolutionary movement and Soviet state promoted radical thinking. There, the strains of Charles Darwin's theories of natural evolution and "struggle of the fittest," together with Karl Marx's theories of historical materialism and class struggle, combined to shape an official scientific atheism open to the boldest new hypotheses about plurality and mediocrity.

By the 1970s, ideas about the origins of life and the possible existence of alien life became more mainstream in the professional and popular media around the world. Science took on questions and posed answers about the ultimate issues, about origins and ends, previously left to theology and metaphysics. Drawing a biblical metaphor, NASA's Chris McKay and spaceflight journalists have called this the search for a "second genesis," the way that life originated or where it may exist beyond the Earth. For the scientist who discovers it, the second genesis will rewrite history and ensure the highest fame.

The Origins of Life on Earth

We intuitively know what life is, better perhaps than we know anything else, because we have always been a part of it. We know that life develops, flowers, reproduces, sickens, and eventually dies. Science properly reduces these processes, and their organic parts, to rigorous classification and explanation. NASA's operative definition holds that "life is a self-sustaining chemical system capable of Darwinian evolution." This formula assumes that life can convert molecules into energy and growth (catalysis), is capable of generating offspring (reproduction), and is able to transform as changing conditions demand (evolution). When we look for life in outer space, not to speak of civilization, we ought to be looking for these markers. Easy enough.

There is an even more fundamental and perplexing question to pose along the way: how does life begin? As we humans search for life beyond Earth, we ought to be looking for analogies to the simplest forms of life here, hoping we might find them on some planet beyond. After all, the theory of evolution, and its growing body of evidence, points backward to a common ancestor or biochemical source for all life on Earth. Knowing how life actually began here offers a clue where to expect it elsewhere. Whether it came from outer space or began on our own planet in some verifiable way, knowing the answers will help us to know what to look for and how to look.

One prominent theory had it that life on Earth might have migrated here from outer space. By the beginning of the twentieth century, leading scientists such as William Thomson (aka Lord Kelvin) and Svante Arrhenius had developed the "panspermia" thesis, that life first came to Earth as organic spores in a kind of frozen animation. Meteors or cosmic dust first seeded life here. This theory followed the logic of the Kant–Laplace hypothesis, an early version of the "Big Bang" theory, according to which the galaxies and solar systems formed from the "original chaos" and "clouds" of stellar material some 13.8 billion years ago. The Earth itself was at one with the universe at large, created out of this common genetic material some 4.5 billion years ago, and from which arose the first organic compounds that eventually arrived, or evolved and thrived here around 3.5 billion years ago. Arrhenius, for example, proposed that the "pressure of radiation" throughout the universe offered a surface upon which the electrical charges of cosmic dust might travel, including microscopic bacterial spores. He even argued that these "traveling germs" were actually launched from planetary surfaces by the power of "electrostatic repulsion," descending back by catching onto larger particles via the "action of attraction." Nature offered humanity its own model of spaceflight launches and returns.

Several German naturalist writers, Max Wilhelm Meyer and Wilhelm Boelsche, whose translated works sold well throughout Europe, North America, and Russia, helped to disseminate the thesis. Boelsche speculated that "certain spores of the simplest living matter" may perhaps have "immigrated and settled on the cooled globe" of our planet. The universe was in motion as a great sea of life. "The milky way of the heavens," so he wrote, "stands opposite to the milky way of life," the oceans on Earth. Here were echoes of that grand metaphor that has so often inspired the rhetoric of the Space Age. As people of the land and air, we humans stand between two oceans: the "ocean of space" upon which we might sail as bold navigators, from whence we originally came, to which we must ascend, just as we once left the oceans of water for land. The cosmos was the ultimate source and end of all life. Here also were the makings of a modern and secular worldview, one that saw the universe as infinite, evolving, and eternal, without the hand of God at work in nature and history.

Panspermia remains a viable theory as to the origins of life on Earth. Scientists continue to experiment and hypothesize about the question, with special attention to Mars, given its proximity to Earth and its age. Because Mars cooled as a planet hundreds of millions of years before Earth, and may have harbored life first, the chances are good that living microbes traveled here, given the right conditions, by Martian meteorite (roughly one ton of Martian meteorites land on the Earth's surface every year). In fact, the very new science of "Exobiology" ("external biology," later known as Astrobiology), "the study of the living universe," had its origins in the panspermia thesis. The geneticist Joshua Lederberg coined the term in 1960, at the First International Space Science Symposium, out of the fear that a space-traveling pathogen might soon contaminate Earth, brought here by comets or meteors or eventually by us, right at the cusp of human spaceflight.

One alternate genesis theory to panspermia is that our planet's biology had a self-starting chemical foundation. This is the "abiogenesis" hypothesis. It means "the process by which a living organism arises naturally from non-living matter, as opposed to biogenesis, which is the creation of living organisms by other living organisms." It circulated widely after T.H. Huxley's propositions in an 1868 lecture, "On the Physical Basis of Life." The origins of life, he posed, were not in Creation, or some mythical force, but in the complex chemical reactions that created the original "protoplasm" of part-inorganic, part-organic matter. Organic life first formed in the Earth's prehistoric "simple, primordial jelly" of chemicals, amino acids and proteins, catalyzed by the energy of volcanoes, or of lightning, or of the sun.

Huxley's point was that Darwinian evolution needed a point of origin, a creation myth all its own: chemical spontaneous generation. By the 1920s, Henry Fairfield Osborn, president of the American Museum of Natural History, was arguing that "the change from the lifeless to the life world" was based on chemical actions and reactions, joining soil, water, and atmosphere with "electric energy and the Sun's heat." Earth, in other words, was an integral part of the living universe, not as a passive receptor of life that came from beyond, but as an active source of life-giving power from within. The museum's spiraling time wheel, most recently reproduced as the Geologic Time Spiral by the US Geological Survey, perfectly illustrated these origins: how the Earth first formed out of a gaseous cloud, gave birth to life out of its primal oceans, and set into motion the inevitable process of natural evolution. Here was unstoppable motion, forward and ascending progress.

The title page of the popular magazine, *Science Wonder Stories* (1931), had cultivated readers with just such imagery, humanity borne along on a trajectory from ape-like creatures standing upright, to looming skyscrapers and rockets launched into space. So did the professional engineering journal, *Astronautics*, in an article by the Peenemünde engineer, Krafft Ehricke, who also mapped the actual rocket trajectories for a human future of interplanetary flight. NASA adapted the map into its "exploration architecture" in the official planning document, *The Vision for Space Exploration* (2014), the spiral of the geologic time scale threading off into the nearby paths for the human colonization of space.

The Geologic Time Spiral also helped inspire a whole new way of doing history, what we call Big or Deep History, history in cosmic terms, from the very beginnings of the universe to its very ends. This is a conceit, of course, as we do not yet know either. But it still encompasses what we know of star and planet formation, the spiral of life on Earth, the Paleolithic and Neolithic ages, the Anthropocene, and what the future may hold. One of the troubles is that Big History leaves so much of the human agency and experience out, the values and freedoms, the vices and blunders, that define us. We are that small step on the spiral without which, after all, there is no story. Nature can be boring without us.

Abiogenesis was a milestone moment in the time spiral. It was already a staple of atheistic Bolshevik ideology both before and after the Russian Revolution of 1917. As the Marxist propagandist, Nikolai Bukharin, wrote: "Organic nature grew out of dead nature; living nature produced a form capable of thought." Human beings were not original; they were not part of some mystical, supernatural order. They were accidental, the by-products of a natural order that claimed them from inorganic matter, if now organized also as mind, as matter in its most developed and perfected form. Tsiolkovsky, too, refuted the theories of Arrhenius that life came from outer space or another planet. Instead, he proposed, life probably originated right here, in some "quiet, spacious, freshwater lake," thanks to just the right chemical solution energized by the power of the sun.

The Soviet biochemist and proponent of scientific atheism, Aleksandr Oparin, gave definitive shape to the classic theory of abiogenesis in his essay, *The Origin of Life* (1924), another Russian first in the history of science and astronautics. Arrhenius was wrong, he argued. The stars were too distant, the spaces between them too lethal, to sustain the interstellar and interplanetary migration of life. Organic life had its origins in evolution from inorganic nature: in the hydrocarbons and myriad chemical reactions forming from them. Once begun in a cycle of conception and reproduction, life could not replicate this moment of natural creation. Yet life was predicated upon simple chemistry. It began as a humble kind of organic slime but with the power to grow and assimilate.

The British Marxist J.B.S. Haldane developed these propositions after 1929, first in a rough sketch, and later along more certain biochemical lines, for which he has shared the title of the "Oparin–Haldane" theory of abiogenesis. It was both a critique of traditional superstitions and the Christian theology of the soul, as well as a reasoned defense of the experimental method and scientific proofs. Haldane offered a study of the boundaries between inorganic and organic matter, speculating that the first organism formed as an anaerobic molecule, our "one ancestor," self-synthesizing and "self-contained" from a combination of water, carbon dioxide, ammonia, and the ultraviolet rays of the sun, all in a "prebiotic soup" and oxygen-free atmosphere. We descended not from a monkey or a worm but from a common virus, back when the "whole sea was a vast chemical laboratory."

Haldane eventually turned these biochemical hypotheses into a grand new utopian science, joining biology with astronomy, a model for visionary scientists and popular science to come. In a variety of essays and science-fiction stories, radio interviews and newspaper articles, he proposed racial eugenics and population control, laboratory-produced babies, and the communal upbringing of children under full socialism. He took the utopia even further, into outer space. In a perfect society set 40 million years to come, humanity is able to domesticate the whole planet: by exploiting the energy of the ocean's tides, by remaking the continents; by engineering a society of individual happiness and fulfillment. The average life expectancy rises to some 3,000 years old. When in this future the planet becomes uninhabitable, an elite corps of humanity sets off for outer space by rocket ship, refashioning itself by means of artificial evolution for the light and heat, gravity and atmosphere, of Venus. There it becomes a near-perfect civilization, a "super-organism" with the power of extrasensory communication with the "complete conscious control of every atom and every quantum of radiation in the universe." Humanity "proves that its destiny is in eternity and infinity."

After the Second World War, laboratory science began experiments to confirm the Oparin–Haldane thesis, and visionaries fortified Haldane's predictions in myriad ways. In the famous Miller–Urey experiment of 1952, Stanley Miller and Harold Urey hypothesized that they had artificially produced organic compounds from a "dead" system. They claimed to have formally replicated abiogenesis by using methane, ammonia, and hydrogen in a series of apparatuses, stimulated by boiling water and an electric spark. They thought they had discovered, through condensation reactions, the consistent production of amino acids, some of the building blocks of life. As Harlow Shapley, the eminent astronomer wrote in 1961, "It is gratifying to be a part of this magnificent evolutionary show, even though we must admit ourselves to be lineal descendants of some rather nauseating gases and sundry streaks of lightning."

Although scientists have identified the limitations of the Miller–Urey experiment, it was still a first, and a foundational and inspiring one. Since 1953, teams of scientists have explored dozens of other related and separate tracks, laboratory versions of the "prebiotic soup," all to prove some variant or alternative of the Oparin–Haldane thesis, all to discover the ultimate genesis for the informational codes of our RNA and DNA. Teams have proposed other places for life's genesis, like hydrothermal vents; or other means, like nucleic acids as the first self-replicating molecules. Many scientists now favor a likely combination of panspermia and abiogenesis at the origins of life on Earth. In one case, As Graham Cooks and his colleagues now presume, some planetary body (a comet, meteorite, or asteroid) likely brought chemicals, water, and amino acids to a young Earth. That mix may have interacted with sea spray, water droplets and air, to form peptides, which are chains of amino acids. Researchers have indeed already synthesized peptides (and DNA and RNA) in laboratory conditions, though they are far from replicating the conditions of that young planet.

Carl Sagan, a Cornell University and NASA astrophysicist, popularized some of Haldane's utopian visions, recalling the work of Camille Flammarion and Konstantin Tsiolkovsky, in a series of popular books and interviews. These culminated in the Public Broadcasting System's 13-part television show, *Cosmos* (1980). It was one of the most widely watched television programs in world history, later updated in a 2014 version. Sagan's approach was to offer a respectful attitude toward religion, though with a presiding claim to supplant it with a modern faith in science. Displaying a unique grace and charisma, he inspired audiences with poetry and music, taught them the wonders of abiogenesis and astronomy, and exhorted us to perfect our technologies not for destructive nuclear war, but for humanity and planet, for the peaceful exploration and colonization of space. We belonged to outer space, he preached. "The nitrogen in our DNA, the calcium in our teeth, the iron in our blood, the carbon in our apple pies were made in the interiors of collapsing stars. We are made of starstuff." Outer space was the "cosmic ocean" from whence we first came, and to which we must return, if now on the wings of science and technology.

The Information Age

The two great "ages" of the twentieth century, the Space Age and the Information Age, are closely interwoven. Ernst Schrödinger's book, *What is Life* (1944), illustrated this. Schrödinger had speculated that the information codes embedded in chemistry and biology amounted to a kind of "negative entropy," an inherent power in life to overcome the Second Law of Thermodynamics, to challenge the degradation of energy into entropy. The book helped inspire Francis Crick and James Watson on their way to discovering some of these very codes within the structure of DNA. They found information in the basic building blocks of life. After the Second World War, advances in silicon chips and computer programs boosted the confidence of some futurists that human beings might someday even fuse their minds with the integrated circuit. Norbert Wiener's, *The Human Use of Human Beings* (1950), proposed that human minds were already machine-like, processors of information by way of feedback loops, regulations, and controls. Information is the "inverse of entropy." It can organize and improve. Isaac Asimov wrote this scenario into his story, "The Last Question" (1956), how humanity will in small steps evolve into computerized information, and thereby conquer entropy and chaos in all of nature, terrestrial, interplanetary, and cosmic.

Gordon Moore's famous "Law" (1965) boosted confidence all the more, predicting the certain rise in computer capacities, measured by the number of transistors in an integrated circuit, to double every year or so, without a corresponding rise in costs, so he surveyed. From computer science, Ray Kurzweil's *The Singularity is Near* (2005) foresaw a corresponding exponential growth in human capabilities, via innovations like computerization, genetic engineering, nanomaterials, neural implants, and AI, until the human body will inevitably join with the computer as a "singularity." No more leaps and plateaus and dips on the technology "S'" curves. All arrows up. From physics, Frank Tipler's *The Physics of Immortality* (1994) expected no obstacles to this fusion, and to our migration as information to the distance reaches of the galaxy and beyond. From history, as former director of the NASA History Office, Steven Dick predicted the bonding of human mind to circuited machine, when we become an interplanetary civilization of electronic loops and signals, dispersed into infinity. From all these fashionable quarters, scholars have prophesied the imminent evolution, by humanity's own brains and hands, into the more perfect forms of information technology. By their count, the Space Age is already the Information Age. Information is inevitably humanity's final form.

The popular culture of the post-Apollo era was attuned to Carl Sagan's cosmic ocean. Atari video games had their start with "Computer Space," featuring a rocket craft pursuing flying saucers through the stars. Atari's famous 1972 logo even imitated the look of a rocket launch. Musicians and their publics tapped into the space vibe, beginning with Billy Preston's electronic funk tune, "Outa Space" (1972); and Sun Ra's Afro-Futurist, "Space is the Place," from *Discipline 27-II* (1972). For the enigmatic Sun Ra, outer space was a place of unlimited opportunity and fulfillment. Disco loved the sounds of spaceflight, as in "Magic Fly," by the French band, Space (1977); and the sights too, with the star-like disco ball hanging above the dance floor. The Soviets offered their own electronic space disco, as in "Hello Music! / *Zdravstvui Pesnya*" by Marathon (1982). But none could beat Michael Jackson, who premiered his classic "Moonwalk," while performing "Billie Jean" for the Motown 25 Special (1983).

Searching for Extraterrestrials

The scientific movement to study life beyond Earth, and to communicate with possible alien life, began at about the same time as the launch of Sputnik. As human beings set off by satellites and probes to outer space, we also began to speculate about alien civilizations doing the same. It took our own outward ventures to entertain the fuller range of possibilities. We had been attempting to signal Mars, and potential alien life, since the late 1800s, proposals in "interplanetary communication" ranging from huge geometric images drawn upon the ground, to lights and mirrors, to wireless radio messages. The new attempts during the 1960s remained within this tradition. Hence, scientists termed this "Communication with Extraterrestrial Intelligence" (CETI), though by the middle 1970s, both university scientists and NASA had settled on the phrase, the "Search for Extraterrestrial Intelligence" (SETI). The objective was to pose and answer the essential questions: where else does life exist, and if it does, might it want to communicate with us too?

The plurality of worlds thesis already had a stronghold in science fiction and scholarship. New data quantified and confirmed it all the more. By 1965, astronomers were calculating that there were at least 100 million stars, in the Milky Way galaxy alone, with the right conditions (dimensions, chemistry, and temperatures) to allow for life. Since there were at least 1 billion galaxies in the known universe, there were likely "billions and billions" of possible worlds, so Carl Sagan famously proclaimed. In 1974, he predicted at least a million civilizations in our own galaxy. We ought to mention several caveats to all of this. These numbers were (and remain, in still greater numbers) "astronomical," staggering in scope. The distances between our solar system and others are staggering too: 4.3 light years to Alpha Centauri, or 11 light years to Epsilon Eridani, and 100 or more light years to still other stars. Please remember that a light year is about 6 trillion miles. Exchanging signals that far will take at least hundreds or thousands of years. This is long-term in the extreme. We build all of this upon possibilities and probabilities, not certainties. The ultimate caveat is a question the physicist, Enrico Fermi, asked in a lunchtime conversation at a Los Alamos dining hall in the summer of 1950. "If there are extraterrestrials, where are they?" The paradox: if there are so many highly advanced alien civilizations in the universe, why have they not contacted us yet?

ET

The word *extraterrestrial* first appeared in the books and journals of the new astronomy during the late 1800s, mainly to describe the map of the solar system, contrasting the realm of the terrestrial or Earth-like planets (Mercury, Venus, Earth, and Mars) and the realms beyond. The reading public understood it to demarcate the cosmos near and far, Earth as one planet among many. *Extraterrestrial* did not yet mean an outer space alien. But the word *zombie* did, what writers borrowed from Haitian Creole folk tales, to mean a monstrous space traveler from another planet. With the advent of the atomic bomb and ballistic missiles after the Second World War, the media used the term *extraterrestrial* to describe the very technologies that threatened Earth's and our existence. Only in the 1960s, with the human mastery of the near reaches of orbit, as the human being became a space traveler, did *extraterrestrial* come to mean alien life or intelligence beyond Earth, the new object of our human desires and interest, the target of first "contact." We set out to search for "ET," perhaps not quite realizing that we were already it.

Several small groups of American astronomers decided that CETI was worth a try by 1959, inspired by the formidable new radio telescopes then in use by the space programs to track spacecraft and satellites, but also to scan outer space for the radio emissions from the planets and star systems. Humanity was also already sending out signals into outer space, what with the ever-multiplying radio and television programs crisscrossing our continents beginning in these years. The pioneers of CETI and SETI proposed scanning the heavens for signals proper, and even sending out some purposeful ones of our own. They assumed that, if humanity had invented rockets, spacecraft and radio telescopes, some among those other plural worlds would have done it too.

Frank Drake, a Cornell astronomer at the National Radio Astronomy Observatory (Green Bank, West Virginia), aimed his radio telescope at the Tau Ceti and Epsilon Eridani star systems in 1960, in search of alien radio signals, though none were found in the initial round, or have been found since. A scientific conference at Green Bank the next year gave a new purpose to the search. Drake put together a series of "meeting points," something people actually did before PowerPoint, and sketched them on a blackboard for discussion. Here were a "string of factors," like the numbers of planets "where intelligent life evolves," to be calculated together in order to estimate the number of "intelligent civilizations in the Milky Way," at least the ones with radio telescopes, rather like our own.

A superbly mathematical rationale for the existence of alien civilizations, the equation is also a perfect description of Earth, of our known facts and visions, of our civilizations and aspirations. It is eminently human. It is the "Frank Drake" equation, after all. It may be more about us than "them," expressing our counts of stars and planets, our somewhat wild ranges of estimates and probabilities, and with our little world as the prime example.

The Greenbush meeting was the beginning of a long and fruitful partnership between Frank Drake and Carl Sagan, on technologies ranging from SETI to the Pioneer and Voyager, and Viking NASA probes. They were not the archetypical closet scientists. As Joel Achenbach has written, "they were not incrementalists. They weren't trying to wriggle down some microtubule of the unknown. They were going for the big kill, the eye-popping discovery, the globe-shattering revelation." When, after a decade of trying, they did not receive the expected signals, they sent out their own. Take, for example, Drake's famous "Arecibo message" (1974). It was sent by the radio telescope at Arecibo, Puerto Rico, to the Hercules Globular Cluster, including 300,000 stars, some 25,000 light years away. It represented basic mathematical and chemical facts about Earth, the structure of DNA, the planets of our solar system, a diagram of the Arecibo telescope, and a stick figure of a human being. In a flash, Drake presumed to speak for the world, one person for all of us. He also assumed that the alien life that might receive it, could also see, hear, or perceive such sensory messages and pictograms as these.

These initial ventures in CETI and SETI prompted a remarkable Cold War partnership between American and Soviet scientists, centered on Carl Sagan and Iosif Shklovsky's joint publication, *Intelligent Life in the Universe* (1966). It was one of the first scientific attempts to survey the possibilities of extraterrestrial life, though it was partially founded on Shklovsky's rather bizarre idea that the Martian moon, Phobos, might actually be a hollowed-out alien spaceship; and it was built upon Sagan's rather imaginative accounts of mythology and astronomical probabilities. Their work continued at the "First International Conference on CETI," held at Byurakan, Armenia (1971). The USSR dedicated significant funds and energies to these tasks, not only to promote science and compete with the US, but also to promote its priority in scientific atheism. US participants, like Philip Morrison and Marvin Minsky, were otherwise focused on communication and cybernetics, as, for example, how to package all of the knowledge from human history (and perhaps someday even us) to be sent off by radio telescope into the cosmos.

Along these lines, one enterprising astronomer, Nikolai S. Kardashev, answered the Drake Equation with its Soviet variant, the Kardashev Civilizations

(1964), a time map of advanced technological civilizations, perhaps millions or even billions of years advanced beyond our own, power civilizations that control whole planets and stars and even whole galaxies. Like the Drake Equation, the Kardashev categories come to us in an idiom of science and mathematics, but they too cannot escape the mystical, reminding us of J.B.S. Haldane's utopian predictions from several decades before. They are perhaps less about what alien civilizations might look like, and more about what our human civilization might or ought to become in the future.

These scientific and technological ventures sparked a minor revolution in the way popular culture began to portray alien life. Such imaginary visions may seem trivial, but they are not irrelevant, because they cultivate expectations in the average reader and television viewer, promoting a bias for the plurality thesis. Like them or not, these images shaped our perceptions. One set of scenarios had humanity traveling to the distant regions of the galaxy. These had a long tradition in the literature, as for example in Olaf Stapledon's classic, *Starmaker* (1937), its main character speeding along through time and space to distant worlds, along the way meeting winged aliens, fish beings, "six-legged" monsters, "mollusk-like" creatures, and plant beings of various sorts. By the 1970s, television and film screened these stories. In the US, *Star Trek* (1966–1969) followed the adventures of the starship USS *Enterprise* through the Milky Way galaxy (roughly during the 2260s), discovering strange alien beings at points beyond, but also telling allegories about racial conflict, nuclear war, and evil empires here on earth. In the USSR, Ivan Efremov's, *Andromeda Nebula: A Space Age Tale* (1959, with a 1967 film), and the television film, *Moscow to Cassiopeia* (1976, about a crew of teenagers speeding light years to an alien planet), were decidedly more upbeat, the stories of the similar voyages of Soviet starships in future times, though also encountering the challenges of vast distances, radio signals from faraway stars, and backward aliens.

Another set of images had to do with the aliens themselves. One kind of alien was a monster or insect, sometimes with technologies of its own or sometimes with just the exoskeleton attached. These were frightening types, at the center of *The War of the Worlds* (1898, with 1953 and 2005 films), *Starship Troopers* (1959 with 1997 film), and of course *Alien* (1979), the start of a whole franchise with its unrelenting killing machine: a large, two-legged monster, with a huge head, several sets of jaws, and acidic blood. A second kind of alien was the human impostor, the alien hidden in plain view here among us. They could be quite scary too, as in the case of *The Day the Earth Stood Still* (1951), and *Plan 9 from Outer Space* (1959), in which human-like aliens have come to warn earthlings against destroying themselves with atomic weapons. In the American television show, *The Invaders* (1967–1968), an alien species masquerades as "us" in order to colonize the planet.

These kinds of movies were all part of a rich tradition in the "imagination of disaster," as Susan Sontag has argued. Americans threaded disaster into moralistic plotlines of their own good and the enemy's evil, filled with machine props to screen their atomic and rocket anxieties. The Russians had no analogs. They built disaster into history itself, threaded by their ideology and propaganda about the violent collapse of capitalism and the inevitable victory of communism. That was scary too.

Friendly and funny aliens with special powers have been a staple of American television too. In the sitcom, *My Favorite Martian* (1963–1966), Uncle Martin looks human, but for the retractable antennae on his head, along with the ability to become invisible, levitate objects, and read people's minds. The third and more lasting type of alien was the bigheaded, round-eyed, frail-limbed diminutive, popularized in Steven Spielberg's, *Close Encounters of the Third Kind* (1977) and *E.T. the Extra-Terrestrial* (1982). These are friendly aliens, of course, almost human. They are precisely the kind we might wish to receive our SETI signals from and even answer back.

Deep-Space Interplanetary Probes

Within only a decade of humanity's first ascent into outer space, beginning with Yuri Gagarin and Alan Shepard's flights in 1961, the US was already launching interplanetary craft to leave the solar system. It was a busy decade, as we reflect, and one of some rather amazing transformations. Spaceflight programs settled into a reliable routine of launch and recovery: of people and machines with Apollo and Soyuz, or of machines alone to the far boundaries of true outer space. Mechanical and robotic probes were important foundations in the history of spaceflight. The Sputnik series and Explorer started it all, humanity's first tentative steps to orbit by machine. They continued with the US Ranger, Orbiter, and Surveyor observations and mappings of the moon, and its Mariner and Pioneer series to interplanetary space; as well as with the USSR's long-standing Luna program and its impressive Venera series (1968–1983), with the first atmospheric probe, surface pictures, and soil analysis of the planet Venus.

For years, spaceflight professionals have debated the advantages of human versus robotic programs, of people versus machines. These have more to do with the competition for scarce government funds and for professional status, rivalries between those who launch astronauts and those who launch probes. There has never been an either-or alternative for spaceflight. Everything about it is mechanical. Everything about living in space demands new and extremely mechanized environments. The human being is nothing there without the technology, without all the small probes that have led the way and without all the component apparatuses that keep us alive and well there.

In terms of the "race" for deep-space interplanetary probes, the US won by default with two extraordinary pairs of spacecraft. NASA's Jet Propulsion Laboratory launched the first pair, Pioneer 10 and 11 in March 1972 and in April 1973, to the outer planets, where they took the first extraordinary photographs and studied the compositions of Jupiter and Saturn. Pioneer 10 then set off to travel through the sun's magneto-tail and the interstellar medium beyond our solar system, away from the center of our galaxy; and Pioneer 11 set off for the interstellar medium toward the galaxy center.

Knowing that the Pioneers would become distant travelers, preserved in outer space like Egyptian tombs, NASA placed some significant human artifacts aboard, perhaps to be found someday by intrepid alien explorers. Carl Sagan headed the committee that chose the materials. Both Pioneer craft carried gold plaques (6 x 9 inches across). They were centered on a map of Earth's location in outer space; on the shape and path of the spacecraft, with an arrow pointed outwards; on some basic mathematical and chemical notions; and on a drawing of a man and woman. The two human figures sparked the most public controversy of all. Here on Earth. They stood mostly hairless and altogether naked, not exactly a true picture of human beings in our civilized state. They were not quite anatomically correct. They were obviously Caucasians. They maintained an altogether sexist and contentious pose, the man dominating the woman. At least the Pioneer craft was silhouetted in the background, illustrating the most eloquent point of all: that we humans have become so dependent on and defined by our machines.

NASA launched a second and even more impressive pair of interplanetary craft with Voyager 1 and 2 (5 September and 20 August 1977), headed on what was called the "Grand Tour," which included both fly-bys and observations of Jupiter and Saturn, with Voyager 2 also observing Uranus and Neptune. Both craft carried packages of scientific instruments that surveyed their atmospheres and radio-emission signals, their various physical properties and geological features, and took the most precise photographs and films of the planets and their moons up to that time. To the surprise of planetary scientists, both craft identified the first known volcanic activity well beyond Earth, on one of Jupiter's moon, Io. They offered precise confirmation of the possible ice crust on another, Europa, under which may be a vast watery ocean. By August 2012, so NASA announced, Voyager 1 was the first human-made craft to leave the solar system and enter interstellar space.

The Gravity Assist

The "gravity assist" was an interplanetary engineering maneuver perfected in the 1960s, designed to use the orbital movement and gravitational force of a planet or planetary body to change the pathway or speed of a spacecraft, mostly in order to save on its all-important reserves of propellant. Two Soviet rocketry pioneers, Yuri Kondratiuk and Fridrikh Tsander, discussed its potential in their theoretical works from the 1920s. The USSR was also first to use it, with the Luna 3 spacecraft (1959), which took photographs of the far side of the moon and then used its gravitational pull to shoot back to earth, this to precisely locate itself in orbit over the USSR for the transmission of its signals. The American mathematician, Michael Minovitch, formally proposed the gravity assist or "slingshot" in 1961, as a navigational maneuver between two planets during a spacecraft's outer trajectory, what he called an "interplanetary free fall." NASA then applied it to the Mariner 10 mission (1973), which used a gravity assist from Venus to head off to Mercury. Thanks to a rare alignment of planets, Voyager 2 (1977) used a gravity assist to skip its way beyond Earth. The Titan III/Centaur booster was only able to launch Voyager to Jupiter, where it would have remained. But it traveled further thanks to a gravity "boost" to Saturn, then again to Uranus, and again to Neptune, and still farther to the edge of the solar system, from where it still transmitted data to NASA.

Like Pioneer, Voyager carried a series of messages for the stars, this time a gold-plated phonograph record, carefully placed in aluminum casings, along with playing instructions. The irony is that most of us today would not even be able to play them, now that we listen to music on our telephones, something few people in the 1970s would have understood. The records contained the sounds, images, and music of the Earth. By the time Voyager reaches the vicinity of a new star system, sealed in its recesses will be a variety of audio offerings. These include the sounds of a kiss from mother to child, and the liftoff of a Saturn V; images like the structure of DNA and a "sunset with birds"; and greetings of "hello" to the universe in 55 world languages The records also include concertos from Bach and Beethoven, drums and choruses from Senegal and Zaire, Native American chants and Peruvian panpipes, and Chuck Berry's rock and roll tune, *Johnny B Goode* (1958), about a guitar prodigy from rural Louisiana. He lives for his mother's promise that someday he will be a star. Such were the melodies sent out into Earth's expanding universe.

After the economic downturn of the 1970s, and amid the first and expensive years of the Space Transportation System, NASA sent off only two interplanetary probes between 1979 and 1991. They confirmed that most of the solar system is uninhabitable, not to speak of uninhabited. So far, there is only evidence of life on Earth, although there is compelling evidence of water, the source of life, elsewhere.

The Galileo mission (1989), which made some gravity assists at Venus and Earth, traveled to the outer planets, meeting asteroids and a comet along the way. Before settling in to explore Jupiter and its moons, it confirmed the presence of a liquid ocean under the moon Europa, and possibly the right conditions for organic life to survive there.

Magellan (1989–1994) traveled to an orbit of Venus, with a successful run of radar mapping and gravity data, culminating three decades of Soviet and

American exploration, discovering more about this planet than ever before. One of the "terrestrial" or rocky planets, Venus has turned out to be very much like a dead Earth, a world suffocated by a massive "greenhouse effect," without a protective ozone layer, the sunlight penetrating to the surface and trapped there. It smothers under an atmosphere of 98% carbon dioxide, at 100 times the Earth's pressure, in heat reaching 800 degrees Fahrenheit.

NASA and its European partners also prepared one of the last great expensive missions, Cassini-Huygens (costing $3.3 billion), launched in 1997 but in development since the late 1970s. It was the first probe to orbit Saturn and the first to land a craft (Huygens) in the outer solar system, on Saturn's moon, Titan. Cassini also discovered what scientists believe may very well be an ocean of water on Enceladus, another of Saturn's moons.

The truth, from all we know so far, from all of these and later probes to and signals from outer space, is that we have found no "second genesis." Not yet. We have only heard many probabilities, predictions, and prophecies. We have never found extraterrestrial life anywhere in our solar system outside of Earth. We have discovered places that might become suitable for human habitation, sites with enough potential supplies of liquid water for survival and civilization. The moon Europa and the planet Mars are two such locations. Our past machines have discovered where our future machines might someday create habitats: for our fragile bodies and all too impressionable minds.

Missions to Mars

Mars remained a great prize in the space race between the USSR and the US. The Lowell canal controversy, along with science fiction in literature and film, had primed generations of enthusiasts to consider Mars as the most suitable abode for life beyond Earth. In the US, Robert Heinlein's *Red Planet* (1949), offered an adventurous plotline about the escapades of young earthling colonizers on the planet, romping about with their Martian compatriots. Ray Bradbury's short-story collection, *The Martian Chronicles* (1950), captivated audiences with darker stories about humans fleeing atomic war, only to find a ferocious struggle to survive on Mars. Soviet political censors allowed both translations onto the Russian reading market, where they became popular books, along with the perennial Soviet classic, *Aelita. Queen of Mars*. The USSR also enjoyed a signal priority in the scientific field: the work of Gavriil A. Tikhov. He was an early supporter of Tsiolkovsky's spaceflight theories in 1913 and claimed to have discovered bands of plant life on Mars by 1946, the foundations for a "new science," one he was soon calling "Astrobiology," yet another first in the Soviet arsenal.

After the Moon, Mars remained the prime target of the superpowers, though it was unyielding at first, with most failures occurring early on in the race. The Soviets attempted to launch nine missions to Mars between 1960 and 1971, though all failed at launch or on their way. The later pair of Mars 2 and 3 spacecraft

(1971), consisting of both orbiters and landers, each with small roving devices, were valiant attempts to beat the US and salvage some of the USSR's spaceflight reputation after the Apollo landings. They saw mixed success, relaying some of the first photographs of the planet, along with physical data and relief maps. The Mars 2 orbiter inserted orbit properly, though the lander crashed onto the planet, along with the Soviet hammer and sickle insignia. Both orbiter and lander worked for Mars 3, but the lander failed to transmit any data in the midst of one of the infamous Martian dust storms, never to recover, though it made a soft landing, by use of aerodynamic braking and parachute.

The Soviets endured successive misfortunes in their Martian ventures. Mars 4 and 5 orbiters and landers (1973–1974) were attempts to beat the impending US Viking missions. A computer failure in Mars 4 sent it beyond the planet; and, although some images were received from Mars 5, the lander failed. The Mars 6 and 7 lander missions (1973–1974) either crashed or flew by the planet, thanks to either electronics or guidance malfunctions. Both of the USSR's Phobos 1 and 2 missions (1988) failed, meant to explore the Martian moon Phobos, though they either flew by the planet or were lost. A popular Russian anecdote, in the form of a rhyme, framed these efforts in bitter terms, given the poverty of most people's lives. "There's talk that we'll be first on Mars [*na Marse*]. It would be better if we were first in meat [*po miasu*]."

The US enjoyed a much better record, with a number of impressive firsts, though there were several failures along the way. Mariner 4 (1964) was the first Mars flyby, with photographs of the planet, followed by the similar missions of Mariner 6 and 7 (1969), the start of what became ever-increasing waves of images and data from what we discovered was a barren planet. Mariner 9 (1971) was the first successful spacecraft to orbit the planet, beating the Soviet Mars 2 by 13 days. Wernher von Braun and assorted guests were at the Jet Propulsion Laboratory for its arrival, anxiously watching television sets of its orbital telemetry, though they received none of the expected transmissions of images, as a dust storm obstructed all views. Yet in time, according to NASA officials, Mariner was

> able to map 85 percent of the Martian surface … in addition to gathering abundant information about the surface and atmosphere. The more than 7,000 images it transmitted included the first detailed views of Olympus Mons, the solar system's largest volcano; Valles Marineris, a vast canyon system that dwarfs the Grand Canyon and was named for this spacecraft; the polar caps; and the moons Phobos and Deimos.

The Viking missions to Mars (1976), a twin mission of orbiters and landers, costing a significant $967 million, offer a culminating moment in the first decades of planetary exploration. They accomplished successful soft landings, and provided the first surface photos of Mars, revealing a sandy and rocky terrain of reds and browns, along with what looked like riverbeds and flood plains, evidence that water may have once flowed on the planet. Viking landed a sophisticated biology package of instruments and devices to find microorganisms, or any sign of life, though by the official judgment it found none. That debate has mostly remained settled.

After the Viking successes, however expensive, NASA faced several financial and engineering challenges through the 1990s. Mars Observer (1992), an orbital platform to explore the climate and planetary characteristics of Mars, failed to make its prescribed orbital insertion, probably due to a mechanical failure onboard. It was an embarrassing $812 million loss. The solution to these budgetary problems, and high risks and costs, was an innovative and mostly successful program called "Discovery," the initiative of NASA administrator, Dan Goldin, and promoted under the slogan of "faster, better, cheaper." The goal was to launch "frequent, highly focused, and cost-effective" missions to study the Earth, Moon, and Sun; interplanetary space and the galaxies; and most of all, Mars.

The Discovery program sent a variety of inexpensive missions to Mars, several successful (Global Surveyor and Pathfinder), several not (Climate Orbiter and Polar Lander).

Premier among them was the Pathfinder lander (1997), whose Sojourner rover explored 300 rocky feet of Mars in three months, and together they provided just over 17,000 photographs of the Martian landscapes. The mission also enjoyed massive public attention, thanks largely to the new phenomenon of the Internet, where enthusiasts could view its images almost as soon as they arrived back to Earth. NASA's news service boasted that its Pathfinder site received 560 million such "hits" in one month, with a record 47 million visitors in one day. An internet archive still maintains them in their original form, as a kind of digital time capsule of virtual spaceflight.

One of Pathfinder's signal legacies was to galvanize several communities of enthusiasts back on Earth, at NASA's centers and at the Mars Society, to mount a credible human mission to Mars. No one represented this passion better than the indefatigable Robert Zubrin. As engineers working for Martin Marietta, Zubrin and his colleagues conceived of a brilliant and influential project, "Mars Direct" (1990–1996) to explore and colonize Mars the old-fashioned way. Unlike the Apollo spacefarers, they would not bring all the components and necessities with them on a quick round trip. Rather, they would send some on ahead, bringing just enough to adapt to the Martian environment and exploit its resources for basic survival, just as our ancestors did on their frontiers in the American West.

It was a powerful pitch, made even more compelling by Zubrin's personal charisma, plying the hallways of college campuses and scientific institutes around the country, selling his books and his mission like an itinerant preacher. He called for $30 billion for Mars, to develop a transportation and habitat system, within a Kennedy-like ten or 20 years, by the beginning of the new Space

Age millennium: 2001. Almost like clockwork, with each successive robotic mission to Mars, like-minded advocates have called for still more billions to reach Mars in that magical decade or two. Like clockwork, the dates pass, and no one remembers, and the human missions languish without full funding. Still, NASA's successes since Pathfinder are getting closer and closer to the mark, the robotic craft acting as advance teams, scouts of the planet for the inevitable human mission. Sometime in the next ten or 20 years.

Pathfinder opened several decades of remarkable achievements, beginning with the Mars Exploration Rover Mission, introducing Spirit and Opportunity (2003), which set out to explore the planet's geography and created a vast archive of data. In 2014, still traveling, Opportunity even beat the record of the Soviet Lunakhod lunar rover, with a total of 25 miles of Martian ground canvassed. The spacecraft Phoenix (2008) landed in the Martian arctic north to search for past or current evidence of water, and to survey for possible colonization sites. Like its fellow robotic craft, it has only found another dead Earth, a planet without life-giving oceans or plate tectonics. Mars has a very thin atmosphere, only 1% of Earth's, made mostly of carbon dioxide. Great dust storms and seasonal extremes of hot and cold cover it. The surface is filled with poisonous soil. For the unprotected human being, the place means immediate death.

The Mars Science Laboratory and its Curiosity Rover (2012) were large and complex platforms, a full laboratory with a robust rover, as of March 2024 still on a search for past or present signs of life (or the potential for it). The package landed not by parachute and balloon cushion, as with earlier probes, but by a rocket-propelled descent stage and an ingenious sky crane, deployed some 25 feet from the Martian surface. It was also the fullest digital mission to date. Malin Space Science Systems, for example, outfitted Curiosity with an advanced digital electronics assembly for superior-quality imaging. The Siemens Corporation provided NASA with the software and hardware in order to model the craft and mission in their digital profiles before the actual machinery was built. Amazon Web Services provided "cloud computing applications" to join the teams of scientists and engineers scattered across the US.

Since Pathfinder, NASA has featured the powers of the digital age to enhance these missions and provide images and explanations for the engaged public. Most recently, the Mars Exploration Program and Mars 2020 have included a new rover, Perseverance (2021) and Ingenuity, the first controlled drone helicopter to fly in another planet's atmosphere. They have captured a serious and smart internet audience.

NASA has not welcomed all of the publicity. The Percival Lowell controversies still cast their shadow from a hundred years before. The Viking 1 orbiter had actually set the stage for one of the more entertaining episodes in the exploration and interpretation of the planet: the infamous "Face on Mars." The whole controversy started with a photograph of the Mars "Cydonia tablelands," discovered in July of 1976 by a member of the imaging team, who was searching for a suitable Viking 2 lander site. NASA initially showed the picture to the press corps as an example of the "tricks of lighting and shadow," although some officials also discussed it as the "Face on Mars," about 1-mile-wide. Others could not help but dramatize it: a tabloid newspaper, the *Weekly World News* (1984); an episode of the television show, *The X Files* (1993); and a popular book by Richard Hoagland, *The Monuments of Mars* (1987). Hoagland, a space publicist, purported to invent a new field of "exo-archeology," and even mapped the ruins of a Martian civilization around the "Face," including a fortress and pyramid, and even teeth for the mouth. Alas, the Mars Global Surveyor (2001) photographed it with a high-resolution camera, revealing nothing more than a hill. By now, the myth had taken on a life of its own, driven by our very human tendency to perceive human faces in otherwise random natural phenomena. Looking for aliens, we found ourselves.

A related controversy in 1995–1996 had to do with something that had come from Mars and actually landed here on Earth: a Martian meteorite, categorized as ALH84001. A team of NASA scientists at the Houston Space Center had used a "field-emission scanning electron microscope," originally designed to find cracks in the parts of the Space Shuttle, to inspect a flake from the meteorite. They produced a nanometer image, about one-billionth of a meter long, of what looked like a Martian "worm," in the shape of a tubular and segmented line. With others on the team, NASA's David S. McKay (a geochemist and specialist on lunar soil) proposed that this might just be the fossilized remains of some Martian bacteria or microbe. Here might be an "extraterrestrial," proof of organic life. The claim did not stand for long, as a team of university scientists from around the country, under the lead of John Bradley, soon argued convincingly that these were not microbes but the edge lines of crystal formations. This was, in essence, merely the optical illusion of a worm created by the very powerful electron microscope. The debate continues unresolved, although it also returns us to an older controversy: Lowell's canals of Mars, when a new and powerful telescope betrayed him. We keep seeing Martian lines in our eyes.

Vertical Horizons: Above and Below Worlds

In the last 30 years, scientists have made some profound discoveries here on Earth, with dramatic implications in the quest for extraterrestrial life. They have found life in the form of "extremophiles," organisms that can live in extreme environments: like inside of rocks, in nuclear reactors, in Antarctic ice lakes, in hydrothermal springs, and in deep-water aquifers and thermal vents. We have learned that microbes are extremely resilient and versatile, able to survive harsh environments that are extraordinarily hot, cold, acidic, salty, or toxic. Some can even flourish in environments without oxygen or light, or under extremely high pressures. We discovered one of the first such extremophiles, the bacterium Thermus Aquaticus, in the hot springs at the Lower Geyer Basin of Yellowstone National Park in 1966, living in temperatures over 158 degrees Fahrenheit. In 2005, researchers in Europe discovered microbes in the anoxic basins or "brine lakes" in the Mediterranean off the coast of Sicily. Still other researchers have located extremophiles inside rocks 1,900 feet below sea level; and encapsulated within ice at least 60 feet beneath the surface of Antarctica. In March 2013, researchers recorded data that suggested microbial life forms existed within the Marianas Trench, the deepest part of the world's oceans.

Because we can use extremophiles for medications, as well as in various industrial and chemical processes, they are becoming something of a new business. Interest in them extends far beyond earthly applications. If life can exist in some of the most forbidding and deep environments on Earth, why should it not be able to exist on planets like Mars? A recent study of the microbes at Yellowstone, for example, asserts that their main energy source in the hot springs is hydrogen, the most abundant element in the universe. You can see where we are heading with this. What is good enough for life under Yellowstone might very well be good enough for any number of places near and far in the cosmos. The microbes found in the Antarctic have also fueled plenty of speculation. If extremophiles can thrive in such extremely cold temperatures on Earth, might they also be found on the frozen terrains of the heavenly bodies within our solar system? Extremophiles have proven that our old assumptions about the kinds of environments needed to sustain life were mistaken; and have provided researchers with a wealth of indications about where and how to search for extraterrestrial life.

Select Bibliography

Achenbach, Joel. *Captured By Aliens: The Search for Life and Truth in a Very Large Universe*. New York: Simon & Schuster, 1999.

Arrhenius, Svante. *Worlds in the Making: The Evolution of the Universe*. New York: Harper, 1908.

Barrow, John D., and Frank J. Tipler. *The Anthropic Cosmological Principle*. New York: Oxford University Press, 1986.

Bernal, J.D. *The Origin of Life*. Cleveland: World, 1967.

Boelsche, Wilhelm. *The Triumph of Life*. Chicago: Kerr, 1906.

Bradley, John P., et al. "No 'Nanofossils' in Martian Meteorite," *Nature* 390 (December 1997): 454–456.

Cameron, A.G.W., ed. *Interstellar Communication*. New York: Benjamin Press, 1963.

Christian, David. *Maps of Time: An Introduction to Big History*. Berkeley: University of California Press. 2004.

Cocconi, Giuseppe, and Philip Morrison. "Searching for Interstellar Communications." *Nature* 184 (1959): 844–846.

Cooks, R. Graham, Dylan Holden, and Nicolas Morato. "Aqueous Microdroplets Enable Abiotic Synthesis and Chain Extension of Unique Peptide Isomers from Free Amino Acids," *Proceedings of the National Academy of Sciences* 119/42 (2022): 1–8.

Darling, David. *Life Everywhere: The Maverick Science of Astrobiology*. New York: Basic Books, 2001.

Dick, Steven J. *Space, Time, and Aliens: Collected Works on Cosmos and Culture*. Cham: Springer International, 2020.

Dick, Steven J., and James Strick. *The Living Universe: NASA and the Development of Astrobiology*. New Brunswick: Rutgers University Press, 2005.

Ehricke, Krafft. "The Anthropology of Astronautics," *Astronautics* (November 1957): 28.

Fimmel, Richard O., James A. Van Allen, and Eric Burgess. *Pioneer: First to Jupiter, Saturn and Beyond*. Washington, DC: NASA, 1980.

Fry, Iris. *Emergence of Life on Earth: A Historical and Scientific Overview*. New Brunswick: Rutgers University Press, 2000.

Haldane, J.B.S. *The Inequality of Man and Other Essays*. Harmondsworth: Penguin, 1932.

Hoagland, Richard C. *The Monuments of Mars: A City on the Edge of Forever*. Berkeley: North Atlantic, 1987.

Hotakainen Markus. *Mars: From Myth and Mystery to Recent Discoveries*. New York: Springer, 2010.

Joseph, Rhawn. *Astrobiology, the Origin of Life, and the Death of Darwinism*. San Jose: University of California Press, 2001.

Kline, Ronald R. *The Cybernetics Moment: Or Why We Call Our Age the Information Age*. Baltimore: Johns Hopkins University Press, 2015.

Koerner, David, and Simon LeVay, *Here Be Dragons: The Scientific Quest for Extraterrestrial Life*. New York: Oxford University Press, 2000.

Kurzweil, Ray. *The Singularity Is Near: When Humans Transcend Biology*. New York: Viking, 2005.

Lederberg, J. "Exobiology. Approaches to Life beyond the Earth." *Science* 132 (1960): 393–400.

McKay, Christopher P. "The Search on Mars for a Second Genesis of Life in the Solar System and the Need for Biologically Reversible Exploration," *Biological Theory* 13/2 2018): 103–110.

McKay, David S., et al. "Search for Past Life on Mars: Possible Relic Biogenic Activity in Martian Meteorite ALH84001," *Science* 273 (1996): 924–930.

Meyer, Max Wilhelm. *The Making of the World*. Chicago: Kerr, 1906.

Morton, Oliver. *Mapping Mars: Science, Imagination and the Birth of the World*. New York: Picador, 2002.

Mudgway, Douglas. *Uplink–Downlink: A History of the Deep Space Network 1957–1997*. Washington, DC: NASA, 2001.

Oliver, Kendrick. *To Touch the Face of God: The Sacred, the Profane, and the American Space Program, 1957–1975*. Baltimore: Johns Hopkins University Press, 2012.

Osborn, Henry Fairfield. *The Origin and Evolution of Life*. New York: Scribner's, 1925.

Pyne, Stephen. *Voyager: Seeking Newer Worlds in the Third Great Age of Discovery*. New York: Viking, 2010.
Roach, Mary. *Packing for Mars: The Curious Science of Life in the Void*. New York: Norton & Co., 2010.
Sagan, Carl. *Cosmos*. New York: Ballantine Books, 1985.
Sagan, Carl. *Murmurs of Earth: The Voyager Interstellar Record*. New York: Random House, 1978.
Sagan, Carl, and Frank Drake. "The Search for Extraterrestrial Intelligence." *Scientific American* 232 (1975): 80–89.
Shklovsky, I.S., and Carl Sagan. *Intelligent Life in the Universe*. New York: Holden-Day, 1966.
Siddiqi, Asif. *Deep Space Chronicle: A Chronology of Deep Space and Planetary Probes 1958–2000*. Washington, DC: NASA, 2002.
Sontag, Susan. "The Imagination of Disaster," *Commentary* (October 1965): 42–48.
Tipler, Frank. *The Physics of Immortality: Modern Cosmology, God and the Resurrection of the Dead*. New York: Doubleday, 1994.
Zubrin, Robert M. *The Case for Mars: The Plan to Settle the Red Planet and Why We Must*. New York: Touchstone, 1996.
Weinersmith, Kelly, and Zach Weinersmith. *A City on Mars: Can We Settle Space, Should We Settle Space, and Have We Really Thought this Through?* New York: Penguin, 2023.
Wiener, Norbert. *The Human Uses of Human Beings: Cybernetics and Society*. Boston: Houghton Mifflin, 1950.

10 Bodies

Living and Cooperating in Space

In the aftermath of Apollo, the US and USSR engaged in one more mostly peaceful spaceflight race. It was a race for Earth-orbit space stations, between the Soviet Salyut and America's Skylab. These hurried space stations turned out quite differently from what spaceflight visionaries had earlier imagined. The classic images of stations, reaching back to the *Collier's* magazine series of the early 1950s, and dramatized in Stanley Kubrick's film, *2001: A Space Odyssey* (1968), were of grand, circular mansions in the heavens. Kubrick's musical score captured the power and majesty of these stations, alternating between Richard Strauss's gravity-breaking, *Also Sprach Zarathustra* (1896) and Johann Strauss's ballet-like, *On the Beautiful Blue Danube* (1867).

With the advice and planning of physicist, Gerard K. O'Neill (Princeton University), NASA entertained this far-sighted "torus" style between 1974 and 1977: a circular tube 4 miles around, with a central bore 400 feet wide, and large enough to accommodate 10,000 people. In another celebrated version of the torus station, the "L-5" Society proposed to build it in Earth orbit, then tow it out to "Lagrange Point Five," that place where the gravitational forces of the Earth and the moon balance each other, allowing for a safer and more efficient orbit and sustained living.

Designers meant these circular stations to be little Earths, tubed for internal living and artificial gravity, rotating around an axis to mimic gravity and provide for normal sensations, an Earth-like up and down, left and right. They were also little utopias, better Earths: Tsiolkovsky-style, filled with pure air and cleansing sunshine, lush gardens, and many happy people. The plans for the stations, with artist Rick Guidice's illustrations, almost look like brochures for the trendy American ranch-style housing development from this era. Yet the "space colonization" movement was also tinged with pessimism, centered on the mortality of planet Earth and our human propensity to destroy it, through war and pollution, urbanization and overpopulation. The hope was that our exploitation of space resources might save the fragile balance of life on Earth.

The actual space stations built through the 1970s, and into the present, have been far less impressive. They ended up being bits and pieces, telescopic sections

DOI: 10.4324/9781003494409-14

that could fit onto a rocket stage, or literally were the conical sections of a multistage rocket, all for easy attachment in space. Our stations looked more like soup cans and television antennae than new worlds. They were not as spacious or populated or visionary as we thought. Rather, they encapsulated us, contained us, like umbrella exoskeletons. Yet they have served as the foundations for what may be humanity's inevitable space future. They have taught us invaluable lessons about long-duration habitation in space, about its costs and benefits for the human body and mind. They are the bridge between those past generations that fought the space race and those future generations that may someday become a new spacefaring human race.

Artificial Gravity

How will human beings live in space if we are to make it our collective future? Will we adapt to microgravity environments, which is especially important if we consider the long durations necessary to traverse the long distances between inhabitable planets? Will we take gravity with us, creating artificial gravity environments that mimic the Earth-bound gravity to which our bodies and minds have become so accustomed? The most popular method for artificial gravity, in both science fiction and in real-life scenarios, is a rotating spacecraft, which uses centripetal force to create gravity-like conditions: the "feel" of gravity on the inner hull of a rotating craft. It pushes against us, and we feel the pull of the hull, essentially turning a wall into a new floor. Since only a few artificial gravity tests have been done in Earth orbit, we can only theorize, but the smartest proposals call for a relatively large spacecraft moving at the slower speeds of two to three revolutions per minute, this to mitigate the effects on the human body of the spinning motion. The trouble with such a proposed craft is its size and speed, as it must be relatively large to create the best conditions and needs some kind of power plant to provide the motion. Size and speed mean more expensive materials to launch into space, of course, which means more money, which has always been the most crucial and problematic aspect of spaceflight.

The Salyut Space Stations

In October of 1969, in a dramatic Kremlin broadcast, Soviet leader Leonid Brezhnev announced a new architecture in the race to space: a competition for Earth-orbiting space stations. He basically recognized the US victory in the Moon race and the USSR defeat, but also raised a new marker for success. Space stations, he said, would be the Soviet Apollo. The USSR would beat the US in a number of new "firsts": long-duration records, microgravity studies, Earth and astronomical observations (as well as all-encompassing military imperatives, too, like military surveillance and satellite defense). Brezhnev was confident based on the USSR's recent missions of the Kosmos, Zond, and Soyuz craft. He was implicitly acknowledging Valentin Glushko's rising power, thanks to the successes of his Proton launcher. He was also subtly hinting that the USSR was still planning to establish a Moon base or Mars mission with the still-viable N-1 rocket. The space race was not over. It had simply entered a new phase.

There were eventually seven Salyut ("Salute") space stations, designed for the changeable crews and re-supply ships of Soyuz. Of the seven, three of the Salyut (2, 3, and 5) were actually military stations of the Almaz ("Diamond") type. All of the Salyut had dual civilian and military uses and shared several basic design features with Almaz, like the hulls and docking mechanisms. The three top-secret Almaz were purely military. They were an answer to the recently canceled US "Manned Orbiting Laboratory" (MOL), a joint civilian–military enterprise of the Gemini program. The Almaz were two-person stations assigned with such tasks as Earth surveillance (photographic and radar observation of US military forces), orbital bombardment, and offensive anti-satellite capabilities. They carried outer-space-style machine guns. Among the Almaz stations were Salyut 2 (1973) which failed soon after launch; Salyut 3 (1974), which held one crew; and Salyut 5 (1976–1977), the most successful of the military stations with two crews. By the late 1970s, the Soviet military canceled the Almaz program in favor of its military reconnaissance satellites, and in the absence of any formidable US threats.

The original plan was to build these varied space stations in gradual and deliberate ways, methodically assembling them piece by piece in Earth orbit. That plan proved far too time-consuming and expensive. So, the early ones were developed and produced rather hastily. Salyut 1, for example, was the hull of an Almaz military space station with the main engines of a Soyuz ferry craft. In the haste to compete with the US, Salyut was plagued by a number of design and construction problems, including several failed dockings and poorly functioning stations, especially early on in the program, although the ground-control teams and crews did learn how to survive and improvise in order to repair the problems. All of the Salyut stations were launched successfully by the USSR's premier Moon rocket, the Proton, with the exception of one failed take-off in 1972.

Salyut 1 (19 April 1971) was the world's first inhabited, Earth-orbiting space station, made up of two simple cylinders, one small and one large, with a propulsion module to keep it steady in orbit. It was serviced by innovative water-recycling and waste-ejection systems, as well as by an environmental platform centered on lithium perchlorate canisters, which cleaned the air of poisonous carbon dioxide. The first crew docked with the station but was unable to enter it. The second crew was scrubbed due to medical complications, and replaced by the backup team of Vladislav Volkov, Georgii Dobrovolskii, and Viktor Patsaev.

Beginning on 6 June 1971, as the Soyuz 11 crew, they established the USSR's first successful long-duration mission on Salyut 1. It was a record-breaking flight of 24 days, beating the 17-day mission of Soyuz 9, achieved in June of 1970. There were a few mishaps, like uncomfortable flight suits and equipment, placed there for the original crew, a faulty exercise treadmill, and a small fire. The achievements were manifest too: television coverage to copy what the US had done with Apollo 11, and numerous medical and scientific tests, including the first plant (flax) to grow in outer space.

All of these achievements and trials made the impending Soyuz 11 accident, during the crew's return back to Earth, even more abrupt and painful. The sequence of events began shortly after the Soyuz descent module separated from the orbital module. The explosive bolts, meant to help with the separation, all exploded at once rather than in sequence, in a jolt that caused a special valve to open too soon. It was supposed to open at a low altitude, in breathable air, to equalize the cabin pressure. When it opened at 106 miles altitude instead, still in the vacuum of outer space, it forced a rapid and lethal decompression of the cabin, letting all the air out. In a frantic attempt to survive, the crew turned off the radio to listen for the leak. When they found it, they simply did not have the time to crank the handle shut (it took 19 turns). They were already incapacitated and suffered a painful decompression for about 20 seconds, were probably unconscious after 30, and were likely dead after 50.

The cosmonauts might have survived had they been wearing spacesuits or had there been an oxygen tank in the cabin, filled with enough air to flood it temporarily. Soviet spaceflight managers, in an effort to save on space and weight factors, had refused both, against repeated cosmonaut complaints about the risks. For the managers, space and weight meant time and money, luxuries they did not have in the Brezhnev race to beat the US. The archival documents also testify to one of their poor judgments. Managers argued that, since a Soyuz-11-type disaster had not happened yet, it would not happen at all. This kind of reasoning and expectation made the discovery of the crew all the more shocking: on opening the hatch, they were found lifeless, and all efforts to resuscitate them failed.

After the disaster, the Soyuz capsule and mission profile went through a number of regular redesigns and improvements, revolving around mandatory spacesuits at launch and return. Soyuz 11 was the last of the Soviet space catastrophes, though there were several near disasters in the program to come, which occurred during ferry missions to the Salyut space stations and spotlighted the dangers inherent in the business of spaceflight. Among the perilous missions was Soyuz 18A (5 April 1975), whose third stage malfunctioned, forcing an abort on the way to orbit, the first time this had ever happened. The craft landed successfully 1,000 miles down range, in the Altai Mountains of Siberia, though it came in at a very steep angle, reaching 10 g-forces for about one minute, and 20 g-forces for several seconds. During this time, the two cosmonauts were unable to breathe, one of them suffering internal injuries. In a scene fit for a Hollywood film, or

Disney cartoon, the craft landed on a mountain slope, rolling down the side and nearly falling off a cliff, but for the parachute that snagged on a rock or tree and caught it just in time.

Soyuz 23 (14 October 1976), failed to dock with a Salyut station due to an antenna problem, resulting in an unplanned return and landing on Lake Tengiz in central Kazakhstan, during a blizzard no less, the parachute dragging the craft across the cracking ice. Expecting to find them dead, the recovery team was happily surprised to discover the crew alive and well. During the dramatic Soyuz T-10A flight (26 September 1983), the Soyuz rocket exploded on the launch pad, the crew was saved by the abort system, and by another 20 g-force ballistic trajectory back to Earth.

The experiment with Salyut 1 did not last long. Its third crew was canceled after the accident; on 11 October 1971, the station deorbited and crashed into the Pacific, its garden still growing. Salyut 4 (1975–1976) replaced it, a basic copy of the first station, and inhabited by two crews. Among its new features was a solar panel for energy supply, a teletype machine for regular communications with Earth, new instruments to measure physiological changes (blood and bone analyses), and the Oasis greenhouse. The first crew inhabited it for 30 days, a failed attempt to match and surpass the US Skylab space station. The second crew achieved 63 days aboard, not enough to beat the US, but plenty of time to test the patience of the two cosmonauts. Each day saw the onboard system of Salyut 4 deteriorate, with increasing humidity, mostly the result of the moisture from the cosmonauts' breathing. Cold plates meant to condense the moisture, and allow for its removal, did not work properly. The windows fogged and green mold grew halfway up the walls. The station began to smell. Ground control rejected the crew's repeated pleas to come home. Partly in protest, partly in frustration, they ate the fresh onions growing in Oasis.

Desert Cosmonauts

One of the classic outer-space movies of all time remains little known to most Americans, and oddly enough has literally nothing to do with space itself. It is *White Sun of the Desert* (1970). This was one of the most popular Soviet movies of its day, a blockbuster in tickets sold. Millions of viewers grew up watching it on television. It was also the favorite film of the Soyuz and Salyut cosmonauts, who established a long-lived tradition of watching it on the eve of their departures for outer space. Why? Here are some clues. The film follows the adventures of an intrepid Bolshevik fighter in the Russian Civil War (1918–1921), Fedor Sukhov. He battles reactionary forces in the distant rimland deserts of Muslim Turkmenistan, a Russian colony on the eastern shores of the Black Sea in Central Asia. This is a bleak and desolate landscape, filled with all varieties of natural and human dangers. Before heading home to Russia, he must wage one final struggle against the bandit, Abdullah. In this, he has an erstwhile ally, Sayeed; and a bungling but loveable sidekick, the soldier Petya. In the meantime, Sukhov often dreams of home, of the lush forests, fields, and gardens of his native Russia, tended to by his beloved Katerina. Comic relief comes in the shape of Abdullah's lost harem, nine women in full-body veils, who follow Sukhov everywhere he goes, and for which he takes full responsibility. Along the way, he shows coolness and courage, a mastery of every situation, with such catchy phrases as "exactly" (*eto tochno*), "don't panic" (*bez paniki*), and "you might as well torture me" (*luchshe, konechno, pomuchit'sia*).. Sukhov was the consummate Soviet man, an archetype of the cool cosmonaut.

The experiences with Salyut 4 were a turning point. Its failures forced the redesigns and improvements that paved the way for the achievements of Salyut stations 6 and 7, which were most notable for their long-duration flights and multiple crews. Salyut 6 (1977–1981) maintained 16 crews over the course of nearly five years, with several firsts: its 96-day record beat the Skylab 84-day record, and it saw the first docking of two crews to one station. The two docking ports, at each end, enabled multiple crews and easier re-supply, including fuel to allow its main engines to hold the station at a proper orbit. The redesign of the Soyuz T-10 ferry spacecraft made the comings and goings much safer. The automated and uninhabited Progress re-supply ship brought fuel, fresh foods, letters, and treats like contraband vodka. The cosmonauts then filled Progress with trash to burn up on re-entry.

These were, after America's Skylab, the world's first true space stations. Salyut 6 achieved further crewed records of 140, then 175, and then 185 days. Cosmonauts inhabited it for 683 days in all. Salyut 7 (1982–1986) maintained ten crews, with a final record of a consecutive 237-day crew in 1984. They also tested a unique heavy-ferry vehicle for the military, inhabited Salyut 7 for a record 800 days, and used it as a platform that helped to supply the next Soviet space station, Mir. Both space stations completed important science studies, among them: Earth observations using a multi-spectral camera and remote sensing; astronomical observations by way of a telescope with infrared wavelength power; and experiments in the microgravity production of optical glass and new alloys and semiconductors.

Most of the work in the stations, it turns out, was in the rather mundane tasks of exercising and conditioning bodies, or in the upkeep and cleaning of the stations. With all its revolving crews, cosmonauts remembered Salyut 6 as if it was an old car, filled with odors and stains. There were plenty of mishaps: radios broke down, the power failed, pipes leaked, and there was another fire, caused by a hot cable burning some insulation. The cosmonauts resolved them. Less quantifiable were conflicts between people. Both of the later Salyuts hosted numerous cosmonauts from the Soviet bloc of occupied satellite states, in Eastern Europe, including Poland and Czechoslovakia, and from its allies around the world, places like Mongolia and Cuba. They were part of an international organization, "Interkosmos," meant to turn Salyut into a global venture. Too many human prejudices and arrogances intervened. Several "guests" later complained that they had been treated as second-rate cosmonauts; bullied by their Soviet-Russian hosts into serving as mere spectators aboard Salyut, rather than as true participants in the mission.

America's Skylab Space Station

Leonid Brezhnev's space-station challenge to the US in 1969 was, in the context of the Apollo successes, rather disingenuous. At least at first, it was really something of a bluff, along the lines of the earlier Voskhod program. We might remember that the Soviet leadership announced Voskhod in response to America's Gemini, and hurriedly put it together before Gemini even flew its first missions, all in a leap for world-recognizing firsts. This time, the USSR did much the same with Salyut 1, which was in fact a response to America's announcement and planning for the Skylab program. Voskhod and Salyut were stop-gaps.

America's first space station, Skylab, had been in the planning stages since 1964, initially part of the Apollo Applications Program, and in the early years was framed with one essential condition: to be completed only after a successful US lunar landing. As the quintessential "plan B," a second priority without proper funding, it was in many ways an improvisation. The US simply did not have the time, money, or personnel to design and build it from scratch. NASA management, so focused on the Kennedy charge to beat the Soviets to the Moon, was hesitant to distract any one part of the organization from the task. Even one Saturn V was not expendable. Thus, NASA only formally announced the Skylab program two days after the Apollo 11 landing, with a scheduled launch for 1972.

In the early planning, Wernher von Braun had initially proposed a "wet tank" approach, to launch a Saturn 1B and adapt its spent hydrogen tank as the main part of the station. The astronauts would have had to install the interior parts slowly and methodically. This proved to be too complex. Instead, von Braun decided to use part of a Saturn V: a third stage of a Saturn V rocket to serve as the station, launched by a Saturn V, to be supplied with crews by way of the Apollo command and service module (CSM), launched by a Saturn 1B. These parameters made Skylab the last hurrah for Apollo, the final use of its "old parts." They also made it quite expensive: costing $2.5 billion. This was a sum, paradoxically, that both empowered Skylab to succeed, and cursed it to cancelation, as both the politicians and taxpayers saw it as too expensive.

NASA launched the Skylab space station on 14 May 1973. It was massive, owing its largesse to Apollo. The orbital workshop was 48 by 21 feet, four times bigger than Salyut in both space and weight, enclosing two main sections, separated by a frame of metal gridwork to allow for the free flow of air. The first half, known as the upper deck, was the largest free interior space ever in orbit, and contained storage cabinets, food freezers, and research airlocks (to move things into outer space). The second half, the lower deck, was reserved as a living area with four rooms. The wardroom had an exercise bike, shower, and emergency medical station. The galley had a large window and a table, to provide for a communal eating experience. As if on Earth, the astronauts were to strap themselves in at the table using footrests and belts. There were no seats. The sleeping quarters comprised three cabins, each with a sleeping bag, and separated by pleated, accordion doors. These doors were popular in the new ranch-style homes throughout the US at the time. The toilet provided one refuge of privacy.

At first, NASA managers were rather naive in the planning process, failing to consider the disorientation of a microgravity environment for astronauts not accustomed to it. As a result, George Mueller, NASA's Associate Administrator for Manned Space Flight, hired the famous designer, Raymond Loewy, to conceive a more habitable human environment. Back in 1939, he had designed an amusement ride for the New York World's Fair, featuring a stratospheric rocket for trans-Atlantic travel. He had also designed the streamlined "Coldspot" refrigerator and the classic Coke bottle. Now he was designing a habitat for real space explorers. In one case, though, it did not quite work. The dining table proved too confining, as the astronauts preferred to float in a fetal position rather than be fastened.

On the outside, Skylab had an exterior hatch for use during a spacewalk, as well as the hatch for the Apollo CSM. The docking module, for transfer from the CSM to the station, was large enough to allow the astronauts to change freely into their spacesuits. Three gyroscopes controlled the roll, pitch, and yaw of the whole craft with precise thrusts and adjustments.

Four solar panels were supposed to provide a supply of energy. But at launch, the thermal and micrometer shield tore away, taking one of the solar panels with it, and damaging the other. This meant that the station lost half of its power supply. As a result, the internal heat of the workshop rose to 130 degrees Fahrenheit, threatening to spoil the stored food, to overheat the instruments, and even to potentially decompose the station's polyurethane insulation, releasing toxic fumes. NASA wisely delayed launching the first crew until it had a viable solution.

On 25 May 1973, the first Skylab crew arrived with a plan for the repair and salvaging of the orbital workshop. "We fix anything," was their motto. Over the course of four hours, in the longest spacewalk to date, Pete Conrad and Joe Kerwin freed one of the solar panels, and draped a parasol of plastic film over the station as a protective shield, together with a collapsible nylon umbrella for shade. The astronauts used telephone repair instruments in their work, including telescopic poles for reaching, and special scissors for cable cutting, several times pushing off the station and flying into space, held on only by the tether that helped pull them back. It was a repairman's ballet.

Skylab saw yet more challenges. There were tenuous moments when docking proved difficult, the drogue and probe not engaging, so the astronauts had to disassemble and reassemble the whole apparatus. The second mission encountered coolant leaks (they were fixed) and a parasol mishap (they were repaired by spacewalk). The labor tasks inside the station were difficult. The astronauts found themselves constantly adjusting their motions in order to float purposefully. They had trouble with up and down, left and right. They were often unable to identify locations and tools, as that microgravity floating constantly changed their perspectives and even the "look" of other floating things. For a short time, one of the Skylab crews rebelled against NASA's strict work regimens, feeling they had been turned into robots. They went on strike and rested.

The crews did enjoy the free flight and acrobatics of microgravity, regaling in some fun jetpack tests in the orbital workshop. Skylab was also the scene of vanguard scientific experiments, like studies of the properties of liquid metals in microgravity, and studies of animal life. Several species made the news, like the minnows who by sight alone swam oriented with their bellies to the side of the station wall; or Arabella and Anita the spiders, who adapted well enough to the new gravity at first, though they spun thinner webs than usual, and eventually died of dehydration.

The docking module had a grand telescope mount with an instrument complex, including an ultraviolet spectrometer and spectrograph, an x-ray telescope and visible light chronograph. These made for an impressive record of achievement in solar astronomy, with 175,000 photographs and images of the sun. Earth observations, including heat-sensitive infrared photographs of Earth, revealed sources of geothermal energy in the western states of the US, cold water in the Caribbean (which influenced storms), water sources in Africa, and oil and copper resources around the world. There was no real-time computerized storage, so many of these discoveries came in the form of some 40 miles of printouts and films of electronic data.

In all, Skylab saw three crews of nine astronauts between 1973 and 1974. They achieved records of 28 days, 59 days, and 84 days in space, the latter two records temporarily beating the Soviets. Of note among the crews were Alan Bean and Pete Conrad, the only human beings so far who both walked on the moon (Apollo 12) and lived on a space station. Astronauts occupied Skylab for a total of 171 days, nearly half a year. NASA abandoned it in February of 1974 because of a lack of funds. It was still inhabitable; with enough oxygen reserves to last another 200 days and enough water for 90 (the orbital workshop could hold 800 gallons of drinking water). As Robert Zimmerman has noted, even at that point Skylab could have flown to Mars. There was one last hope for it, a plan to complete the forthcoming Space Transportation System (the Space Shuttle) by 1978 in order to reach Skylab. Alas, the Shuttle did not arrive in time, so Skylab fell back to Earth, burning up over the Indian Ocean and Australia.

The Apollo–Soyuz Test Project

Amid the lingering USSR–US tensions, through this race for space stations, something remarkable happened on 15 July 1975. The world's cosmonauts and astronauts did not make first contact with some exotic band of extraterrestrials, but with each other. We found not some alien race in outer space, but our own human race: the American and Russian rivals in the Cold War. This was the beginning of the short-lived but historic "Apollo–Soyuz Test Project" (ASTP). The Soviets called it "Soyuz–Apollo," thank you very much. The two spacecraft joined to create perhaps the most important space station of all. This was the first of a series of fully shared firsts. It was the first intense exchange of spaceflight knowledge and technology between the USSR and the US. The first public and live launch of a Soyuz rocket and craft. The first time there were seven international "spacefarers" in orbit at one time: those aboard Apollo–Soyuz and two in a separate Soyuz craft also orbiting. This term, *spacefarers*, seems an appropriate description of this new kind of crew, dedicated to a common purpose, and humanity.

The conjunction of several historical forces made this alliance possible. American–Soviet cooperation in space had been in preparation for some time, reaching back to conversations in the early years (1959–1962 and 1964–1965), when both sides reached out to each other about the possibilities of sharing scientific information. A breakthrough came in 1971, when the two countries agreed to exchange what innocuous data they could afford to share: about weather satellites, interplanetary science, and space medicine, for example. These opportunities truly opened after the initial Apollo landings, once the tension of the Moon race broke. The culminating moment of cooperation, the ASTP, happened during the short era of "detente," when the USSR and the US signed a series of nuclear-arms limitations agreements (1972–1974), Richard Nixon becoming the first US president to visit Russia.

Two impressive crews served the ASTP. Representing the US was Deke Slayton, an original Mercury 7 astronaut though grounded for medical reasons. He went on to an important career as head of the Astronaut Office at NASA, directing and assigning crews and missions. He was cleared and returned to flight status by 1972. Thomas Stafford was a veteran of Gemini 6 and 9 and Apollo 10. Vance Brand was a Marine Corps jet fighter pilot who later flew three Space Shuttle missions. On the Russian side was Aleksei Leonov, a former Red Air Force pilot and the first person to conduct a spacewalk (on Voskhod 2); and Valerii Kubasov, a civilian engineer who also flew on Soyuz 6 and Soyuz 36.

The two crews exchanged visits to each other's spaceflight centers. The astronauts went to "Star City," the cosmonaut training center outside of Moscow;

to Kaluga, Tsiolkovsky's home city; and to the launch base at Baikonur. They suffered many Russian banquets and vodka toasts, they investigated what they could of Soviet spaceflight expertise, and they even partook in space rituals, watching a favorite cartoon of the cosmonauts, *I'll Get you Yet* (*Nu Pogodi*), a lighthearted version of the *Tom and Jerry* series. In an era of male chauvinism, they often exchanged ribald jokes. At his visit to Houston, Leonov made a silly toast in broken English, to a "sex-full [successful] life in the future." Stafford handed out photographs of a Houston stripper he renamed, Miss Apollona Soyuzovna, with three mission patches covering her sensitive parts. It is hard to say if these episodes were included in the CIA and KGB debriefings each side made after their visits.

The most challenging technological aspects of the "test project" were the docking adapters and airlock module to join the two craft, given that they were so different. The docking adapter was a petal-like instrument that opened on Apollo to meet the one on Soyuz, in order to align and hard lock. The Soviets designed the docking system, and Rockwell International Division built it in California. The airlock module was crucial in order to equalize the pressure between the two capsules. In fact, the process of opening and connecting the two hatches took three hours, given the delicate process of calibrating and equalizing the air pressure.

Once rendezvous, docking, and equalization were complete, the crews made their way through to the connecting module and each other's cabins, where they made several visits over the next two days. "Ah, good to see you," said Leonov at first contact. *Ochen' rad* ("happy to see you"), responded Stafford. The two jokesters could not help themselves. Leonov put vodka labels on one of the servings of borscht soup; and after one undocking, Stafford played a tape for the Soyuz cosmonauts to hear in the background, the voices of giggling women, with his comment that they were all resting comfortably. There were serious tasks to perform, as in the joint scientific experiments: on the effects of microgravity, on the analysis of Earth's geology and meteorology, and on the photography of the sun's corona. The craft linked for 44 hours. They docked and undocked twice. The crews also performed their political functions, talking with President Gerald Ford, and exchanging gifts like flags, pins, and a book by Konstantin Tsiolkovsky.

On the third day, Apollo undocked from Soyuz and placed itself between it and the sun to create an artificial eclipse. A second docking test, coupled with a wrong firing of the Apollo side-roll thrusters, depressurized and damaged the docking module and nearly "crippled" Soyuz, according to Leonov. The project did not end that well.

It ended worse for Apollo. During its re-entry sequence, one of the astronauts failed to hit a switch to stop the reaction-control thrusters from firing. He may have been distracted by a communication problem and loud noise. This caused the guidance system to start up, rolling the cabin about. Stafford shut off the valves to the thrusters, but it was too late. When the parachutes deployed at 24,000 feet, a vent valve properly opened to let in fresh air, sucking poisonous nitrogen tetroxide (as a yellow-brown mist) into the cabin. Here was a strange "repeat" of the Soyuz 11 accident, if with the opposite effect. The craft also hit the water hard, at 10 g-forces, the astronauts hanging downward in the air by their shoulder harnesses, struggling to put on their oxygen masks. Their faces and mouths and throats were burned from the fumes, and they had to spend two weeks in rehabilitation from the swelling of the lungs and pneumonia symptoms.

ASTP was America's last human spaceflight before the inaugural flight of the Space Shuttle in 1981. Still, as one of the last great R&D projects of the post-war era, it was a promise of cooperation to come, especially in the later flights of the Shuttle–Mir program. Project Apollo's more immediate legacies came in the long formation of the European Space Agency (ESA, 1975) and the National Space Development Agency of Japan (NASDA, 1969; later the Japanese Aerospace Exploration Agency, JAXA, 2003). Both applied the best of their earlier national models, and America's Apollo program management and systems engineering, to form a new worldwide partnership for civilian, democratic spaceflight. To these, the USSR had no analog.

The Human Body in Space

From the first days of human spaceflight, cosmonauts and astronauts have been humanity's perfect specimens. They have been the chosen ones, originally selected from the elite ranks of military jet fighter or test pilots, now coming mostly from the military, engineering, and science professions. They have been subjects of the most demanding physical and mental tests for their acumen, their readiness to endure the extreme g-forces of rocket travel and the extreme rigors of Earth orbit. We test them in centrifuges and spinning seats and in isolation chambers. We choose the best human specimens. Then we put them into space to destroy them, at least in part, given what always happens to their bodies and sometimes to their minds, so we have learned from Salyut to the International Space Station.

Many spacefarers experience nausea from launch, or from being unaccustomed to the constant motion in space. The Skylab 2 crew suffered extreme bouts of vomiting. Even Skylab 3's "iron belly" Bill Pogue, a veteran of 43 combat missions in Korea, succumbed. The condition usually passes. Much more severe are the initial changes to the human body from microgravity, at least while the mission lasts, and often for some time afterwards. We are, after all, creatures of gravity. Evolution has adjusted our bodies to its forces. We stand and sit, walk, and breathe adapted to it, automatically and unconsciously. Space changes all that, when we get there, and on our return.

The circulatory system undergoes a redistribution of fluids to the chest and head. The feet (and even the legs, arms and hands) shrink. The head swells as fluids pool there, especially in the face and neck, making for flu-like symptoms, itchy eyes and a stuffy nose. The fluid shifts also cause changes in the hormonal system, raising aldosterone levels that promote frequent urination. The excess fluid in the head, by pressing on the optic nerve, might also be the cause of the poor or blurred vision experienced by many astronauts, especially those from long-duration missions. Lethargy and depression are also secondary effects of the swelling.

These changes in the circulatory system led to changes in the skeletal system, as the blood stops depositing calcium in the bone, and the bone gives back calcium into the bloodstream. The result is a kind of osteoporosis, a weakening of the bones, a condition compounded since they do not need to be strong to support weight. Data from combined Russian and American studies show a loss of bone mass from the lower vertebrae, hips, and upper femur. The high calcium levels in the blood can cause kidney troubles as well, either anemia, or calcification of the soft tissues, or kidney stones. Without gravity and the need to walk, the spinal column also loosens and lengthens by one or two inches as the disks relax. The body tends to form into a fetal position; the soles of the feet shed their callouses and become soft and pink.

The muscles atrophy and suffer a loss of mass, especially in the legs, from lack of use. There is no need, after all, to stand erect. The relaxation of the abdominal muscles also causes flatulence. This, combined with the difficulty of toileting, can create uninviting odors and atmospheres. One Salyut cosmonaut remembered the toilet as a "witch's broom," painful at best and unforgiving of any mistakes, much to the dismay of the other crew. One Skylab astronaut recalled, "we have to pass so much gas. I think passing gas five hundred times a day is not a good way to go. The only redeeming feature is that everybody is passing the same amount."

Table 10.1 The physiology of spaceflight (in microgravity). Author.

The Physiology of Spaceflight (in microgravity)
Circulatory:
Nausea and Motion Sickness
Head and Neck Swell
Eye Edema
Skeletal:
Spinal Column Lengthens
Loss of Bone Mass – Osteoporosis
Ease of Fracturing
Muscular:
Atrophy
Relaxation
Loss of Muscle Mass
Psychology:
Overwork and Stress
Monotony and Isolation
Sleeplessness

Space stations, in sum, look and feel less like cosmic sanctuaries, and more like sick rooms. Studies have also found that microbes tend to grow and multiply in space stations at dangerous rates, leading to minor infections of the eyes, mouths, and sinuses; these are compounded by compromised immune systems, brought on by the physiological changes, by the absence of a diverse microbiome, and by the ineffectiveness of antibiotics in space. There are a host of other distractions and frustrations. Fans, air conditioners, and other instruments create noise. Floating crumbs and dust, even metal shavings and debris, can infiltrate the lungs and cause serious illnesses.

These physiological and environmental factors cause changes in the sinuses and taste buds, rendering even the tastiest foods bland and unappealing. Even eating becomes a chore rather than a joy. Appetites were not enhanced by pouring hot water into powdered food, or squeezing dehydrated food from tubes, or sucking liquids from a straw. Skylab spacefarers ate lobster, filet mignon, and butterscotch pudding, but still had to salt or spice it all for taste. Salyut spacefarers had some scrumptious meals, like lamb and rice soup, or varieties of breads and cheeses, but still preferred to eat them with lots of salt and onion preparations.

For all of these challenges, the human body adapts fairly well to the conditions of microgravity. The main trouble is in coming back, when the spacefarer's body must reacclimatize to Earth's conditions. Among the early long-duration flights, different persons and crews experienced varying effects both in space and in rehabilitation back on Earth. As a rule, the later missions learned important lessons from the earlier ones. Exercise helps, mainly in keeping one's muscle strength. With intense exercise and conditioning on treadmills, the Skylab 3 crew were able to walk normally, without assistance, within 24 hours. On the other hand, one Salyut 7 crew, though they exercised every day, were not able to walk on their own for a whole week and needed several more weeks of "intense rehabilitation." Bone loss is a more serious and unavoidable problem. The Soviets created the first innovative body suits to alleviate it, by forcing blood into the lower limbs to stabilize circulation. The trouble was that the suits were uncomfortable to wear and partially immobilized the cosmonaut for a time.

There are joys and wonders to behold in space. Cosmonauts and astronauts testified universally to the exhilaration of seeing the Earth from orbit, in 90-minute blocks; 20 minutes of this was over the USSR alone. Part of the orbit is day, part night, the lines separating them drawing out the colors of the Earth, its oceans and continents, against the background of outer space. Spacefarers remembered several scenes with special force. The grayness, not blackness, of outer space, what with all of the brilliance of the stars. The effect of the moon shining on the Earth, to reveal the intimate details all the better. The dazzle of a lightning storm on the planet, what one cosmonaut described as a "luminous dancing sea of lights within the clouds."

These scenes of Earth were delightful, yet thoughts of family and home could also depress the spacefarers, especially those on long-duration missions, living in such close quarters without much human touch or affection. One cosmonaut, Valentin Lebedev, found an interesting way to cope. He used the "flashes" of light that many cosmonauts and astronauts see in space, and that linger when they close their eyes, to illuminate memories of home and family. These sound like hallucinations, likely brought on by the sleep and sense-deprivations of spaceflight. But they were also really very human flashbacks, cinematic memories of the sights and sounds, smells and tastes of the vivid, living Earth; all the more powerful and memorable within the hard and cold, metal, and plastic surroundings of the space cabin. Andrei Tarkovsky captured some of this disorientation and imagination, in his film *Solaris* (1972), its distant interplanetary crews suffering from isolation, seeing things upon the visual screen of an alien planet. Space became a place where we lose ourselves.

All of these slow but rather stunning transformations in the human body and mind, brought on by prolonged space travel, teach us that we may evolve as a species in space. We ought to ask, evolve into what? Popular culture may offer a clue. The physiological changes of spaceflight transform us into new kinds of beings, a new kind of space species: with large heads and eyes; bloated bodies with soft and pale torsos; with thin and long limbs and less muscle mass. In space, we become slower and more graceful in our microgravity movements. It all sounds vaguely familiar: the very image of the loveable alien extraterrestrial that became so popular with the Steven Spielberg movies of the late 1970s. We look for the alien, the "other." We find ourselves. The "aliens" may be us, the image of us in space. We may be the very travelers our spacecraft and probes travel off to find.

This kind of extraterrestrial, the smallish humanoid with large heads and eyes, has a long provenance in science fiction and popular culture. A century ago, the media referred to them as "Greys." Typically, they have gray skin; and small, slender bodies with slim, elongated arms and legs. Their bodies are smooth, sexless, seemingly without muscle mass or skeletal structure. The most obvious characteristic of the Greys are their disproportionately large, long heads and gigantic black eyes. These are the same characteristics that psychologists tell us we human beings are "hard-wired" to love and cherish, for we first see them in the human infant, the new person we must protect and nurture in order to allow our species to survive.

H.G. Wells imagined humanity itself as eventually transforming into little gray men with big heads, as in his article, "Man of the Year Million," first published in *Pall Mall Gazette* (6 November 1893). Wells also described the invading Martians, from *The War of the Worlds* (1898), in a similar fashion. "Greys" later became a part of mainstream American culture after the infamous Betty and Barney Hill abduction case of 1961, retold in a made-for-television movie in 1975. The Hills reported that aliens kidnapped them in rural New Hampshire; they allegedly took the couple aboard a saucer-shaped spacecraft, where smallish men with gray skin, wearing military-type uniforms, conducted experiments on their bodies. The story became the prototypical narrative for many subsequent cases of reported alien abductions (or repressed memories). Greys were also associated with the alien beings supposedly recovered, and autopsied, after the imagined crash landing of an alien craft in Roswell, New Mexico in 1947.

Love them or hate them, suffer their abuse or be abused by them, "aliens" seem to appeal to the average human being. They fascinate us. They may serve some kind of deep psychological or spiritual need. Our emotions toward them are constantly mixed and confused. Maybe this is because we see ourselves in them, both our innocence and our guilt, all of the alienation in the "aliens" in us.

But the alien extraterrestrial is not our only fantasy. As the writer J.G. Ballard has mused, "science fiction is the body's dream of becoming a machine." To overcome our Earthly limitations and even survive in outer space, we dream of enhancing our bodies by turning them into metallic monsters, or heroes, or villains. A long line of movie franchises proves the point, from *Terminator* to *X-Men* to *Iron Man*.

Vertical Horizons: Above and Below Worlds

One of NASA's signal achievements during and after the Apollo program, as part of its Technology Utilization Plan (1962), were its appeals to "spin-offs" as a way to justify the high costs of spaceflight to the American taxpayer and consumer. The expensive technologies that took us into outer space were supposed to come back to Earth with greater returns for the average person. The political patrons of Apollo turned to hyperbole to sell the program to the American public. As Senator Robert Kerr said in one boast, "In my judgment, one of the benefits that will come from the program will be an increased life span of at least ten years for each American under 50 years of age today." The NASA Public Affairs Office argued that the Apollo "brand" translated into many new minor brands for ordinary Americans.

Early on, NASA publicists claimed spinoffs where there were none. NASA did not invent a number of the claimed items, but popular myth presumed this, and NASA officials were happy to oblige and take the credit. For example, General Foods first developed the powdered orange drink, Tang, in 1957. Not for NASA but for the general public, though the astronauts did use it. Dupont invented Teflon in 1938 and began using it on frying pans, though NASA eventually applied it to heat shields and space suits. The Swiss invented Velcro in the 1940s, while NASA used it beginning with the Apollo missions in order to hold equipment in place in microgravity.

Since 1976, NASA began seriously publicizing the various types of technologies and consumer items that were products of its research. Its magazine, *Spinoff: Technology Transfer Program*, eventually claimed nearly 2,000 items in such areas as transportation, safety, the environment, and health. These included industrial-type goods. Goodyear Tire developed a "fibrous material that was five times stronger than steel, for NASA to use in parachute shrouds to soft-land the Viking Lander spacecraft on the Martian surface." It then expanded the technology and went on to produce a new, more durable radial tire. NASA scientists created an "aluminum composite material" to use on rocket casings. It was transferred to create the "lightweight materials" (for facemasks, body suits, and air tanks) now used in firefighting. NASA engineers, along with its private contractors and research universities, studied how to keep water cleaner for extended periods of time, an imperative on space stations. We use the carbon filters they developed in home water filters today, and the purification systems help underdeveloped regions of the world create drinkable water.

A number of other spinoff items filled niches in the mass-production consumer market. NASA developed a "diamond-like lubricating coat over plastic lenses" to protect astronaut visors. This became the Scratch Resistant Lens, which we now use for sunglasses and safety glasses. NASA's "infrared astronomy technology," used to measure the temperatures of the planets and stars, became the ear thermometer, used to measure the amount of energy emitted by the eardrum, of special use for children and incapacitated patients. Under a NASA contract, the Ames Research Center created Memory Foam (or temper foam) to "improve seat cushioning and crash protection for airline pilots and passengers." Today it is used in mattresses and pillows, military and civilian aircraft, automobiles and motorcycles, sports safety equipment, amusement park rides, furniture, and human prostheses. Other spinoff-related products include Black and Decker cordless tools and the Cochlear ear implant. In all these ways, what was good for space was good for Earth as well. What was good for NASA was good for America and the world too.

Select Bibliography

Ballard, J.G. *A User's Guide to the Millennium: Essays and Reviews*. New York: Picador, 1997.

Belew, Leland. *Skylab. Our First Space Station*. Washington, DC: NASA, 1977.

Clancy, Susan A. *Abducted: How People Come to Believe They Were Kidnapped by Aliens*. Cambridge: Harvard University Press, 2005.

Conn, Phyllis. *Cooperation In Space: The Soviet Space Program and International Science, 1957–1972*. Ph.D. Dissertation, Indiana University, 1994.

Dean, Jodi. *Aliens in America: Conspiracy Cultures from Outerspace to Cyberspace*. Ithaca: Cornell University Press, 1998.

Denzler, Brenda. *The Lure of the Edge: Scientific Passions, Religious Beliefs, and the Pursuit of UFOs*. Berkeley: University of California Press, 2001.

Ezell, Edward Clinton, and Linda Neuman. *The Partnership: A History of the Apollo-Soyuz Test Project*. Washington, DC: NASA, 1978.

Fendley, Caitlin. *Countdown to Zero: A History of Grassroots Population Activism in the United States, 1968–1991*. Ph.D. dissertation, Purdue University, 2023.

Foss, Richard. *Food in the Air and Space: The Surprising History of Food and Drink in the Skies*. Lanham: Rowman & Littlefield, 2015.

Hall, Rex, and David Shayler. *Soyuz: A Universal Spacecraft*. New York: Springer, 2003.

Hunter, Maxwell. *Thrust into Space*. New York: Holt, Rinehart, Winston, 1966.

Ivanovich, Grujica S. *Salyut: The First Space Station: Triumph and Tragedy*. New York: Springer, 2008.

Krige, John, et al., ed. *NASA in the World: Fifty Years of International Collaboration in Space*. New York: Palgrave Macmillan, 2013.

Lardier, Christian, et al. *The Soyuz Launch Vehicle: The Two Lives of an Engineering Triumph*. New York: Springer, 2013.

Lebedev, Valentin. *Diary of a Cosmonaut: 211 Days in Space*. New York: Bantam, 1990.

Leonov, A.A., and V.I. Lebedev. *Perception of Space and Time in Outer Space*. Washington, DC: NASA, 1969.

McCray, Patrick. *The Visioneers: How a Group of Elite Scientists Pursued Space Colonies, Nanotechnologies, and a Limited Future*. Princeton: Princeton University Press, 2012.

Oberg, James. *Pioneering Space: Living on the Next Frontier*. New York: McGraw-Hill, 1986.

O'Neill, Gerard K. *The High Frontier: Human Colonies in Space*. New York: William Morrow, 1976.

Piantadosi, Claude. *The Biology of Human Survival: Life and Death in Extreme Environments*. New York: Oxford University Press, 2003.

Rabinowitch, Eugene, and Richard Lewis. *Man on the Moon: The Impact on Science, Technology, and International Cooperation*. New York: Basic Books, 1969.

Sagdeev, R.Z., and Susan Eisenhower. *The Making of a Soviet Scientist: My Adventures in Nuclear Fusion and Space from Stalin to Star Wars*. New York: Wiley, 1994.

Scott, David, and A.A. Leonov. *Two Sides of the Moon*. New York: Thomas Dunne Books, 2004.

Stafford, Thomas, and Michael Cassutt. *We have Capture: Tom Stafford and the Space Race*. Washington, DC: Smithsonian, 2002.

Tumminia, Diana G. *Alien Worlds: Social and Religious Dimensions of Extraterrestrial Contact*. Syracuse: Syracuse University Press, 2007.

Vakoch, Douglas, ed. *Psychology of Space Exploration*. Washington, DC: NASA, 2011.

Zimmerman, Robert. *Leaving Earth: Space Stations, Rival Superpowers and the Quest for Interplanetary Travel*. Washington, DC: Joseph Henry, 2003.

Zukowsky, John, ed. *2001: Building for Space Travel*. Boston: Harry N. Abrams, 2001.

11 Orbits

Planet Earth from Outer Space

In the fast-paced years of the Moon race, the US Lunar Orbiter 1 took a picture of a "crescent" Earth on 23 August 1966, a photograph of the Earth from the Moon. This was the first time the world saw our home planet from another planetary body, alone and exposed in the cosmos. It was a rare look back. Lunar Orbiter focused on mapping in preparation for the Apollo landings. As such, the photograph was not in the news very long. The US had other priorities. It took the crew of Apollo 8 in 1968 to actually see and perceive our planet from a similar perspective, in the famous photograph of "Earthrise," and force the rest of us to perceive it too. A robotic craft had photographed the crescent Earth, but only the human touch, eye, ear, and memory, could really frame and value it and popularize it. We call this the "overview effect," how the human perspective, on seeing Earth from outer space, humanizes us, draws out our compassion for each other and our common homeland.

Since that moment, the Earthrise photograph has become one of the emblems of the environmental movement. So has the famous "Blue Marble" photograph of the whole Earth from Apollo 17, some 18,000 miles away. Images of Earth from space have become commonplace: satellite views of our weather, our oceans and landscapes, our electrified civilizations. Earth's orbital stations seem almost to envelop the planet in a kind of protective visual cocoon, offering us views of the living planet, the "biosphere," that contrast sharply with the dead Earths all around us, near and far. The farthest and most dramatic of perspectives we have up to now is a stunning image of Earth as a planet, the famous "Pale Blue Dot," a photograph taken in 1990 from humanity's Voyager robotic spacecraft, already 4 billion miles away, though still less than one-thousandth of a light year in distance.

All of these images have literally spotlighted the fragile Earth as the focus of humanity's interests in the Space Age. They have served as symbols of either the principle of "exclusivity" (rare Earths), or the principle of "mediocrity" (plural Earths). Depending on our preferred view of the world, we are either one of one, or one of many. In either case, we are always of this Earth, fragile planetary beings locked into one small part of the cosmic order. Yet these very photographs

DOI: 10.4324/9781003494409-15

also remind us that we are an interplanetary species, given how far we have traveled with Apollo and Voyager. The photographs become avatars of the human quest for spaceflight. They remind us that for many people, spaceflight is more often about optics than technics. The Space Age, after all, has run concurrent with the Information or Digital Age. Space telescopes and digital cameras, computer analyses and the internet, have all revolutionized what we see and how we perceive ourselves and the planets and stars. They figure our expectations, with either hope or dread of the spacefaring future.

The Earth as Biosphere

Our contemporary perspectives of the Earth as a living planet actually have a longer provenance than the Apollo era of the 1960s. The Austrian geologist, Edward Suess, first introduced the term, *biosphere*, in his classic work, *The Face of the Earth* (1883), imagining the planet as if an alien being was approaching it "from outer space," peering through the veils of the clouds to discover the lacework detail of nature and culture on the thin envelope of a livable planet. The Russian geochemist, Vladimir Vernadsky, popularized the term in his masterpiece, *The Biosphere* (1926), which defined the Earth in integral terms: as a spherical world within the solar system, graced by its life-providing rocks, fauna, animals, atmospheres, and humans. Such was the "cosmic or common organism" of the biosphere, what he called "the face of the Earth viewed from celestial space."

Vernadsky painted a canvas of the delicate colors and lines, shapes and forms, natures and cultures, which make our planet so distinctive. Here was one of the foundations for the "rare Earth" hypothesis and the principle of exclusivity. He also imparted to the biosphere a unique material force: that envelope or layer of plant and animal life, of ground and air, which captured the energy of the sun and gave life to an otherwise dead planet. Human history was a brief and temporary chapter in the natural history of the living Earth: the habitable moment that we now live in, of only some 200,000 years. Yet Vernadsky also entertained a willful confidence. "Man alone violates the established order," he wrote. We enjoy unique powers and promises. "It is man who ascends to the greatest heights of the stratosphere, unwittingly taking with him the forms of life that accompany his body or his products." We were destined to venture into cosmic space, taking part of the biosphere with us.

A devout Christian, Vernadsky did not embrace the communist regime. But it embraced him. Marxist political patrons and academics found his storyline too enticing: of a living planet embedded in the material forces of evolution and civilization, in their terms without God. They called on the authority of Karl Marx and Friedrich Engels to raise the human being as master of the plant and animal world, of climate and globe, as an ultimate "geochemical agent." As early as 1926, inspired by Vernadsky and Tsiolkovsky, the Soviet astronomer Vladimir Kostitsyn promised the human conquest of space by parabolic rocket, along with orbiting satellite for Earth and interplanetary probes. A map of the biosphere, in the leading Soviet journal, *Nature / Priroda* (1936), one of the first in the academic literature, charted the physical realms of the Earth and atmosphere, up to the ionosphere, and all the oceans and land masses and altitudes of air in between, including the human artifices of balloons and airplanes and rockets traversing them.

The concept of the biosphere inspired a whole new school of adherents. The Jesuit priest and paleontologist, Pierre Tielhard de Chardin, adapted the living Earth as biosphere, into the thinking Earth as noosphere. As he wrote in *The Phenomenon of Man* (1955), we humans are the realm of thought (*nous*), whose "physical curvature of the mind" matches and dominates the "spherical geometry of the Earth." We are, like Earth, both solid and liquid, if standing upright and walking and talking upon it. We are "radial energy," ever improving on a journey to a threshold "mega-synthesis," to a "hyper-personal" apex, an "omega point" of human evolution. The zoologist, Adolf Portmann, in his essay on "Earth as the Abode of Life" (1953), took a much more practical and experiential approach. He spotlighted the rich array of animal life on the planet, centered on us, adapted by our physiology and psychology to the gravity of Earth and the moon, to the light of the sun and its seasons. We are "planetary phenomena," living by its rhythms and notes, but we are also the "I" that speaks and knows, masters of calculation and imagination, still becoming more fully Copernican, more fully aware of the orbiting Earth and potential spaces beyond.

R. Buckminster Fuller popularized this very notion of "spaceship Earth" in a short booklet at the start of the Apollo missions, focusing on the vibrant Earth and our human power to save both it and us. We had already mastered the planet with engineering projects like the industries and railroads and steamships of old, and now had new hope to improve it with the potentials of automation and computerization in the present, and wind and solar and cosmic energies to come. "We are all astronauts," Dyson wrote, worthy of a "metaphysical metabolic regeneration" of mind over matter. Human ingenuity and energy would conquer division and entropy.

James Lovelock adapted the theory of the biosphere into his colorful metaphor of Earth as Gaia, named after the primeval Greek deity who was the great mother of all living things, and from whom all the gods of the Greek pantheon were born through her unions with the sky and the sea. He wrote it while working at NASA's Jet Propulsion Lab and thinking about the possibility of life on Mars. His book, *Gaia: A New Look at Life on Earth* (1979), argued that inorganic and organic matter have evolved together upon our mother Earth to create one holistic, self-regulating, cybernetic system, the sum of its interconnected and dependent parts. The Earth itself controls its own global temperatures as well as the salinity of the oceans and content of the atmosphere. Earth is an energy creator and entropy reducer. Lovelock's theory has intrigued scientists, like microbiologist Lynn Margulis (at the University of Massachusetts), who demonstrated how microbes influence the health of the Earth's atmosphere, one element within the planet's vast symbiotic ecosystem.

NASA scientists, with European and international partners, took up the charge of defining the biosphere in more exact scientific terms, based on the advanced instruments and techniques of the Space Age, our satellites in orbit and sensors upon the Earth. By the 1970s, these machineries began to archive a trove of big data about our natural world and the very unnatural harms threatening it from civilization. In response, Frank Bretherton at NASA helped pioneer the new school of Earth System Science (1986), especially with his famous diagram, an informational grid of feedback and control boxes, representing the life-giving and interactive forces of the sun; the atmosphere, oceans and land masses; Earth's chemistry and biology. He also introduced the concept of "climate change" as a crucial human factor amidst it all, our pollution and overuse of the land, our excessive production of carbon dioxide into the delicate equilibrium of the biosphere. One of Bretherton's inspirations was also a 1977 NASA study on how to design a space station. Its models, illustrating the building blocks for an artificial ecosystem and habitat, helped scientists explain the bigger natural habitat of our own planet, the actual life cycles of plants and animals, air and water, soil and us. Space station Earth clearly needed some further study and fine-tuning.

Some of us earthlings tried to replicate the natural biosphere in a planned artificial community, known as Biodome 2 (Biodome 1 was the Earth itself), established in the Arizona desert in 1991 with an initial investment of $200 million. Biodome 2 was a series of glass domes, sealed from the outside, meant to sustain itself and its human occupants with all the substitute air and elements of the planet. A first team of eight paired men and women lived in the dome for two years, though because of faulty construction materials, oxygen levels depleted, and the ecosystem fell out of a healthy balance. The experiment devolved into internecine disputes between several ingoing and outgoing teams, with charges of conspiracy and sabotage. Replicating the Earth on earth was not easy.

Gaia theory also inspired several trends in a new-age spiritualism. Artists like C. Bangs, in her *Gaia & NGC 4414* (1992), married mythic figures with cosmic nebulae and galaxies, celebrating the beauties and mysteries of the planet, our delicate and sensitive mother nature. The feminist theorists, Mette Bryld and Nina Lykke, applied Gaia to critique the US–USSR space race itself: as a male-dominated, sky-driven discourse of power and exploitation, a narrative only the winsome dolphins can remedy, emblems of a more nurturing and maternal oceanic world to come.

Satellites and Geospace

As modest a satellite as it was, Sputnik changed everything. We already know that it was a spur to the missile and space races of the Cold War. It was also a kind of quantum leap in the wider span of human history, transforming how we see our climate and landscapes, how we communicate with each other, and how we relate as a species on a global scale. Before Sputnik and Explorer, natural "satellites" were celestial bodies that orbited around other celestial bodies: the Moon around Earth, or the Earth around the Sun. "Constellations" were clusters of stars and planetary bodies that framed our nighttime skies, to which we attached myths and astrological powers. They were passive objects of awe and wonder, drawing our eyes outwards. After 1957, "constellations of satellites" described human-made objects, technologies in space that we now aimed inwards to serve human needs.

Sputnik was the first communications satellite, though it had a rather specific and peculiar task, to relay its alluring and terrifying beeps to a cowered American public. Within 50 years, James Motz counted some 1,000 satellites in operation, with thousands more decommissioned but still in orbit. Some were inhabited craft, designed to win victories in the space race; or, as with the International Space Station, to conquer near-Earth orbit for humanity. Most were simpler machines designed to magnify and multiply what human beings can see, and speak, and hear between each other. In these ways alone, their results may not have made the headlines or history books, but they certainly have been dramatic in the breadth and depth of their results.

After the Corona spy satellites of the Cold War, the US designed the first civilian satellites to monitor the weather, as was the case with the Television Infrared Observation Satellite (TIROS), launched by NASA on 1 April 1960, with nine more following by 1965. These were a joint program of NASA, the Department of Commerce (its National Oceanic and Atmospheric Administration), and the National Weather Bureau. At first, they were simple television cameras pointed to Earth, the first ones revealing stunning global atmospheric conditions, thousands of miles of intricate cloud patterns covering the continents. TIROS 3 (1961) alone tracked five hurricanes, and discovered one new one in formation, several days before conventional tracking approaches. The US, in turn, sponsored new venues for international cooperation, in the World Weather Watch and Global Atmospheric Research Program.

Successors to TIROS have included a series of Geostationary Operational Environmental Satellites (GOES) after 1975. They orbit with the planet's movements, at a high 22,000 mile-altitude, in order to provide continuous viewing of the continental US (and the western hemisphere), for real-time use by the National Weather Service. In tandem with Doppler radar stations and other ground observations systems, they provide for precise local and national forecasting, as well as "now-casting" of severe weather dangers like tornadoes and hurricanes. With TIROS, meteorologists invented the reliable evening weather report, and the five-day forecast, soon expanded to ten. We have learned to appreciate the planet's life in the short term.

Further innovations after 1978 have included the Polar Operational Environmental Satellites (POES) that orbit the whole Earth around the poles, dedicated to the work of the environmental and planetary sciences. They have photographed mountain snow-packs, ocean currents, forests, farmland crops, geographical fault-lines and formations; using infrared imaging for nighttime pictures, and ultraviolet sensors to measure ozone levels.

Both types of satellites monitor long-term global conditions, providing a range of images and data that measure changes in temperature, radiation, and moisture; that cover the oceans and coastline; that monitor pollution, natural disasters, and the human transformation of plant and animal habitats. They feed computers with hundreds of millions of pieces of information to create weather and environmental forecasting models. GOES became one part of the US initiative, Mission to Planet Earth (1989), soon joining Europe, Canada, Japan, and Russia with NASA's "Earth Science Enterprise." The program first exchanged its data and images by tape, microfiche, and CDs; NASA made it available for open access on the internet by 2002, via the Earth Observing System Data and Information System (EOSDIS), what it called "as demanding a project as sending the satellites to space," and predicting that the satellites "will send enough data back to Earth during its first 15 years of operation to fill nearly 6.5 million books per day."

Along similar lines but focused on remote-sensing technologies, the US government launched Landsat 1 in 1972. It was the first in a series of satellites that created a vibrant commercial market for images of Earth resources and landscapes. Millions of images and data sets became available on the "Earth Explorer" website of the United States Geological Service. The early Landsat satellites found some evidence of global warming, with glaciers melting in Alaska and ice shelves cracking in Antarctica, the Sahara spreading into the south, and the Amazon rain forest being cleared by human intervention. France joined this Earth-observation project with its SPOT satellite (*Satellite pour l'observation de la terre*) in 1986; with Landsat, it crafted three-dimensional digital landscapes to map territories and locate boundaries.

Communications satellites soon followed upon these Earth-observation platforms, easily becoming the most profitable ventures in early commercial spaceflight. In one especially dramatic moment, the launch of the Signal Communication by Orbiting Relay (SCORE) satellite in December of 1958, President Eisenhower answered Sputnik's beeps with some choice words of his own, the first to be broadcast from space, thanks to an Atlas launcher and a backup tape recorder.

> Through the marvels of scientific advance, my voice is coming to you via a satellite circling in outer space. My message is a simple one: through this unique means I convey to you and all mankind, America's wish for peace on Earth and goodwill toward men everywhere.

The pop-rock group, Angels and Airwaves, later honored Eisenhower's words in a preface to their heartfelt song, "Love Like Rockets" (2007).

Telstar (1962–1963) was the first practical communications satellite, a joint venture between AT&T, NASA, Bell Laboratories, and the US, British, and French governments. It relayed television images and telephone calls across the world, including the first live television pictures over the Atlantic. They included an American news show, pictures of the Statue of Liberty and Eiffel Tower, and a baseball game at Chicago's Wrigley Field. Telstar was a treasured technological artifact in its day. A Ford car, video game, and soccer ball were named after it. The Tornadoes released a fast-paced instrumental tune devoted to it, "Telstar," on their album *Jungle Fever* (1962).

Telstar opened a vibrant new commercial market for satellite television and telephone coverage during the 1960s. Space was open for business. NASA, for example, began a joint venture with Hughes Aircraft in the Synchronous Communications Satellite (Syncom). Western Union's "Westar" became the first commercially launched geosynchronous satellite. It provided continuous voice, video, and data communications to the US. Most significant was the formidable International Telecommunications Satellite Organization (Intelsat), a global consortium directed by the US that came to dominate the international telephone and communications services into the 1990s, and in membership and influence even became something of a rival to the United Nations.

The 1980s and 1990s witnessed transformations in the satellite industry, spurred in part by the deregulation of the US economy, and which encompassed the transportation and energy, aviation and communications industries, freeing market forces for competition and innovation. The new business partnerships, along with the profits to be made with new internet services (and such innovations as video-on-demand, digital recording, and high-definition television), led to a dramatic turnabout by the turn of the century. Riding a boom of tens of billions of dollars in satellite investments, commercial launches of spacecraft surpassed government ones for the first time in history. This led to the development of new launch gateways around the world. Besides the established sites at Cape Canaveral and Vandenberg Air Force Base (in the US), as well as Plesetsk and Baikonur (in the former USSR), the new commercial firms now launched from Kourou spaceflight center in French Guyana. It was home of the European Space Agency's "Ariane Space" system, founded in 1980 as the "first commercial space transportation company"; and from Sea Launch, first tested in 1999, a mobile platform to launch from varying prime latitudes throughout the Pacific Ocean. Spaceflight was diversifying, breaking away from government controls and from US dominance.

Satellite firms like DirecTV, nicknamed "Death Star" for what analysts thought it would do to its cable competitors, helped to drive these trends. Yet commercial satellite ventures also had their failings and scandals, compounded by a devastating bust in the communications satellite market in the early 2000s. Doomsday predictions for the telephone and cable companies never came true. Their hard-wired networks, fiber-topic cables, and transmitting towers that crisscross the world, not to speak of the vast underwater cable systems that unite the continents, have proven to be far more durable than first thought. While DirecTV stumbled through market ups and downs, and a variety of owners, its satellite dishes became the object of spirited derision in advertising, more like relics of a failed neanderthal past rather than exemplars of a bright future. The below world of vertical horizons still mattered.

The Iridium bankruptcy scandal of 1999 offers another cautionary tale. Drawing on an international fleet of launch systems (American Delta rockets, the Russian Proton, and the Chinese Long March rockets), the company had placed an impressive 66 communications satellites in low-Earth orbit beginning in 1997, offering the most total global coverage of any system, and able to communicate with each other and with an extensive network of ground relay stations. Yet a weak business model, along with mobile telephones that were consumer unfriendly, forced the company into bankruptcy on its whopping $5 billion in investments and loans, from which it only partially recovered, though the orbital platforms remained in working order.

One more satellite system is worthy of mention, the Global-Positioning System (GPS), originally a joint program of the US Air Force and Navy, designed for military air-traffic control and navigation. Perfected in the late 1970s, it was a communications network of satellites, synchronized with each other, and with ground-based tracking stations, by precise atomic clocks. President Ronald Reagan declassified the system at a poignant moment in the Cold War, just after the infamous Korean Airlines disaster on 1 September 1983, when a Soviet Sukhoi jet fighter shot down a civilian passenger jet that had strayed into Soviet airspace, killing all of the 269 people aboard. Originally flying a route from New York City to Seoul, a GPS system would have saved it by locating it precisely in space and time. In response, the US government offered its services to the world. Within a decade, the government's original GPS became one of the twentieth century's great commercial success stories, receivers costing only $199 to provide a fixed geographical position (within 50 feet in three directions), using the simultaneous reading from several different satellites. Now they are much cheaper and seemingly everywhere: in telephones and watches, computers and cars, ankle bracelets and smart bombs. They catch speeding cars and predict the arrival of your bus, locate your place on the map, and give you instructions on where to go.

The history of these varied weather, communications, and global-positioning satellite systems may lack some of the luster of other spaceflight ventures. Part of their magic is in just how invisible they all seem; just how natural it is to predict the arrival of a snowstorm by clicking on the weather channel on your home computer; or to telephone a loved one halfway around the world within seconds; or to predict the fastest way home through rush-hour traffic using Google Maps. Serving all these needs are the thousands of small machines, latticed above our globe, pointed downward to direct our lives. In these purely practical terms, satellites have no doubt been some of the most revolutionary and transformative of modern technologies: the ultimate "spinoffs" of the Space Age.

Outer Space and Cyberspace

As we weigh the impact of the Space Age on human history, we ought to remember that it alone cannot define us. Other ways of marking and measuring time rival it, for example in what we call the Information or Digital Age, that sweeping revolution in such digital technologies as the computer, cellular telephone, and internet. By their capacities to create, manipulate, and share copies of images and texts, movies and sound files, these technologies have radically transformed the ways we perceive and understand our world, and other possible worlds. One of the important tasks for the historian is not to isolate these two ages but to see how they are intertwined.

We have seen several global computer networks already at work: in the early warning detectors (radar and satellites) and SAGE computers begun in the 1950s, and in the NASA-guided Earth surveillance sensors and satellites after the 1970s. Paul Edwards has called them part of a "vast machine" of global coverage. The arrival of the internet bridged both of these wired worlds. It had its origins in a Cold War and Space Age moment: the work of J.C.R. Licklider, a behavioral scientist at the Advanced Research Projects Agency (ARPA) of the US Department of Defense. Building on ARPA's charter to create the new satellite and missile technologies of the Space Age, Licklider envisioned an "Intergalactic Computer Network" in 1963, so as to network computers around the world, and someday beyond. His immediate goal was to securely transmit and share data and computer programs.

These ideas inspired the development of "Arpanet" in 1969, a small network of computers, joining several of ARPA's partner universities and enterprises by way of email communications and file transfer. Within only 20 years, it became the model for the public World Wide Web (1989) and the global internet (1991). Both platforms, with advances in digital miniaturization and computation, have created astounding leaps in the creation, storage, and transmission of information. Ours has become a wired, signaled, information-driven world.

Outer space also helped define a new term, first popularized by William Gibson in his science-fiction tale, *Neuromancer* (1984): cyberspace, now commonly understood as the computerized realm of virtual reality on our screens. To match our knowledge of the expanding universe beyond, humanity created a new global information web for us here below. It has entwined us and many of our needs for quick and easy (and ever more monetized) access to texts, telephone, television, videos, and music, news and opinion. Unmoored from traditional communities and personal contacts, we rely instead on the superficial knowledge and discourses, attractions, and temptations, of the digital realm. What we once used to search for public knowledge or goods, now searches us, with the algorithms that thread our private information into big data sets for market manipulation. Or in the case of China's "social credit" system, punishes us for political or cultural self-expression.

Varieties of Planetary Defense

One of the most remarkable moments in the space race, during the heyday of the Apollo missions, happened not in outer space but here on the planet: Earth Day, 22 April 1970. Centered in New York City and Washington, DC, it was a manifestly happy mass celebration of clean air, water and land. In perfect weather of moderate temperatures, blue skies, and light winds, people joined teach-ins, played frisbee on the capitol Mall, and walked down Manhattan's 5th avenue, closed for the day. With some 20 million people joining, this was one of the largest simultaneous civic events in US history. It was inspired by the social protest movements of the 1960s, now targeting the abuse and pollution of the environment and our bodies. Protesters focused on the defoliation campaigns in the Vietnam war, and on the infamous oil spill off the coast of Santa Barbara, California. Rachel Carson had spotlighted the pesticide poisons in our foods, especially DDT in her *Silent Spring* (1962). In his *Unsafe at Any Speed* (1965), Ralph Nader warned us about the design dangers and pollution from our own automobiles. Barry Commoner, in his *The Closing Circle* (1971), argued for a new high-tech approach to a cleaner and more sustainable ecology.

The Blue Marble photograph of Earth, made from Apollo 17, soon became one of the key emblems of Earth Day. We humans had to become detached observers of the Earth before we became more responsible stewards of it. The libertarian and environmentalist, Stewart Brand, had featured the Blue Marble on the cover of his *Whole Earth Catalog* (1968–1972), a compendium of self-help products and techniques to empower individuals, help them save the Earth and themselves by breaking away from the destructive cycles of production and consumption. Information was the key to a new survival. Brand was also a computer pioneer, who joined the event we now remember as the Silicon Valley "Mother of All Demos" (1968), which introduced the world to the coming tools of the computer mouse, electronic mail, and the wonders of the personal computer (like screen windows, hyperlinks, and video meetings). NASA played its own role in the transformation, helping to fund the early design of the computer mouse, and creating the space camera that morphed into the cell phone camera.

The space-based portrait of Earth has enjoyed quite a legacy, as in the computer-enhanced "Blue Marble: The Next Generation," the default screen for Apple; or the myriad images of Google Earth. Some of the more stunning adaptations are the computer-generated images of space debris orbiting Earth, first publicized by NASA in 1999, and in the ensuing years multiplying so as to visually obscure the planet, as well as endanger spacecraft, space stations, and operational satellites.

Earth Day was the symbolic beginning of the environmental movement. The US Congress legislated the Environmental Protection Agency (EPA) and the National Oceanic and Atmospheric Administration (NOAA) that year, to consolidate and coordinate enforcement of new standards for cleaner air and water, and for the protection of animal and plant species. On the lighter notes of popular music, "Nature's Way" (1970), by the rock group Spirit, warned about industrial pollution in American rivers. The Soviets had their own informal versions of Earth Day, not in any one event, but in grassroots environmental protection groups and in nature tourism. The Soviet pop-rock group, Earthlings / *Zemliane*, even offered a love song about Earth, "The Grass of Home / *Trava u doma*" (1983), which eventually became the official hymn of the Russian space program.

Critics warned humanity that these were all small measures to match intractable, doomsday scenarios. Science and technology were helping humanity to grow beyond the capacity of the planet's resources to sustain us. The Club of Rome's, the *Limits to Growth* (1972) and Paul Ehrlich's, *The Population Bomb* (1968), feared the worst about overpopulation, predicting global catastrophe and mass starvation. Human interventions soon forestalled the crises, at least so far. The Green Revolution in agriculture created a series of high-yield crops, like dwarf wheat and rice, that have fed millions. State family-planning policies in China and India, along with new interpersonal norms about marriage and family, have curtailed population growth, in exchange for immediate national and individual prosperity. Total Fertility Rates have fallen, especially in East Asia. The planet may no longer be in immediate peril. But given the results of falling population numbers, labor forces and economic development now are.

Thanks to our new knowledge about the planet, spurred by the networks of land and air-based monitoring sensors, and orbiting satellites, humanity found new threats to Earth. Scientists discovered the ozone hole over the Antarctic after 1985, where the Earth's protective ozone layer was most depleted, making the planet more vulnerable to the ultraviolet radiation from the sun. NASA publicized the dilemma over the coming decades, in vivid satellite-based images of the globe in peril. Chlorofluorocarbons were to blame, widely used in refrigeration and aerosol technologies, and which leaked their poisons into the air (as we actually learned from the work of James Lovelock). The nations of the world united, in a rare unanimous agreement, to save the ozone by way of the Montreal Protocol (1987).

The dawn of the Space Age also spotlighted the threats of climate change, in the form of the high carbon emissions from the burning of fossil fuels. Climate change had its own emblems of warning. From the Scripps Oceanographic Institute, Charles Keeling published his famous Keeling curve, charting the rising carbon dioxide in the Earth's atmosphere and contributing to the "Greenhouse Effect" of trapped gases poisoning and heating the air. From NASA, James Hansen offered his own chart of the resulting rise in global temperatures, submitted as testimony to the US Congress in 1988, with later warnings that the world was at a "tipping point" of collapse with melting glaciers and rising oceans.

These dire predictions helped inspire the Green movement at the turn of the twenty-first century. Gaia went Green. In the tradition of biosphere and noosphere, the Noble Prize winner in Chemistry and a specialist on ozone depletion, Paul Crutzen popularized the term Anthropocene in 2000, to mean the latest geological epoch in Earth's history, the one we created by the revolutions in agriculture and industry, and that have both mastered and harmed the Earth itself.

American politicians followed the trend. Former senator and presidential candidate, Al Gore, made climate change the centerpiece of his public movement, the Alliance for Climate Protection and the Climate Project. He first honed the pitch in a series of PowerPoint shows that he presented at college campuses, then published in a profitable book and film, *An Inconvenient Truth* (2006), and web page (www.wecansolveit.org). In his electoral campaigns and two terms, President Barack Obama drew from a ready mix of space race metaphors to define the problem, calling for a new 2008 "Apollo Project" to build sources and technologies for clean energy; and in his State of the Union speech, naming 2011 as America's new "Sputnik moment," now to "invest in biomedical research, information technology, and especially clean energy technology." His administration translated these progressive, anti-growth policy positions into hundreds of millions of dollars in federal subsidies for solar and wind power, and electric cars, to defeat the grip of fossil fuels on the American economy.

In all these ways, the Apollo portrait of "Earthrise" trumped some of the allure of rockets and spaceflight. Earth mattered more. The real-world consequences have been provisional international agreements to reduce global temperatures, in the Kyoto Protocols (1997) and Paris Climate Accords (2015). Equally significant have been the shifts in public consciousness in favor of these measures, thanks to the media and popular culture. The movie, *Wall-E* (2008) captured the spirit well, the story of humanity escaping a dying Earth via spacefaring arks, though bound to return "down to the ground," as Peter Gabriel's theme song put it, to a landscape reborn. A series of IMAX high resolution documentaries, so often screened in science museums around the world, have also spotlighted the wonders and perils of Earth, and in the case of *A Beautiful Planet* (2016), filmed from within the confines of the International Space Station.

From beyond Earth, thanks to our space telescopes and space probes, we have also identified the threat of possible extinction events, at least the ones we have not started ourselves. Asteroids figure prominently among them. Impact theory shows that bodies falling from space (either an asteroid or comet or meteor) led to a several extinctions in Earth's past. They offer a Big History countdown of cataclysm: 3.5 billion years ago, then 251 million years ago, then 65 million years ago, the one that probably killed off the dinosaurs. All are of recent "Space Age" discovery. Near-Earth Objects are also actionable. Several thousand of the 33,000 asteroids that approach Earth have the potential for collisions. In 1998, Congress charged NASA to monitor them, and NASA created a Planetary Defense Coordination Office in 2016.

NASA and its partners have also sent out small probes and landers to meet and greet some of these objects. The Near-Earth Asteroid Rendezvous (NEAR Shoemaker) was the first spacecraft to orbit an asteroid and land on it successfully (2001). The Deep Impact probe (2005), following up on earlier comet flybys, reached the Comet Tempel 1, with an impactor probe to study its composition. The European Space Agency's Rosetta (2014) was the first rendezvous, orbit, and landing on a Comet (67P), some 250 million miles from Earth. In 2020, the Japanese Hayabusa 2 returned a soil sample from the asteroid Ryugu. And NASA achieved a threshold moment in 2022, when the Double Asteroid Direction Test probe (Dart), speeding at 15,000 mph some 7 million miles away, successfully altered the orbit of the asteroid, Dimorphos. Hayabusa 2 and Dart were also among the first deep-space probes to be powered by an ion (electric propulsion) thruster. Thus, the comets and asteroids that had once portended doom, as we saw in Chapter 1, have now become destinations, and even engineering projects.

For all these achievements in distance and contact, we human beings remain a fragile species on a fragile planet. We are all mortals. By our own mortal lives (120 years may be the maximum a few of us might be able to reach). And by the Second Law of Thermodynamics and the scourge of entropy. Peter Ward's and Donald Brownlee's *The Life and Death of Planet Earth* (2004) has surveyed the lifelines and breaking points. How life on Earth depends on so many finely balanced factors like its magnetic core, the gravitational pull of the moon, and the water and gases stored in the land and oceans. How we live in the rather brief Age of Animals, only a billion years in scale. How the planet and species are bound to doom with the inevitable heat death of the sun.

Looking for Other Earths

One turning point in the history of the Space Age was rather sedate as turning points go, especially given some of the more spectacular spaceflight firsts and media events that we have covered, like Sputnik and Gagarin, Apollo and Viking. This turning point was the foundation of the NASA Astrobiology Institute in 1998, joining a group of 150 scientists and their home institutions, centered at the Ames Research Center in California, and soon networked by way of cutting-edge video conferencing and internet links. They were on a mission to define a whole new discipline: "Astrobiology." It replaced the original term, Exobiology, as the new meta-science "study of the living universe," encompassing chemistry, biology, geology, paleontology, astronomy, and the planetary sciences, to name a few. This is a cosmic science that searches for "biomarkers" beyond Earth. For any signs of the chemical building blocks of life, like oxygen and nitrogen; or for the essential medium for life, water; or for any of the more significant signs of organic and intelligent life. Biomarkers can be dead, as in the chemical traces within planetary soil, or the remains in fossils, or the data from spectrographs. They can be living, too, as in any actual microorganisms, or a coded radio signal perhaps.

The literature on Astrobiology has exploded since then, curious for a field that has so far found no life beyond Earth, only tantalizing probabilities and possibilities, based on what we are constantly learning about the vast breadth and depth of the universe. True, its sources and archives have grown significantly every year, what with the new data provided by the improved technologies of spaceflight: for example, all of the interplanetary probes and space telescopes that have charted the chemical and physical formations of the solar system, the galaxy, and parts of the universe beyond. The Hubble Telescope has been especially dramatic in its results, with photographs and commentary often featured on the NASA web site, "Astronomy Picture of the Day." One of the most famous Hubble photographs was the composite, Ultra Deep Field (1996), a small frame of the nighttime sky, just blackness to the human eye but containing 10,000 galaxies, each with perhaps a trillion stars, some possibly with habitable planets. They were at the time the farthest points ever seen from Earth, over 13 billion light years beyond.

As images go, a visualization of the Opte Project (2005) offered a close second to the Hubble revelations, in a color-coded representation of a part of the internet, showing the dizzying synapses, connections and routes that direct us, as if the internet was a distant galaxy all its own, but an artificial one, right here between us on the planet.

Astrobiology is an exploratory science, searching for what might be out there in space, where life may be "elsewhere." Part of the attraction to those who join it, and to the public that reads about it, is its transformational potential. We await, in other words, the most dramatic of discoveries, maybe the biggest "first" ever in human history: the confirmed fact of life beyond Earth, Second Genesis. Beyond the thrill of discovery, the search for extraterrestrial life also has much to do with values and meanings, even with theology and metaphysics. What would "life elsewhere" mean for our traditional religious and philosophical conceptions about God and humanity?

Some scientists claim it would finally consolidate the modern worldview that began with Copernicus's sun-centered universe and peaked with Darwin's theory of evolution. As Ray Jayawardhana has argued, the discovery of life beyond Earth would finally complete the Enlightenment project of replacing God with science, faith with reason. Yet given our complexity as human beings, it also might lead some people to a dark pessimism, as if we had suddenly become very small and insignificant in a multi-variant universe. Others might find the discovery invigorating, a moment of maturity for us humans as a species, now the embracers of new worlds. In both cases, the discovery may very well fortify traditional religions rather than eviscerate them. We may experience the need for sacred meanings in our lives, for creation and salvation, all the more strongly.

The Search for Extraterrestrial Intelligence (SETI) remained a centerpiece of the new Astrobiology, one of its earliest foundations. SETI has even become something of a mass movement, what with the digital and internet revolutions of the 1990s. As computer chip technology and computing speeds improved dramatically, the organization drew its enthusiasts and young people into the radio quest with "seti@home," a screen saver that tapped the energy of the home computer (when not in use) to search for radio signals in a piece of raw data from the Arecibo telescope. By 2003, it had 4 million users across the world.

Digital entrepreneurship offered another boon to SETI, in the form of the Allen Telescope Array, a series of 42 radio telescopes located some 300 miles northeast of San Francisco, established in 2007 with substantial funding (approaching $30 million) by Paul Allen, one of the founders of Microsoft. Unlike the one movable radio telescope at Green Bank, or the stationary one at Arecibo, the Allen Array can sweep wide and varying segments of the universe for basic astronomical observations, as well as for the radio bands of SETI. The range expands SETI's coverage, but also spotlights just how utterly daunting its tasks remain. SETI receivers bring in hundreds of millions of signals that computer hardware and software programs must sift in order to detect a meaningful sign through all the static. Given the size and distances of the Milky Way Galaxy alone, and the rather slow pace of mapping, it will take thousands of years to canvass only a small percentage of the galaxy.

The US government maintains its own investments in radio telescopes, as in the Very Large Array, New Mexico (operating since 1980), and which surveys the universe for the signatures of stars and black holes, planets and quasars; for radio signals from any possible advanced alien civilizations; and even for the radio signals from part of us, the Voyager 2 probe as it swept past Neptune. NASA's Deep Space Network of antennas and communications systems, with locations in California, Australia, and Spain, keeps in primary contact with all of NASA's space probes and does radio astronomy of its own. In 2008, NASA sent out a peculiar interstellar message from the Madrid location of the Deep Space Network, aimed at Polaris, a star 431 light years away. The message was a song, "Across the Universe" by the Beatles (1968), a signal of "limitless undying love, which shines around me like a million suns, / It calls me on and on across the universe."

Amid all the noise and routine, SETI scientists continue to astound, as, for example, with their most fascinating new field of astral semantics, a new linguistics of virtual spaceflight. On the initiatives of Douglas Vakoch, SETI's "Director of Interstellar Communications," the group has entertained a variety of "star messages." For example, transmitting three-dimensional pictograms of human beings in motion, doing acts of kindness and compassion with their fellow human beings; and initiating "Earth Speaks," an online program to collect messages for the far constellations from average people around the world, this to illustrate humanity's instinctive urge to equality and democracy. Both kinds of dispatches represent the best in us, of course, against the background clamor of human violence and hatred also committed and transmitted around the world and into space. SETI keeps beaming our messages up, though we have yet to hear anything back.

Most astrophysicists, in the meantime, concentrate on finding not civilized extraterrestrial worlds but simply habitable ones. They are the "exoplaneteers," hunting the galaxy for the "exoplanets" that orbit stars like our own sun (the very first of which, 51 Pegasi b, was discovered and confirmed only in 1995). Doppler spectroscopy and imaging spectrographs, for example, have helped scientists to measure the light wavelengths and gravitational forces of select stars and their planets, all to map the "habitable zones" of solar systems. These are the zones at just the right distance from a sun in order to allow water to pool on the surface without evaporating, or boiling away, or freezing. Since liquid water is the essential foundation for carbon-based life, at least the only kind of life that we know, the habitable zone is a potential zone of life. Since the location of these planets must be "just right," that is neither too hot nor cold, neither too small nor large, scientists have dubbed them "Goldilocks planets." They are like the golden-haired girl in the fable who searches for the porridge, then the chair, and finally the bed that is "just right" for her. It is a wistful term to apply to science, of course, as the tale does not end well. In some versions, the three bears eat Goldilocks, revenge for invading their home. In others, she simply runs away. In either case, "Goldilocks" is a cautionary tale, about going where one does not belong, taking what is not yours.

The Kepler Space Observatory has provided some of the most provocative data about exoplanets and habitable zones in recent years. NASA launched the platform in 2009 on an eight-year mission to search for Earth-like planets in the Milky Way galaxy. Kepler served as something like our Earth's own eyes, constantly pointing its sophisticated digital camera to a small corner of the galaxy, the constellation Cygnus and its 156,000 stars, some as far as 3,000 light years away. It used an extra-sensitive photometer to visualize and measure "star-blinks," the quantity of a star's diminishing light as its planets moved through its foreground. The "dimming" of the star's brightness, together with other factors, offered a range of data: about the size of the planet and its orbit, about its temperature, and about the size and temperature of the star. Kepler had its limits, though, as it could not offer any details about the planet's mass, its "densities and compositions," or what organisms might live or think there.

Over its life span, Kepler found many unexpected results: like a huge planet, twice the diameter of Earth, boiling at its surface; or planets that orbit binary stars (stars orbiting each other); or planets in strange orbits that take them first near and then far from their sun, enough to alternately burn and freeze them. Yet it also found thousands of recognizable and seemingly normal exoplanets: already 2,740 of them by January 2013, although the Harvard-Smithsonian Center for Astrophysics estimated "at least 17 billion Earth-size exoplanets in our galaxy," and the University of California at Berkeley claimed 40 billion. We were riding a wave of optimism that exoplanets are everywhere. With these kinds of odds, and with each year bringing in more discoveries, scientists are now boasting that it is only a matter of time before we find that one (or more) habitable planet, that perfect other Earth.

There was a heady dose of anthropomorphism in all of this. William Borucki, scientific director of the Kepler telescope, described the twin planets, Kepler-62-e and Kepler-62-f (discovered in 2013), which are closer to each other than Earth is to Mars, as very much like Hawaii and Alaska in temperature. That kind of comparison certainly humanizes them. On their discovery, NASA has also commissioned artistic renderings of these exoplanets, baby pictures of a kind, to join our family portraits of the sun and moon and our own nearby planets. Like them, they are all recognizable and admirable in their colors and roundness. Cornell University's "Institute for Pale Blue Dots" (2014) began to offer comprehensive studies and models of Earth-like planets, as if mirror images of the original Pale Blue Dot. All of this is the best we can do for now, to search for ourselves out there in space, for life as we know it.

Eco-Living

The Space Age has even influenced how we live in our homes here on the planet. Spaceflight inspired Finnish architect, Matti Suuronen, to design his Futuro home, built in the shape of rotund flying saucers standing on quadrupeds. He originally developed them as portable, easy-to-install ski chalets, not exactly an environmentally friendly or democratic medium. But buyers got creative and used them to streamline and downsize their lives. Other even more unique 1960s Space Age homes include the futuristic Garcia House of Los Angeles, sitting 60 feet high on the edge of a cliff like an inviting alien spaceship; or the more recent Blue Dream home on Long Island at East Hampton, New York, of softer parabolic curves, embedded naturally in the dunes. More popular and widespread have been homes built as geodesic domes. First proposed by the architect, R. Buckminster Fuller in 1954, these were to be real living spaces, in domes eventually made up of tensioned triangles or polygons, mass produced for easy installation and comfortable interior space, an answer to the presumed crises in population and resources after 1968. With the Green movement, entrepreneurs began to design and market eco-friendly "Earthship Homes," built with clean materials for clean energies and responsible living. They are built of stone and mud and grass, like the prairie sod houses of old. Or as tree houses in the woods. Or adobe in the desert US southwest. Thus, in our immediate surroundings, the comforts of home and earth, we can dream of spaceflight and even help protect the spacefaring Earth.

Already by 2015, with a bit more modesty, NASA confirmed the first nine planets in the Milky Way galaxy most identifiable as "Earth twins," although most of them are too large. Only one was both in the habitable zone of another star and approximately the same size as Earth: Kepler-186f. As exciting as the discovery was, planetary scientists were not sure if it was rocky enough to be a twin, though they suspected it was. They were even more divided as to whether it was warm and wet enough to sustain life. It may be more like Venus and Mars than like Earth. These kinds of open questions and debates remind us how frail our knowledge is about these places. Space defeats us. Kepler-186f is roughly 459 light years away from Earth. Even if it is like us in some way, shape, or form, it is very far away. Time defeats us too. SETI radio telescopes can now send one of our human messages there. If our evolution and history follow the same rules on these other planets, their civilizations are likely to have risen and fallen already, or not to have appeared yet at all. There may be no one there to hear us.

Kepler's visions also highlight a special privilege of the moment, something like the special privilege of living on the Earth at our own most habitable moment. As Caleb Scharf has argued, all of what we now see in the universe beyond is fleeting in its own way. Since the universe formed with the Big Bang some 14 billion years ago, it has been expanding into ever wider and wider spirals. In time, our ability to see much of its vast expanses will fade. "We exist during what may be the only cosmic period when the universe's nature can be correctly inferred by observing what is around us." The moment will last a few more billion years. That's history in the long term. Though science teaches us that even such a long moment will not last. Given the distances and time spans between the stars, what lights we are seeing are already past, a screen of cosmic wonders filled with as much destruction and chaos as creation and order. They are more records of time rather than maps of space; places coming and going on their own scales, rather than places where we might venture.

In a curious and even doleful way, the Kepler discoveries confirm what we already know about the biosphere, our mortal Earth. What other Earths that we have found in that one corner of the galaxy may very well be young ones, or already dead ones well beyond their prime of life, or they may not be other Earths at all. Kepler's discoveries remind us that life may possibly be elsewhere, but from all the hard evidence, it is most certainly here. Our searches of the stars yield new knowledge and pride about the universe, and new wisdom and modesty about Earth's fragile and dynamic place within it.

We remain, in other words, fluctuating somewhere between the models of plurality or singularity, the mediocre or the rare Earth. Hubris trends us toward plurality, as when we presume that we can create new worlds ourselves, a twist on Second Genesis, according to the model of terraforming ("earth shaping"). Kim Stanley Robinson popularized the notion in three novels, *Red Mars* (1993), *Green Mars* (1994), and *Blue Mars* (1996), about creating a new Earth on that otherwise hostile planet.

Vertical Horizons: Above and Below Worlds

As spacefaring nations set their sights on the eventual human exploration and colonization of Mars, they find themselves first drawn to extreme climates and environments on Earth. In order to prepare for its hostile conditions in space, we need to learn how to survive in comparable conditions down here first. Mars will kill any human being outright. Its barren landscape, freezing temperatures, dust storms, and toxic atmosphere make it extremely inhospitable. When we look to duplicate such an environment here on Earth, we look for desolation, for places like ice sheets and deserts where no humans have ever lived for long or would ever care to live.

The Mars Analog Research Station program (MARS), a multinational project initiated by the Mars Society in 2000, sought to cultivate just such places, prototype research centers in forbidding climates, all to discover how the human body and mind, and our equipment, might survive on Mars. The Mars Society has operated the Flashline Mars Arctic Research Station at Devon Island, located in the Canadian Arctic, a pseudo-Mars facility on the edge of a 14-mile-wide crater in the middle of a polar desert. With its sub-zero temperatures, it is one of the most Mars-like environments on Earth. The Mars Society has also founded the Mars Desert Research Station in southern Utah, situated on the San Rafael Swell, a large anticline composed of sandstone, shale, and limestone that pushed up 60–40 million years ago. Many of the Swell's geographic features resemble Mars, including plant life that does not exist anywhere else in the world.

NASA has managed several analog stations, including the Haughton Mars Project, also on Devon Island; and the NASA Extreme Environment Mission Operations (NEEMO) project, underwater near Key Largo, Florida. Its Hawaii Space Exploration Analog and Simulation (HI-SEAS) is located on the Mauna Loa volcano, where Martians-to-be can spend 4-, 8-, or 12-month journeys. They wear Hexoskin monitors in their clothes (not yet implanted in their bodies) to gather data about their physiology. And they simulate communication delays with the outside world for up to 25 minutes, as if from Mars.

Analog stations are about what the human body can tolerate, and also the human mind. They are experiments in what we now call Astrosociology. Yes, more "Astros." There is now an Astropolitics to get us to space, and Astroethics to justify it, and an Astrosociology to keep us there. None of this will be easy. Even if humans could survive on Mars in some kind of fabricated habitat, the question of whether they could even survive the journey to Mars in the first place remains problematic. Mars is 140 million miles away and the journey will take up to nine months, through the darkness and radiation of interstellar space.

To confront the dilemmas of such a journey, in 2009 the Institute of Biomedical Problems in Moscow, as part of its Mars 500 project, kept an international team in a Soviet-era isolation chamber for 500 days, hoping to learn about the effects of prolonged isolation and monotony on the human psyche. Talk about nostalgia for the old communist regime. There are also all the problems associated with medical issues, such as bodily injuries. What if someone gets sick or breaks a bone while living on Mars? We already know that weightlessness causes the loss of bone and muscle mass. Cosmic radiation going there and being there greatly increases the likelihood of cancer. Add to this the Martian toxic dust, which will corrode equipment and spacecraft systems, and be harmful to humans if inhaled. Finally, if humans survive the journey and can learn to cope with the challenges, what will they eat and drink? The Mars Analog Station in Utah maintains a "greenhab," hoping to learn how to transform Martian soil into something that can sustain plant growth. It envisions a self-sustaining environment for future Martian colonies. They will be fully enclosed habitats in which all waste is recycled and purified. Plants will provide food and oxygen for humans, and microbes will break down waste. NASA expects that most foods on Mars will be of the plainer and tasteless kind: wheat, rice, potatoes, beans, and soy. Our Martian explorers can expect a menu of wheat cereal, scrambled tofu, lentil loaf, sloppy joe tempeh, soybean loaf, and vanilla soy pudding. Enjoy.

Select Bibliography

Benedick, Richard Elliot. *Ozone Diplomacy: New Directions in Safeguarding the Planet*. Cambridge: Harvard University Press, 1991.

Borowitz, Mariel. *Open Space: The Global Effort for Open Access to Environmental Satellite Data*. Cambridge: MIT Press, 2017.

Bruno, Andy. *Tunguska: A Siberian Mystery and Its Environmental Legacy*. New York: Cambridge University Press, 2022.

Butrica, Andrew. *Beyond the Ionosphere: Fifty Years of Satellite Communication*. Washington, DC: NASA, 1997.

Bryld, M., and N. Lykke. *Cosmodolphins: Feminist Cultural Studies of Technology, Animals and the Sacred*. New York: Zed Books, 2000.

Cirac-Claveras, Gemma, "Re-imagining The Space Age: Early Satellite Development from Earthly Fieldwork Practice," *Science as Culture* 31/2 (2022), 163–186.

Connelly, Matthew J. *Fatal Misconception: The Struggle to Control World Population*. Cambridge: Harvard University Press, 2008

Doel, Ronald E. "Constituting the Postwar Earth Sciences: The Military's Influence on the Environmental Sciences in the USA after 1945," *Social Studies of Science* 33/5 (2003): 635–666.

Edwards, Paul. *A Vast Machine: Computer Models, Climate Data, and the Politics of Global Warming*. Cambridge: MIT Press, 2013.
Fuller, R. Buckminster. *Operating Manual for Spaceship Earth*. New York: Arkana, 1969.
Gabrys, Jennifer. *Program Earth: Environmental Sensing Technology and the Making of a Computational Planet*. Minneapolis: University of Minnesota Press, 2016.
Gribbin, John. *Alone in the Universe: Why Our Planet Is Unique*. New York: Wiley, 2011.
Grinspoon, David. *Lonely Planets: The Natural Philosophy of Alien* Life. New York: Harper Collins, 2003.
Holbrow, Charles, and Richard D. Johnson, ed. *Space Settlements: A Design Study*. Washington, DC: NASA, 1977.
Jayawardhana, Ray. *Strange New Worlds: The Search for Alien Planets and Life Beyond our Solar System*. Princeton: Princeton University Press, 2011.
Kirk, Andrew. *Counterculture Green*. Lawrence: University of Kansas Press, 2007.
Kolbert, Elizabeth. *The Sixth Extinction: An Unnatural History*. New York: Henry Holt, 2014.
Lemonick, Michael. *Mirror Earth: The Search for our Planet's Twin*. New York: Walker, 2012.
Lowman, Paul. *Exploring Space, Exploring Earth: New Understanding of the Earth from Space Research*. New York: Cambridge University Press, 2002
Lovelock, James. *Gaia: A New Look at Life on Earth*. New York: Oxford University Press, 2000 [1979].
Mack, Pamela. *Viewing the Earth: The Social Construction of the Landsat Satellite System*. Cambridge: MIT Press, 1990.
Margulis, Lynn. *The Symbiotic Planet: A New Look at Evolution*. New York: Basic Books, 1998.
Messeri, Lisa. *Placing Outer Space: An Earthly Ethnography of Other Worlds*. Durham: Duke University Press, 2016.
Moltz, James Clay. *Crowded Orbits: Conflict and Cooperation in Space*. New York: Columbia University Press, 2014.
Pool, Ithiel de Solla. *Technologies without Boundaries: On Telecommunications in a Global Age*. Cambridge: Harvard University Press, 1990.
Scharf, Caleb. *The Copernicus Complex: Our Cosmic Significance in a Universe of Planets and Probabilities*. New York: Farrar, 2015.
Selcer, Perrin. *The Postwar Origins of the Global Environment: How the United Nations Built Spaceship Earth*. New York: Columbia University Press, 2018.
Slotten, Hugh Richard. *Beyond Sputnik and the Space Race: The Origins of Global Satellite Communications*. Baltimore: Johns Hopkins University Press, 2022.
Tielhard de Chardin, Pierre. *The Phenomenon of Man*. New York: Harper, 1955.
Turchetti, Simone, and Peder Roberts, ed. *The Surveillance Imperative: Geosciences during the Cold War and Beyond*. New York: Palgrave Macmillan, 2014.
Turner, Fred. *From Counterculture to Cyberculture: Stewart Brand, the Whole Earth Network, and the Rise of Digital Utopianism*. Chicago: University of Chicago Press, 2006.
Vakoch, Douglas, ed. *Between Worlds: The Art and Science of Interstellar Message Composition*. Cambridge: MIT Press, 2004.
Vakoch, Douglas. *Archaeology, Anthropology, and Interstellar Communication*. Washington, DC: NASA, 2014.

Vernadsky, Vladimir. *The Biosphere*. New York: Copernicus, 1998 [1926].

Ward, Peter, and Donald Brownlee. *Rare Earth: Why Complex Life Is Uncommon in the Universe*. New York: Springer, 2003.

Ward, Peter, and Donald Brownlee. *The Life and Death of Planet Earth: How the New Science of Astrobiology Charts the Ultimate Fate of Our World*. New York: Henry Holt, 2004.

Weart, Spencer. *The Discovery of Global Warming*. Cambridge: Harvard University Press, 2008.

Westwick, Peter. *Into the Black: JPL and the American Space Program, 1976–2004*. New Haven: Yale University Press, 2000.

White, Frank. *The Overview Effect: Space Exploration and Human Evolution*. Reston: AIAA, 2014.

Although the environmental challenges facing Earth are formidable, a number of authors have argued, based on all the historical evidence so far, that human intelligence and innovation will surmount them. See the works by Matt Ridley, *The Rational Optimist: How Prosperity Evolves*. New York: Harper Perennial, 2011. Peter Diamandis and Steven Kotler, *Abundance: The Future is Better Than You Think*. New York: Free Press, 2012. Christopher Preston, *The Synthetic Age: Outdesigning Evolution, Resurrecting Species, and Reengineering our World*. Cambridge: MIT Press, 2019. Marian Tupy and Gale Pooley, *Superabundance: The Story of Population Growth, Innovation, and Human Flourishing on an Infinitely Bountiful Planet*. Washington, DC: Cato Institute, 2022.

12 Thresholds

Shuttles, Star Wars, and Stations

This final chapter brings us to a culminating moment in the history of the Space Age, when the US and USSR began to construct or plan vast new space architectures in Earth orbit. These assets included the Space Transportation System and the Strategic Defense Initiative, the Mir Space Station and International Space Station. They were assets for both war and peace, still beholden to the old ways of the missile race and space race, but also previsioning something new, an unprecedented era of international cooperation in space. A series of crisis moments in the 1980s made this new era possible, centered on President Ronald Reagan's military buildup, one that intensified the Cold War rivalries, and that spotlighted space as a field of futuristic machines and weaponry to break through the logic of Mutual Assured Destruction.

Outer space, in other words, became a tipping point for dramatic developments here on Earth, influencing the collapse of the USSR in 1991. It fell for many reasons, including its corrupt political system and command-administrative economy. Among them were also its investments in nuclear and conventional forces, helping to bankrupt the country through the 1970s and 1980s, all at the expense of people's needs and wants. This imbalance was compounded by a severe fall in world oil prices between 1980 and 1986, on whose exports the USSR economy depended for sustenance. This perfect storm of influences shaped a rare event in the twentieth century: the relatively peaceful transition of power from the Communist Party and USSR to a more democratic Russian Federation in 1991.

A second rare event followed the collapse of the USSR, as its former enemy and rival, the US, reached out to the new Russia with political and financial aid, offering significant support to the Russian Space Agency in difficult times. The US Space Transportation System (Space Shuttle) was the keynote technology through all of these extraordinary historical turns.

The Space Transportation System

The Space Transportation System (1981–2011) was the only element of the Space Task Group proposals of 1969 to survive the Nixon-era budget cuts. Gone

DOI: 10.4324/9781003494409-16

were the orbital space station, the Moon colony, and the mission to Mars. The Space Shuttle remained, with the promise of inexpensive and frequent access to space. Critics called it all means and no ends, a Shuttle to nowhere; a "technology in search of a mission," as the editors of the *New York Times* complained in 1981. Yet this judgment only counts against the grand vision of the Apollo era. In its lifetime, the Space Shuttle served a number of essential national, commercial, and international goals. At least originally, the costs of research, development, and production, at roughly $10 billion, were quite reasonable for the various pieces: the Enterprise sub-orbital glide test vehicle, and a fleet of five orbital craft: Columbia, Challenger, Discovery, Atlantis, and Endeavor (which replaced the lost Challenger in 1991).

The Shuttle had its moments of glamor. Columbia, for example, took its maiden voyage on 12 April 1981, only 20 years to the day of Yuri Gagarin's flight into space. The fleet of Shuttles also owed their original design to the creativity and forward-thinking of the Apollo era (1968–1972). North American-Rockwell built it, having also made the Apollo Command and Service Modules. NASA assembled and prepared it for flight at the massive Vehicle Assembly Building, and launched it at the Saturn V pads at Cape Canaveral. Mission Control at the Cape and at Houston managed the flights. It was crewed by the long-standing Astronaut Corps, whose techniques and personnel had served the Mercury, Gemini, and Apollo missions. There were important and symbolic moments, too. Like America's first woman in space, Sally Ride, and the first African American, Guy Bluford (both in 1983). The rock group Rush even captured the thrill of a Shuttle launch in the song, "Countdown" (1982).

Among the Shuttle's signal achievements were the design and engineering of its technology. The aerodynamics were based on the achievements of the NACA X-plane program, namely the hypersonic X-15 and the experimental X-20 Dynasoar; as well as the design genius of Max Faget and his teams, who had also overseen the creation of the Mercury, Gemini, and Apollo spacecraft. The Shuttle launched into space by rocket, gliding back to Earth like an airplane. It introduced cutting-edge computerized flight controls, airframe design, electrical power system, thermal protection system, and main propulsion unit. It was also a partially re-usable launch system, with two Solid-Rocket Boosters (SRB), built by Morton Thiokol, the largest ever flown. Once spent with their solid propellants (an ammonium perchlorate mixture), they disengaged and were parachuted back to Earth for an ocean landing and recovery and re-use. The External Tank, built by Martin Marietta, filled with liquid oxygen and liquid hydrogen that fed the Orbital Vehicle's re-usable main engines, and which disengaged just as the Shuttle reached orbital insertion: at 17,500 miles per hour. The Shuttle's main engines were remarkable achievements, three of the most efficient chemical rocket engines ever built. Like the Saturn engines of Apollo, they enjoyed a 100% success rate.

There were multiple orbiters for multiple flights, one of the guiding ideals and practical values of the program. The most controversial design elements of the Shuttle were its delta wings. They allowed it to glide back to Earth through the atmosphere. As a rule, the Shuttle computers guided re-entry by using its thrusters, atmospheric drag, and wide banking turns; as well as aerodynamic braking and parachute deployment as it came to a pilot-controlled landing (in some circumstances). The Shuttle's landings at Edwards Air Force Base and the Kennedy Space Center landing facilities were always an impressive sight, soaring down like a jet plane, the crew eventually standing nobly on the tarmac. This was quite a contrast to the traditional blunt-force ballistic reentries, rescue parties pulling the cosmonauts and astronauts out of the re-entry modules on land or sea. Yet these wings were also a significant expenditure in weight, as that launch systems had to carry them up to orbit in the first place. New designs and thermal tiles had to protect from the intense heat of re-entry. This was a complex venture from the start.

In orbit, the Shuttle was a space station, serving as a platform from which to deliver, position, repair, and retrieve satellites. It was an international living space and orbital laboratory, and a service module for the Mir space station. It was also the building site for the International Space Station. It had 135 missions and 1,334 days in Earth orbit, made 21,152 orbits of Earth, and traveled over 542 million miles; 852 people flew on the Shuttle. Over the years, these kinds of activities became less and less newsworthy, more and more routine, reducing what had been a cadre of celebrity astronauts to a mere list of rather anonymous pilots and flight engineers, mission specialists and payload specialists.

The Shuttle could carry up to 60,000 pounds of payload into low Earth orbit. It was also able to bring satellites and other space assets back down to the Earth. One of its crowning satellite achievements was much more than a satellite, of course: the Hubble Space Telescope, which the Shuttle delivered and positioned into Earth orbit in 1990, and was repaired by Shuttle astronauts in 1993, with four major repair and replacement missions that followed. The Shuttle sent off the Galileo and Magellan interplanetary probes using the launch platform of the Inertial Upper Stage. Shuttle astronauts also conducted thousands of large and small microgravity and human-physiology experiments. Twenty-four Shuttle missions between 1983 and 1998 raised the international Spacelab, with partners in the European Space Agency, Germany and Japan. This science laboratory, part pressurized and part not, with special pallets that allowed astronauts to place items in open space, achieved remarkable studies of plant growth and cellular development, materials science and bodily changes, in microgravity.

One of the most dramatic and poignant moments in the history of the Space Age happened on 7 February 1984, when astronaut Bruce McCandless became the first person to fly in space on his own accord, untethered to the spacecraft. The "manned maneuvering unit" or jetpack he was wearing allowed him to travel 300 feet from the Space Shuttle *Challenger*, a moment captured in a historic photograph. This was the first of four such spacewalks on this mission, and four future astronauts experimented with the pack that year, servicing and retrieving several disabled satellites, though it was discontinued in favor of smaller emergency jetpacks, and reliance on the Shuttle's robotic arm. McCandless was the first "human satellite" on his 7 February test, flying in orbit at 17,500 miles per hour, the fastest any lone, unattached human being has ever flown. He was also the first human rocket: the jetpack included two tanks of gaseous nitrogen with 24 thrusters, operated by fingertip controls and powered by two silver–zinc batteries. Working time: six hours. Range: 450 feet. Weight: 340 pounds (but weightless in space). Cost: $10 million each.

Two politicians also flew into space aboard the Shuttle. Republican Senator from Utah, Jake Garn, flew on Discovery in 1985 as a payload specialist. He was also chair of the Senate sub-committee for NASA appropriations, and a former US Navy pilot and National Guard member. Democratic Senator from Ohio, John Glenn flew in 1998, his second flight after his 1962 Mercury orbit of Earth. Only this time he was 77 years old and the oldest person ever to fly in space. The justification was science, to study the effects of spaceflight (which in many cases mimics aging) on an old body. At the time, Glenn was also a member of the special Senate committee on aging.

From start to finish, the Shuttle was an expensive, high-technology proposition. Each of the five Shuttles was a complex machine of some 2.5 million parts, carrying a high-risk factor were any one of them to fail. Early on, the thermal tiles that protected the delta wings from the heat of re-entry posed significant challenges. On its maiden voyage, Columbia lost 16 tiles with damage to over 100. Other Shuttles experienced such problems as cracks in the main engines, ruptured tubes, corroded bolts and hoses, and short circuits of wiring. After the investigations and safety inspections following the Challenger disaster, engineers found hundreds of serious faults in the Shuttle's parts and mechanisms, as in the sensors and components of the Orbital Vehicle.

Through the 1990s, NASA modernized each of the Shuttle's computer systems to accommodate more memory and faster speeds. The Shuttle also became the first spacecraft with full "fly by wire" controls, an electronic interface and computerization between the pilot's controls and the craft's flight system; and with the glass cockpit, the "full-color, flat-panel display screens" that took the place of the old analog flight instrumentation. In 1999, Columbia was totally overhauled, through the course of 17 months and $145 million, twice as long and twice as expensive as originally thought. The cockpit was upgraded to modern standards and 1,000 pounds of wiring stripped. The work also found fragile thermal tiles, unusual drag on the left wing, weak airframe bonds between the wings and the hull, and 3,500 problem issues with the wiring overall.

In 2002, NASA grounded the whole fleet of Shuttles after discovering fuel-line cracks in all four of the remaining orbiters, subsequently repaired. In their last decade of flight, the Shuttles required around 2,200 contractors and over 20,000 people for each launch, costing between $300,000 and $500,000 each time. The expense was mostly due to the age of the Shuttles, as every major component was replaced, repaired, or refurbished after each flight. The Shuttle had come a long and winding way from the original promises of cheap and available transport to outer space. NASA no doubt had a high standard of safety and oversight, though this did not prevent the major disasters that befell the program.

By 2003, three generations of Americans had witnessed a space disaster: Apollo 1 in 1967, Challenger in 1986, and Columbia in 2003. In each case, human error was the ultimate cause; in each case, NASA corrected the mistakes. Both of the Shuttle disasters were eminently avoidable, more results of political pressure and bureaucratic oversights than any fault in the engineering or mechanics. The Challenger disaster, for example, happened at a tense moment in the Cold War, just as the Space Shuttle was proving its central role in the Strategic Defense Initiative. The Reagan administration was applying intense pressure on NASA administrators and managers to perform, especially given their recent pattern of schedule delays. Doubled upon this military and geopolitical imperative was a public relations initiative, the first schoolteacher in space, Christy McAuliffe, who was to direct one week of experiments for hundreds of thousands of school children either watching or listening to the mission.

The accident resulted from a failure in the O-rings (flexible gaskets), joining two sections of the Solid-Rocket Boosters, which had hardened in the extreme cold of that morning, as low as 30 degrees Fahrenheit, 15 degrees colder than any other launch. They then ruptured from the heat and force of the launch, the hot gases and flames burning through, causing a series of explosions only 73 seconds into the launch, at 46,000 feet, killing the crew of seven. Engineers at Morton Thiokol, the makers of the O-rings, had warned of their potential failure in excessive cold. Groupthink also played its part. Some NASA managers rejected the warnings, under political and bureaucratic pressures, and fearing a delay until the next window for a launch (in April). Instead, much of the country watched the disaster (and deaths) live on television. NASA was subjected to an intense investigation, and the Shuttle program was put on hold for two and a half years as fixes were implemented: new O-rings for the boosters, an escape system for certain conditions, new weather standards for launch, and stringent oversight within and over NASA, by an Office of Safety Reliability and Quality Assurance.

The second great Shuttle disaster happened on 1 February 2003, with the 28th flight of Columbia, already 22 years old, and the 112th flight of the Space Shuttle. Fifteen minutes from its scheduled touchdown, flying some 200,000 feet over Texas, the Orbital Vehicle broke up and disintegrated, killing another seven of the crew. The technical cause, once again, was something simple: foam insulation on the fuel tank had come loose at launch and made a small hole in the Shuttle wing, just enough to create a breach that imperiled the whole craft when exposed to the intense heat and flames of re-entry. The integrity of the foam insulation and potential for hazards were long-standing issues at NASA. Even after the initial incident, engineers called for Department of Defense imaging, or a

Columbia spacewalk, in order to identify the exact nature and threat of the damage. NASA administrators refused, along with any chance for repair. Columbia attempted to re-enter the atmosphere, but the delta wing did not hold.

NASA investigations and new safety protocols followed again, as the Shuttle program was on hiatus for the next several years. Yet the Shuttle Discovery was nearly subjected to the same accident when foam insulation flew off once again in July of 2005, grounding the fleet, but with a crucial difference. On Discovery, the insulation broke off at 120 seconds into the flight, where the air was already less thick, and the insulation simply floated away. On Columbia, the insulation had broken away at 82 seconds into the flight, where the air was thicker and forced the foam to make a deadly impact on the wing. This was a difference of 38 seconds. Such were the powerful variables at play that helped make a Shuttle mission succeed … or fail.

The Strategic Defense Initiative

Much like the Atlas and Titan missiles before it, which served the US as both ICBMs and as launch systems for the Mercury and Gemini programs, the Space Shuttle represented a dual military and civilian program in space, if now combined as both launcher and orbiter. From the Shuttle's earliest designs, the US Air Force dictated the size and characteristics of the payload bay, necessary to fit the dimensions of the country's newest Keyhole spy satellites and other secret military assets in space. One of the reasons for the delta wings, too, was the military's imperative that the Shuttle be able to make precise emergency landings at Vandenberg Air Force base from Earth orbit, in case it was carrying a military payload. The Space Shuttle was, in its original purposes and some of its essential missions, partially a military spacecraft, introducing a radically new element of threat and surprise into the Cold War.

Up until the Shuttle, the operating trend in American and Soviet relations had largely been for the peaceful uses of outer space and for parity in their nuclear forces, if always within the broader context of Cold War competition. There were signs of hope and cooperation, as in the US sponsorship of both the International Geophysical Year (1952), and the United Nations Committee on the Peaceful Uses of Outer Space (COPUOS, 1959). In the wake of the menaces of the Cuban Missile Crisis, the US and USSR agreed on a number of nuclear limitations. The Partial Test Ban Treaty (1963) banned all atmospheric and space-based tests of nuclear weapons (though still allowing underground tests). The Outer Space Treaty (1967) prohibited nuclear weapons or weapons of mass destruction in outer space or on any celestial body, providing for the "free exploration and use" of outer space "by all States." The Non-Proliferation Treaty (1968) banned signatories from selling or providing, or buying or receiving, nuclear weaponry, although this did not prevent Israel, South Africa, Pakistan, India, and North Korea from becoming nuclear powers.

These positive developments occurred against the backdrop of the USSR's massive military buildup in the later 1960s, centered on a new fleet of improved ICBMs, an investment that diverted state funds from consumer markets (and civilian spaceflight) to military production. In the words of one of its architects, Dmitrii Ustinov (Minister of Defense between 1967 and 1984), "the only thing the people need is bread and defense." Beginning in 1965, he deployed the UR-100 in force, the most widely deployed ICBM in history, with 720 in place by 1970. To these were added 240 of the R-36 ICBMs. The USSR had achieved a rough ICBM parity with the US.

Parity brought the US and USSR to the Strategic Arms Limitations Talks agreement (SALT) of 1972, which officially locked the two powers into the doctrine and reality of Mutual Assured Destruction (MAD), each of its civilian populations essentially held hostage to the nuclear balance of power. By its terms, the US limited its launchers of nuclear weapons to 1,054 ICBMs and 710 Submarine-Launched Ballistic Missiles (SLBMs). The USSR to 1,618 ICBMs and 950 SLBMs. An essential codicil to the agreement was the Anti-Ballistic Missile (ABM) Treaty of 1972, which banned a nationwide missile defense system for each country, allowing only two such sites, with no more than 100 ground-based interceptors each. The parties amended the treaty in 1974 to allow for only one site each: one for an ICBM site in North Dakota, and one for Moscow. In this way, both sides conceded the balance of terror, their reciprocal power to destroy the other. This was the basis of the late Cold War peace.

This new equilibrium remained unsteady through the 1970s and 1980s. The SALT and ABM agreements set standards on ICBM numbers, but they could not hold back the continuing pull of technological advances. Both sides built up their stockpiles of nuclear weapons, well beyond the SALT limitations on launchers. Each developed stealth techniques and technologies to hide their missile-equipped nuclear submarines. Through the 1970s, both sides set out to break through the quantitative limits with qualitative improvements in missile technology. In terms of accuracy, the US advanced far ahead of Soviet

Table 12.1 The milestones of US–USSR détente. Author.

The Milestones of US-USSR Détente

UN Committee on the Peaceful Uses of Outer Space (COPUOS, 1959)
Partial Test Ban Treaty (1963)
Outer Space Treaty (1967)
Non-Proliferation Treaty (1968)
Strategic Arms Limitation Treaty or **SALT I** (1972)
Anti-Ballistic Missile Treaty (1972-1974)

	US	**USSR**
ICBMs	1054	1618
SLBMs	710	950

SALT I limitations on delivery systems

technology, with its application of electronic miniaturization and computerization (the integrated circuit) in the Minuteman III and MX missiles. The Multiple Independently Targetable Re-entry Vehicle (MIRV) gave each missile multiple warheads. These deployed at the height of the midcourse phase of flight, along with decoys, improving the chances for city and silo strikes.

For its part, the USSR also developed missile accuracy and MIRV capacity, though it was roughly a decade behind on both. It also answered American developments with a system of "Phased-Array Radars," an initiative in ballistic missile defense located in Krasnoyarsk, Siberia, to allow for an early warning of a US attack and likely to coordinate the management of an illegal ABM system. In space, the Almaz space stations (Salyut numbers 2, 3 and 5) tracked US military forces with telescopes and radars in real time, and threatened US space assets with anti-satellite weapons. With an R-36 variant, the USSR also experimented with a "fractional orbital bombardment system," essentially an ICBM that orbited the Earth and offered a surprise "first-strike" capacity, theoretically able to hit targets in the US within three to four minutes rather than 10 to 30.

Ronald Reagan's Republican victory in the 1980 presidential election transformed this landscape of both balanced and dynamic competition. Over the next few years, the conservative and fiercely anti-communist president began an enlargement and modernization of the US military costing $32 billion, introducing 3,000 new combat aircraft, 10,000 new tanks, and 3,700 new strategic missiles. By 1988, his administration increased defense spending from 4.8% of the Gross Domestic Product to 6.0% ($116 billion to $290 billion). In 1983, he deployed Pershing II and Cruise IRBMs to US allies in western Europe to offset a massive Soviet conventional military buildup in Eastern Europe, recently fortified by Soviet IRBMs (1976–1988) that threatened to destroy NATO forces in the case of a ground war. The president also intensified the rhetoric of the Cold War, naming the USSR as an "evil empire," driven by an ideology of party-state dictatorship. It was "the focus of evil in the modern world," to which the US would not accept a position of "military and moral inferiority."

President Reagan capped these measures with a prime-time television speech to the nation on 23 March 1983, announcing the Strategic Defense Initiative (SDI). In a radical departure from previous US military policies, he proposed to break the logic of the balance of terror and the arms race, the stranglehold of reciprocal offensive capacities, in order to "render nuclear weapons impotent and obsolete." In defiance of agreements like the Outer Space Treaty and ABM treaty, he offered a space-based satellite system, and eventually with ground-based weapons, to shield the American population from a Soviet ICBM attack. This was a powerful symbolic moment, a "defense initiative" that was really a strategic offensive and radical turning point in the Cold War arms race.

The Strategic Defense Initiative Organization (SDIO), created in 1984 to implement the new policies, experimented with some rather fantastic proposals over the next ten years, centered on America's premier military laboratories,

Table 12.2 Fundamentals of the Strategic Defense Initiative. Author.

Fundamentals of the Strategic Defense Initiative (1984-1993)
Objectives: Population Defense ICBM Defense
Means: Directed Energy Weapons (X-Ray Lasers, Chemical Lasers, Particle Beams) Kinetic Energy Weapons, and High Velocity Rail Gun Satellite Surveillance and Tracking Systems Conventional Ballistic Missile Defense
Targets: ICBMS (primarily in the Boost Phase) Space Assets (Satellites)
Costs: $25 billion projected (up to an estimated $40 billion in eventual costs over ten years)

like Lawrence Livermore and Los Alamos. These projects ranged from directed energy weapons (X-ray and chemical lasers) to high-energy acceleration guns. They were meant to be satellite-based weapons that mainly targeted Soviet ICBMs in their boost phase (the first three to five minutes of a launch), or Soviet satellites.

ICBM Phases

By 1972, the Intercontinental Ballistic Missiles (ICBM) in the American and Soviet arsenals required only about 30 minutes to reach their targets: aimed at other ICBM siloes and the largest cities, in either the USSR or US. Each followed the basic flight path of the original V-2 rocket, though now coasting through the shallows of outer space. In the boost phase, the first three to five minutes of an ICBM launch, the missiles fire and release a powerful infrared signal easily detected by satellites, with the missile moving slowly and its warheads not yet deployed. This was the favored window of opportunity for the SDI program. In the midcourse phase, the missile coasts above the atmosphere, in space, at around 4 miles per second, for 20 minutes or so, as the warhead deploys (with decoys and chaff). In the terminal phase, the last one minute of flight, as the warhead and decoys re-enter the atmosphere, the decoys burn up but the warhead speeds to its target. In order to avoid damage to a city, any ground-based ABM interceptor must destroy the warhead at about 20 miles altitude (or for a missile silo, at about a 1-mile altitude).

Most of these systems were just possibilities. As the president admitted, they were likely decades away from deployment, if even possible at all. Their practical results were either meager or delayed. The Reagan administration touted them in terms of the remarkable historical record of American industrial and technological achievement. Advocates even likened SDI to the Apollo program or the Space Shuttle. Of the two, though, the Shuttle program gave SDI its firmest foundation in reality. NASA worked closely, if at first not always cooperatively and successfully, with the SDIO. Its first director, Air Force Lt. General James Alan Abrahamson, was a former director of the Space Shuttle program; its Deputy Director for Technology, Michael Griffin, later became NASA administrator.

Political opponents called SDI "Star Wars," after the popular movie franchise. The video arcade game, Space Invaders, was also hugely popular at this time, an entertaining and challenging battle of lasers in outer space, much like SDI. The punk rock group, the Ramones, had already made some lighthearted fun with the nuclear "balance of terror," featuring a Moscow-bound love rocket on their album, *Rocket to Russia* (1977) along with the very rocket-sounding song, "Rockaway Beach." American punk rock was also a weapon in the Cold War.

Other critics were much more serious. Carl Sagan and some of his colleagues at Cornell and NASA warned that SDI was moving the world closer to an all-out nuclear war, which would smother the Earth in a cloud cover of ashes and poisons, a nuclear winter of planetary catastrophe. The television movie, *The Day After* (1983), watched by some 100 million viewers, and based in part on Jonathan Schell's riveting, *The Fate of the Earth*, offered graphic images and storylines of human devastation.

SDI was one of the most expensive military initiatives in US history, eventually costing some $40 billion up to 1993. Early on, projections were even for a possible trillion-dollar investment. The government never imposed funding limits or timetables on it. The Reagan administration publicized the costs and open schedule. One of the curious elements of the proposal was Reagan's offer to share SDI technologies with the USSR, a point of leverage in arms control negotiations. All of this was planned and purposeful, spotlighting SDI as a dare, the ultimate in brinksmanship, a bookend to the Sputnik and missile gap bluff from two decades earlier. Only now the US was forcing the Soviet hand. The USSR, as a CIA National Intelligence Estimate (12 September 1983) predicted, would have to match US investments with its own, risking bankruptcy and economic collapse.

Thanks to these investments and intersections, and to Reagan's ongoing military transformations, Soviet policymakers took SDI seriously. By 1983, the US had a proven launch system in the Shuttle to deploy the full range of SDI platforms. Soviet negotiators consistently revealed their frustration and fears of the Shuttle, demanding that the US stand down from militarizing it as an anti-satellite or ABM platform. Soviet military analysts also feared it as a possible fractional orbital bomber with a potential first-strike capacity. Thus, in the logic of the Cold War competition, the USSR had no choice but to increase Soviet military spending up to 1988, research its own potentials for SDI, and create its own version of the Shuttle: the Energia–Buran system.

Valentin Glushko had lobbied successfully for the Energia launch system in 1974, as a replacement for the failed N-1 moon rocket. It was a return to his earlier successes with cryogenic engines, based on a core module and booster rockets. The USSR ordered its Shuttle variant, Buran (Blizzard) by 1976, just two years after the Space Shuttle was approved. Mark Wade has called it a "straight aerodynamic copy," although also a "blend of mature American technology and Soviet innovation." It was the last Soviet copycat of Space Age technology, after the V-2 rocket, the US atomic bomb, and the long-range B-29 bomber. The Energia launcher was one of the most powerful systems ever built. It had to be. Unlike the Shuttle system, whose external tank disengaged as the Shuttle inserted itself into orbit, Energia carried the Buran Shuttle directly into orbit. Buran had no main engines. Soviet engineers simply did not have the expertise or the time to build a perfect American copy.

Energia–Buran was a serious and well-funded venture. The USSR planned to use it as a system to launch heavy satellites and military payloads, and for projected missions to the Moon and Mars. The USSR prepared a cosmonaut contingent beginning in 1978. It built a runway at Baikonur by 1983. It constructed six Shuttles for Earth-bound tests by 1985. The USSR Ministry of Defense also planned to use Buran in order to build an SDI-type system: the Polyus orbital military platform, with a carbon-dioxide laser. Only it had to be cheaper and simpler than the US versions. The USSR could simply not afford and did not have the engineering potential to shoot down American ICBMs in the boost phase, so it planned instead to aim the lasers at SDI assets in space.

The whole Energia–Buran program was beset by paralyzing delays and problems. Energia only lifted off twice, working extraordinarily well each time: once to launch Polyus in 1987, and once to launch Buran in 1988. Polyus failed to reach orbit due to its own design and production flaws (in the guidance system) which sent it crashing back to Earth. Buran flew around the Earth in automatic mode, though without manual avionics, life support, fuel cells, or a crew. All of these were far too expensive for a Soviet economy already in collapse. Moreover, the Soviets never realized their plan for an improved version of Buran, called the Uragan (Hurricane) system.

The new Russian Federation mothballed Buran in 1993. The whole program collapsed during the greater crisis in the Soviet political and economic system. In 2002, the Buran flight test vehicle was destroyed, and eight people were killed, when a poorly maintained hangar collapsed at Baikonur. Other Buran models littered the base in ruins. In a twist of poetic justice, Glushko's Buran met the same fate as he had once assigned to Korolev's N-1 rocket.

The Collapse of the USSR and Shuttle–Mir

Faced with the Reagan military buildup and SDI challenge, with falling economic growth rates and national income at home, the new communist party leader, Mikhail Gorbachev, attempted a series of partial democratic reforms between 1985 and 1991, promoting freer cultural debate, a controlled parliamentary system, and chaotic free-market economic reforms.

These initiatives began to spiral out of control in 1989–1990. The USSR lost its last "satellites," the occupied states in Eastern Europe, as they each declared full independence from Moscow. Protesters helped to topple the Berlin Wall; Germany soon reunited its capitalist West with its formerly communist East. Soviet troops retreated. They went home. The very constituent "republics" within the USSR state began to voice and struggle for independence and national self-determination as well, especially in the Baltic (Estonia, Latvia, and Lithuania), Ukraine, and Azerbaijan.

With worsening political and economic conditions, Gorbachev's regime relied on targeted acts of violence and intimidation, especially against Lithuania and Azerbaijan, as well as rule by a "presidential dictatorship." When he did not take these measures far enough, hardliners within his own government conspired to overthrow him by a coup in August of 1991. The Russian army and police forces, the people of Moscow and democratic opposition, rose against the conspirators. Newly and democratically elected president of the Russian Federation, Boris Yeltsin, gradually seized the military, political, and financial resources of the country over the course of the fall and winter. By December, the USSR had ceased to exist. The communist system that had sponsored all those Soviet space firsts went bankrupt and imploded.

Amid all of this, at four crucial summits between 1985 and 1988 (Reykjavik, Geneva, Washington, and Moscow), Gorbachev attempted to force Reagan to give up the SDI program in favor of massive nuclear-arms reductions. In each case, Reagan refused. In the end, Gorbachev capitulated, resulting in several later arms-reduction treaties. He had found himself locked into the last "gap" of the Cold War: an ultimate technological and financial gap of the USSR's own making, and from which it was never able to extricate itself. It was never able to catch or surpass again.

One of the last significant events of the space race, a point of challenge from the USSR, was the launch of the core block of the Soviet space station Mir (Peace) on 19 February 1986, boosted into orbit by a Proton rocket. The Soviets originally planned it for a five-year civilian and military mission, to last until 1991, when it was to be replaced by Mir 2. With the collapse of the USSR in that very year, it lasted until 2001, and became the first long-term International Space Station during the transformative Shuttle–Mir program, a bridge of cooperation and understanding between the former enemies in the Cold War.

In its size, achievements, and longevity, Mir was a foundational event in the history of the Space Age. This was still not the living wheel or torus of early space-engineering dreams, more like an ungainly spider with its various pieces and essential solar-power arrays. The core block included a multiple docking adapter to attach the modules to come. The two remaining axial adapters were for the Soyuz TM ferries, which brought the hosting and visiting crews (two Soyuz could dock at any one time) and for the automated Progress resupply ships.

Two Kvant (Quantum) modules were raised between 1987 and 1989, with a unique wheel mechanism to provide the station with proper orientation and attitude control, as well as an astrophysics laboratory with telescopes and spectrometers, along with added environmental regenerating equipment and a shower unit. Kristall (Crystal), an Earth-observation apparatus, was attached in 1990. The remaining modules were launched and attached after the collapse of the USSR, beginning in 1995: Spektr (Spectrum), a platform for the remote sensing of Earth, and in time for European and US experimental microgravity laboratories; and Priroda (Nature), also for remote sensing, but of Earth's climate and oceans. On completion, all these modules took on a massive, multi-dimensional cruciform shape.

Mir was in space for 15 years, continuously occupied for nine. It was a platform for 16,000 experiments and 86,331 Earth orbits. Designed to house three cosmonauts, it could also accommodate six for short periods of time. In all, 104 cosmonauts lived on Mir. They originally came from Soviet-occupied states and allies: Eastern Europe, Afghanistan, Cuba, Vietnam, and Syria; but eventually from former enemies like France, Germany, Italy, and the US. Soviet cosmonauts broke all the records: five cosmonauts spent over consecutive 212 days on Mir, and three cosmonauts spent over 325 days. By 1995, cosmonaut Valerii Poliakov spent a historic 437 days on Mir.

These spacefarers were able to live and thrive on Mir thanks to the earlier accomplishments (and trials and errors) of the Salyut space stations. Soviet engineers improved the life-support and atmospheric conditions, applying innovative mechanisms for the recycling of water, including atmospheric moisture and human waste products. An oxygen-generating "candle" helped to diminish the buildup of carbon dioxide. The Progress resupply ship brought fresh oxygen and nitrogen. Soviet psychologists also integrated Earth-orienting markers throughout the living and working spaces: the representation of up and down with white ceilings, green walls, and carpeted floors. One docking apparatus was "ahead" and the engine was "behind." There were two personal compartments with sleeping bags. There were also foldout food preparation tables, with a writing desk, complete with a porthole, perhaps for gazing when writing home. The toilet was enclosed with a door. Some of these earth analogs were easy; others were hard. To maintain physical fitness for the return to Earth, each cosmonaut had to cycle on the stationary bicycle five miles a day and walk on the treadmill for three.

Cosmonauts occupied the Mir space station for five years before the USSR collapsed in 1991. In one of its most dramatic moments, Sergei Krikalev spent

313 days aboard Mir during the collapse, only to return to a whole new country (the USSR became the Russian Federation), a whole new flag (the red flag with hammer and sickle became the tricolor red, white, and blue), and a whole new hometown (Leningrad became St. Petersburg). After returning, his substantial pre-flight salary of 500 rubles a month only amounted to a few US dollars.

Through these years of breakdown and trauma, the Soviet spaceflight program also gave up its secrets. The world learned about the R-16 disaster and N-1 moon program, both kept hidden for all those years. In one remarkable exposé, the investigative journalist Nina Chugunova disclosed how the Institute of Medical and Biological Problems had even prepared a whole cadre of human test subjects (guinea pigs), for physical and psychological abuse, in order to assess the effects of spaceflight. Doctors referred to them as the "mechanics" or "non-cosmonauts," subjecting them to catapult tests, pressure chambers, and isolation compartments, often under powerful narcotics and at great personal suffering, all in the name of Soviet state achievement. Maciej J. Drygas spotlighted some of these practices in his documentary, *State of Weightlessness* (1994).

Not all of these revelations were so stunning and incriminatory. Through memoirs and news stories, we also discovered some of the strange customs of the Soviet spaceflight program. Aleksei Leonov, the first person to conduct a spacewalk, shared how on the eve of his flight, for good luck, he slept in the small cabin at the Baikonur cosmodrome where Gagarin had also slept. He also revealed how the Soviet state rewarded him for his feat: a 15,000-ruble bonus to his 80-ruble-a-month salary, a Volga car, and a deluxe 45-day vacation. We also learned about some of the strange and often sexist cosmonaut rituals, reflecting the traditional rural values of Soviet society. Cosmonauts did not allow their wives to come with them to Baikonur, as it was "bad luck to see any woman" not associated with the space program. They took a sip of Soviet champagne, and then all briefly sat down in a ceremonial moment of rest before leaving. They also urinated on the right rear wheel of the bus that took them to the launch pad. For good luck.

The new freedom of speech in the USSR and its successor state, the Russian Federation, also translated into a series of parodies of Soviet spaceflight, poking fun at its obsession with cosmic firsts and its drive for priority and superiority, all at the expense of real people. The avant-garde artist, Ilya Kabakov, staged his exhibition, "The Man Who Flew into Space from His Apartment" (1988). It was the scene of a cluttered Soviet living room from which a man has just catapulted into outer space … by slingshot. Viktor Pelevin's novel, *Omon Ra* (1992), whose title was a play on the name of the Soviet police forces and the Egyptian sun god, pilloried a fictional Soviet space program that mangled cosmonaut bodies, to the point of dismembering their legs, just to fit them into an efficient moon lander. The punk rock group "Hairy Glass" (*Volosatoe steklo*) sang a vulgar tune, "About Gagarin /*Pro Gagarina*" (1995), to parody his personality cult. On a lighter and cleverer note, Aleksei Fedorchenko directed the black comedy, *The First on the Moon* (2005), splicing scenes from Soviet science fiction and documentary films with a pretend story about how Stalin's government first landed a man on the moon … in 1939.

The forces of the free market intervened to upset the new Russian space program as well. The Sotheby's auction house sold off Russian historic space memorabilia in 1993, at prices well above the pre-sale estimates. One of the biggest buyers was the Texas billionaire and former presidential candidate, Ross Perot, who paid $1,652,500 for a Soyuz space capsule; $600,000 for three space suits; and $189,500 for the mannequin, "Ivan Ivanovich," that preceded Yuri Gagarin into space. The Russians soon learned that they could make more money from space advertising. In 1996, the Mir cosmonauts took video footage of a giant Pepsi can and a banner reading, "Even in space … Pepsi is changing the script." Pepsi Cola reportedly paid $5 million for it.

In 1997, an Israeli company filmed a milk commercial aboard Mir; and in 1998, two cosmonauts appeared live on the QVC shopping channel, hawking Fisher Space Pens for $32.75 and 11 tiny pieces of Mars meteorites, priced from $90 to $2,500. One of the Soviet Burans found a second life in these years. A joint-stock company owned and operated by the pioneer cosmonaut, German Titov (the second Russian in space), brought a Buran model to Gorky Park in central Moscow, where for $20 each, 48 visitors could take a pretend two-hour space voyage in it, eat space food, and even fend off a meteor shower.

These peculiar events masked deeper problems and crises within the new Russian Space Agency (*Roskosmos*), which by 1995 received only one-fifth of its 1989 funding. With the breakup of the USSR, and the declaration of a newly independent Kazakhstan, the Agency suddenly had to pay hundreds of millions of dollars to rent the use of its own launch facilities at Baikonur. The city lost half of its population of 100,000 (mostly Russian). Those who remained were left with unemployment and shortages. NASA came to the rescue with $475 million, and the Space Shuttle–Mir program (1994–1998), when the US and the Russian Federation began a dramatically new era of cooperation and coordination in space.

This moment recalls earlier American financial and material assistance to Russia. As during the famine of 1922, and in the lend-lease program of the Second World War, the US now came to Russia's rescue between 1992 and 2000, with financing of $1.5 billion to dismantle nuclear weapons, $2 billion to support the Russian economy, agricultural aid of $2.5 billion, and support of $8 billion in World Bank loans. US companies like Proctor-Gamble and Mcdonald's invested broadly in the Russian economy at this time. Aerojet and Orbital Sciences purchased the Kuznetsov engines for the N-1, for use in the Antares program. General Dynamics (Lockheed Martin) bought the Glushko Energia engines for use in the Atlas III program.

In the Shuttle–Mir program, Sergei Krikalev was the first cosmonaut to fly on the Space Shuttle. Norm Thagard was the first astronaut to fly on Soyuz.

Cosmonauts conducted spacewalks in Shuttle suits, astronauts in Soyuz suits. The Space Shuttle also rescued stranded Mir, bringing people and supplies, fresh oxygen, and power to extend its proper orbit. Seven Americans flew on the Mir space station between 1995 and 1998. They became, in a word, America's "first cosmonauts," charged with adapting both to the long-duration flights through microgravity and to the vicissitudes and peculiarities of Soviet-Russian culture. With minimal assistance from NASA, most of them had to learn Russian on their own and had to live in daunting and impoverished circumstances during their training in Moscow and at Baikonur. They found that it was easy to fall in love with Russians, normal human beings filled with good humor and warmth, patience and hospitality. It was not always easy to accept how they acted, what they thought, and how they treated each other. The Shuttle–Mir program was a study of the culture shock at the end of the Cold War. The program bridged two systems: NASA's high technology, and a culture of purity and cleanliness; and the Russian Space Agency, a culture of workable technologies and a repair-as-needed mentality. A society of wealth and consumption encountered a society of poverty and scarcity.

The American flights on Mir were trials. Norm Thagard spent 115 days on the station and lost 17 pounds. Shannon Lucid completed 188 days, conceding that the lack of oxygen in the cabin had incapacitated the crew at times. Jerry Linenger experienced the first of a series of life-threatening situations in 1997, when a flash fire broke out in the station due to a ruptured oxygen generator (February). The main oxygen and coolant systems failed next, the interior temperature rising to 86 degrees Fahrenheit (March); then the air purifying system failed (April). Finally, the Progress resupply ship smashed into the Spektr module and punctured its hull (June), causing a loss of life-support systems. This was probably the most dangerous moment, as the collision damaged a solar panel, knocking the station out of alignment with the sun, shutting down the climate control, oxygen generators, and carbon-dioxide scrubbers. Yet in a short time, the cosmonauts (and Michael Foale) fixed the alignment and restored power.

Despite all of the equipment failures, computer breakdowns, and general mishaps, the crews and Mission Control teams of Mir faced the challenges and overcame them. One of the space station's significant lessons was that human beings could learn to survive in an old and failing orbital technology. Just like on Earth. They all learned how to handle the disrepair and dysfunction. The cosmonauts spent 80% of their time at the end just fixing the broken parts of Mir. It was an important lesson. Mir proved to be a perfect school for future spaceflight improvisations. It taught us that space is not some utopia of perfect environments and total happiness. It is a place with all of the frustrations and challenges, achievements and failures, which any human endeavor brings. This series of small and large victories did not stop the critics. The American media turned Mir into an object of derision, calling it a "junkyard car," or "flying death trap," or scene of "terror" in space. Dan Rather's CBS evening news reports were especially scathing, as was Jerry Linenger's righteous anger in his memoir, *Off the Planet* (2000), although Linenger was at least fortunate not to have flown on either the doomed Challenger or Columbia.

The Mir space station reentered the atmosphere on 23 March 2001, parts of it burning up as it cascaded through the atmosphere, parts of it falling into the Pacific Ocean. The United States Space Command, one of the very agencies tasked with defending the US from a Russian ICBM attack, monitored this descent and provided the Russian Space Agency with telemetry and much-needed information. An old chapter in the history of the Space Age was closing as a new chapter was just beginning.

Meanwhile, as the US and Russia were cooperating in space, the Yeltsin and Putin regimes were rebuilding some of the old empire by way of discord and violence, staging a series of civil wars and frozen conflicts on Russia's own territories and rimlands, in places like Azerbaijan, Moldova, Georgia. Chechnya, Dagestan, Syria, and Ukraine. The two venues, space and Earth, were related. Evgeny S. Troitsky and his Association for the Study of the Russian Nation, partnering with the Slavic International Union of Aviation and Aeronautics, found wide support in the official Russian media. They celebrated the space station Mir, and the whole heritage of Tsiolkovsky and Russian cosmism, as engineering extensions and high-tech culminations of the Russian empire. Mixing a mystical nationalism and cosmic internationalism, they refashioned the Russian Federation as a boastful guarantor of Eurasian multinational comity, against the presumed offenses of NATO expansion and US globalism. Russian achievements in space justified imperial Russian conquests here on Earth.

The International Space Station

The crowning moment of the US Space Shuttle fleet was the launching and building of the most significant parts of the International Space Station (ISS), the first station without a name, only a description. Like the various space stations before it, the ISS had its early origins in the intense competition of the Cold War. President Ronald Reagan initially conceived of it as a response to the Soviet space station, Mir, calling on NASA to build it within a decade, so he declared in a Kennedy-style, Moonshot challenge, in his State of the Union speech on 25 January 1984). He proclaimed, in stagey terms:

> Our progress in space, taking giant steps for all mankind, is a tribute to American teamwork and excellence. Our finest minds in government, industry and academia have all pulled together. And we can be proud to say: We are first; we are the best; and we are so because we're free.

Echoes of Apollo

President Kennedy's Apollo decadal directive made spaceflight achievements a political prize of high value. Several more of his successors tried to replicate the magic of the grand Moonshot. On 20 July 1989, President George H.W. Bush, for one, announced a Space Exploration Initiative on the steps of the National Air and Space Museum, on the 20th anniversary of Apollo 11, calling for a vast project to build a space station, a return to the Moon, and the exploration of Mars, with estimates approaching $500 billion over time. "Like Columbus, we dream of distant shores we've not yet seen," he announced. "It is humanity's destiny to strive, to seek, to find. And because it is America's destiny to lead." President George W. Bush revived something of his father's plan in 2004, at NASA headquarters: the Vision for Space Exploration, defining the Constellation program, apportioning $100 billion to send the US back to the Moon by way of the Ares heavy-lift rocket and Orion spacecraft. As the younger President Bush exclaimed, "we do not know where this journey will end, yet we know this, human beings are headed into the cosmos." Stirring words all around, yet both ventures failed. Space policy can be exuberant, but also expensive, and therefore expendable. President Barack Obama, to the consternation of many NASA officials and aerospace executives, people deeply invested in Constellation, Ares and Orion, canceled the programs in April of 2010, in a cautionary speech at the Kennedy Space Center. He replaced it with several long-term, distant enterprises to take Americans to an asteroid (by 2025) or even to Mars (by 2040), in what became the Space Launch System (SLS) and a revived Orion Multi-Purpose Crew Vehicle.

Still, in the media culture, the Moonshot metaphor retains its allure, in the exaggerated promises to cure cancer, end poverty, or solve global warning. Conversely, a cadre of conspiracy theorists have created the Apollo "hoax" thesis, that the Moonshot never really happened at all.

Canada, Japan, and the European Space Agency joined space station "Freedom" in 1988, promising to contribute billions of dollars in hardware and support. The true surprise came in 1993, just after the collapse of the USSR, when the new Russian Federation agreed to join the project. The administration of President Bill Clinton offered to pay the Russian Space Agency $1 billion for its machinery and expertise rather than launch its own new Mir. Thus, the Soviet-era Proton rockets launched the "Zaria" (Dawn) initial propulsion unit (now storage), in November of 1998, an American purchase; and the "Zvezda" (Star) service and living module, in July of 2000. The ISS effectively became something of a "Mir 2" for the world. This was an appropriate and auspicious start, as the Russians had been the first to pioneer and master space station engineering.

The Russian modules were only the beginning. Over the coming years, along with six Proton and Soyuz payloads, some 34 missions of the American Space Shuttle built the ISS, using its payload bay and robotic arm, assisted by over 260 astronaut and cosmonaut spacewalks, all of which assembled the station with multiple trusses (over 300 feet in all), airlocks, thermal radiators, and some 240 feet of eight solar arrays. The Canadians added a robotic arm for the ISS itself, one of several to come. Among the 14 pressurized living and working modules (167.3 feet of them in all) are four major research modules: the Russian Nauka, the American "Destiny," the European-American "Columbus," and the Japanese-American "Kibō." Bigelow Aerospace tested a new kind of module beginning in 2016, an inflated one, in an experiment to study its structural integrity and inhabitability.

The ISS is truly international. With eight docking ports, eight visiting spacecraft can dock at any one time. Cargo and crew have been ferried, besides the Space Shuttle, via the Russian Progress and Soyuz, the Japanese H-II Transfer Vehicle (HTV), the European Automated Transfer Vehicle (ATV), the Northrop–Grumman Cygnus, the Boeing Starliner, and the SpaceX Dragon.

The ISS is the largest human-made structure in outer space and one of the most expensive structures ever built in human history. NASA publicists like to advertise the ISS as the size of a football field, no better a point of reference for American audiences. In truth, it is more like one of those new apartment-style campus residence suites, a dormitory pod of sorts with six sleep stations. An international venture by its spirit and letter, as for example the international protocols that define it, the US is the majority shareholder in the ISS "corporation," having invested over $120 billion in the building of the structure itself, which included the cost of maintaining the Space Shuttle missions that were so crucial to building the ISS.

Between 2000 and 2023, the ISS has been continuously occupied. 273 people visited, from 21 countries. It circles the Earth at 250 miles altitude every 90 minutes, at 5 miles per second. Fifty-two computers manage and control the operating systems. All of its power comes from the solar panels. The living environment is kept clean by air filters, "micro-impurity reactors," and "carbon-dioxide scrubbers." Purifying resin beds collect and filter water products. Life has improved remarkably since the days of the early stations. The spacefarers have their own sleep cabins and wireless access for entertainment and communications with home. A seven-window cupola module offers them a sunroom of sorts, partially designed and assembled in Italy. But it does not bring in the sun or stars. Like a bay window, it brings in the Earth, the glorious portrait views of Earth.

The spacefarers on the ISS experienced several milestones, as they made a wrench using a 3D printer; when they were able to eat a meal out of salad grown right on the station; and when they were able to drink potable water recycled from their own urine. NASA mission controllers raised a toast about the new drink, though the news reports did not mention what was in their glasses.

The food has only marginally improved: it is still either canned or bagged and still rather tasteless. Peanut butter and cheese whiz seem to be favorites.

As for life aboard, the American rule holds that at least two hours of strenuous exercise each day (on a treadmill and stationary bike) will keep bone and muscular loss to a minimum. Exercise is essential not just for the eventual return to Earth, but also for the proper functioning of the station itself. The spacewalks that repair the older and dysfunctional parts of the station can be laborious and exhausting. Space may be gravity-free, but it is not labor free. Working in microgravity environments, especially in bulky space suits in outer space, is always difficult and tiring. Since 2013, the station has seen air leaks, computer crashes, solar-array failures, and electrical shutdowns. The ISS, in other words, is becoming more and more like good old Mir as it ages.

Personal space on the ISS is abbreviated to be sure. Sleeping quarters amount to the boxlike dimensions of a small closet. Spacefarers can bring about three pounds of personal belongings. That is the sum total of privacy. There is no place to walk away or hide. Almost everything about life and work is scheduled and planned.

Some of the ISS spacefarers have been paid customers. In April 2001, the Los Angeles multi-millionaire executive and engineer, Dennis Tito, blasted into space aboard a Russian Soyuz, making him the first space tourist (if we do not count the Interkosmos tourists on Salyut). Tito spent nearly eight days in orbit during a mission to the ISS. He reportedly paid $20 million for his trip. The Soyuz rocket that delivered Tito also delivered vacuum-packed Pizza Hut personal pan pizzas to the crew. As the company's spokesperson put it,

> As a leader in the pizza delivery business, we're determined to give customers what they want, when they want it and where they want it, even if they are in space. Wherever there is life, there will be Pizza Hut pizza.

This also followed an earlier advertising stunt when Pizza Hut paid to have its logo painted on the Russian Proton rocket that delivered the Zvezda module in 2000.

Other space tourists have followed, people of extreme wealth and privilege, including a software developer, a computer game pioneer, and the inventor of Microsoft Office. The cost of a vacation on the ISS, courtesy of a private company, Space Adventures Orbital Spaceflight, and the Soyuz ferry, rose to $60 million and more.

The ISS was undoubtedly one of the trophy accomplishments of the US Space Shuttle, which NASA retired in 2011. At the time, each of the final flights of the Space Shuttle seemed all the more poignant, a seeming funereal drumbeat of America in decline, but really just another pause, another gap. For the next nine years, the Russian Soyuz booster was America's only access point to space and to the ISS.

Vertical Horizons: Above and Below Worlds

On 11 September 2001, 19 terrorists hijacked four passenger jets and turned three of America's ordinary airliners into deadly ballistic missiles. Two American Airlines Boeing 767s struck the North and South Towers of the World Trade Center in New York City and a third American Airlines Flight 77 slammed into the west side of the Pentagon in Arlington, Virginia. A fourth plane, United Flight 93, intended for the White House or US Capitol, crashed into a field in Pennsylvania after the passengers and flight attendants fought and overcame their hijackers. In a single day, 2,996 people were killed on American soil (125 military personnel and civilians at the Pentagon; 246 passengers and crew on the four planes; 2,606 people at the World Trade Center, including 400 first responders and the 19 terrorists). In the immediate aftermath of that day, Americans felt a sense of shared grief and unity. The events of "September 11" also had profound effects on the history of the Space Age.

One American had a unique perspective on these events, NASA astronaut, Frank Culbertson, who witnessed the attacks from the International Space Station. From 250 miles above Earth, Culbertson and two Russian cosmonauts viewed a huge column of smoke billowing out of New York City at the site of the World Trade Center. "I didn't know exactly what was happening, but I knew it was really bad because there was a big cloud of debris covering Manhattan." When he learned the news from Mission Control, he felt particularly isolated. "That's when it really became painful because it was like seeing a wound in the side of your country, of your family, your friends." Culbertson was able to film video and take photographs of the attack from space. He also saw the effects of the Federal Aviation Administration's prohibition of commercial aviation across the US for three days, starting on September 11. Culbertson was startled when he noticed that

> all of the contrails that are normally crisscrossing the United States has disappeared … except for one airplane that was leaving a contrail from the central US toward Washington, DC, and that was Air Force One headed back to DC with President Bush. It was a very sobering time for us.

The large plume of smoke and debris witnessed from space came from the collapse of the Twin Towers, of course, each at 1,360 feet with 110 floors. They released an enormous amount of energy when they came crashing down. The Towers contained more than 200,000 tons of steel, 425,000 cubic yards of concrete, and 600,000 square feet of glass (in 43,000 windows).

Each story had a floor pad weighing around 4.8 million pounds. Their collapse was a result of the intense heat generated from thousands of gallons of burning jet fuel, which weakened and buckled the building's steel framework, causing the upper floors to collapse, igniting a chain reaction, borne down by the force of gravity. The combined weight of each floor pancaking on to the one below grew so great that the steel and concrete became increasingly pulverized. The total energy released from the destruction, considering the weight of the planes, their speed at impact, the jet fuel and the mass of the building, was equivalent to 600 tons of TNT, or one-twentieth of that released by the 10-kiloton atom bomb dropped on Hiroshima.

Most of the energy came from the gravitational collapse. The large plume of dust and debris was pushed up, by both the air released from inside the buildings and by a powerful downdraft as the debris hit the ground, forcing more air up. This dust cloud spread over Manhattan at 50 miles an hour, depositing significant debris more than a half-mile away and smaller amounts up to 2 miles away. The destruction of the Towers also left an unsettling gap in the cityscape, and in the economic fabric of the city itself; as well as gaping holes in the ground where the two towers once stood. For a time, starting in 2002, the space in the cityscape has been filled by two ghostly beams of light, known as "The Tribute in Light," an installation of 88 searchlights, creating two vertical columns reaching into space, representing the Twin Towers. In 2005, two memorial pools and waterfalls (known as "Reflecting Absence," each 50 feet deep) were constructed within the gaping footprints. And in May 2014, the 9/11 Memorial Museum opened at Manhattan's bedrock, with 10,000 objects, 23,000 images, and 500 hours of film; taking visitors 70 feet down on gentle ramps to the very epicenter of what used to be "the pit."

The events of "September 11" irrevocably transformed America's national priorities, away from a potential mission to Mars, toward a series of painful wars in the Middle East, with high human and material costs. As once in the 1990s enthusiasts had called for $30 billion to take the US to Mars (by 2001) now the US invested $437 billion for the Afghanistan and Iraq wars between 2001 and 2006 alone. Total US spending on war between 2001 and 2019 approached $5 trillion. The Pentagon justified these wars under the new geopolitical rubric of the "Core and Gap," meaning the defense needs of the industrial trans-Atlantic economies against the unstable countries of the mostly equatorial gap, rich with energy resources, but isolated with debt and politically disenfranchised. This rubric was not really that new, just a version of the older Great Game and Cold War containment policies, of heartlands and rimlands, still focused on the projection of power on Earth, if now fortified by the missile and satellite assets of space power.

Select Bibliography

Barnett, Thomas P.M. *The Pentagon's New Map: War and Peace in the Twenty-First Century*. New York: Putnam, 2004.

Baucom, Donald R. *The Origins of SDI, 1944–1983*. Lawrence: University of Kansas, 1992.

Benjamin, Marina. *Rocket Dreams: How the Space Age Shaped Our World Beyond*. New York: Free Press, 2003.

Brannigan, Michael. "America's First Cosmonauts: Reflections on the Human Cost of Shuttle-Mir," *Astropolitics* 15/1 (January 2017): 27–50.

Burrough, Bryan. *Dragonfly: NASA and the Crisis aboard Mir*. New York: Harper Collins, 1998.

Chugunova, Nina. "Cosmonauts Number Zero," *Bulletin of the Atomic Scientists* 50/3 (1994): 16–23.

Eisenhower, Susan. *Partners in Space: US–Russian Cooperation after the Cold War*. Washington, DC: Eisenhower Institute, 2004.

Geppert, Alexander C.T., et al., ed. *Militarizing Outer Space: Astroculture, Dystopia and the Cold War*. London: Palgrave Macmillan, 2021

Hadfield, Chris. *An Astronaut's Guide to Life on Earth*. New York: Back Bay Books, 2015.

Harland, David. *The Story of the Space Shuttle*. New York: Springer, 2004.

Harland, David. *The Story of Space Station Mir*. Chichester: Praxis, 2005.

Harvey, Brian. *China in Space: The Great Leap Forward*. New York: Springer-Praxis, 2013.

Hendrickx, Bart, and Bert Vis. *Energiya-Buran: The Soviet Space Shuttle*. New York: Springer, 2007.

Heppenheimer, T.A. *The Space Shuttle Decision*. Washington, DC: NASA, 1999.

Houston, Rick. *Wheels Stop: The Tragedies and Triumphs of the Space Shuttle Program, 1986–2011*. Lincoln: University of Nebraska Press, 2013

Inboden, William. *The Peacemaker: Ronald Reagan, the Cold War, and the World on the Brink*. New York: Dutton, 2022

Klerkx, Greg. *Lost in Space: The Fall of NASA and the Dream of a New Space Age*. New York: Vintage, 2005.

Linenger, Jerry M. *Off the Planet: Surviving Five Perilous Months aboard the Space Station Mir*. New York: McGraw-Hill, 2000.

Logsdon, John. *Ronald Reagan and the Space Frontier*. New York: Palgrave, 2019.

MacKenzie, Donald. *Inventing Accuracy: A Historical Sociology of Nuclear Missile Guidance*. Cambridge: MIT Press, 1993.

Neal, Valerie. *Spaceflight in the Shuttle Era and Beyond: Redefining Humanity's Purpose in Space*. New Haven: Yale University Press, 2017.

Oberg, James. *Star-Crossed Orbits: Inside the US-Russian Space Alliance*. New York: McGraw-Hill, 2002.

Pakkanen, Saadia. *In Defense of Japan: From the Market to the Military in Space Policy*. Stanford: Stanford University Press, 2010.

Pelevin, Victor. *Omon Ra*. Trans. Andrew Bromfield. New York: New Directions, 1994.

Portree, David. *Mir Hardware Heritage*. Houston: Lyndon B. Johnson Space Center, 1995.

Reiss, Edward. *The Strategic Defense Initiative*. New York: Cambridge University Press, 1992.
Schell, Jonathan. *The Fate of the Earth*. New York: Avon, 1982.
Schweizer, Peter. *Reagan's War: The Epic Story of His Forty-Year Struggle and Final Triumph over Communism*. New York: Doubleday, 2002.
Seedhouse, Erik. *Astronauts for Hire: The Emergence of a Commercial Astronaut Corps*. New York: Springer, 2012.
Suzuki, Kazuto. *Policy Logics and Institutions of European Space Collaboration*. London: Routledge, 2003.
Wanjek, Christopher. *Spacefarers: How Humans Will Settle the Moon, Mars, and Beyond*. Cambridge: Harvard University Press, 2020.
Zabusky, Stacia E. *Launching Europe: An Ethnography of European Cooperation in Space Science*. Princeton: Princeton University Press, 1995.

For more on the collapse of the USSR, see Mark Harrison, "How Much Did the Soviets Really Spend on Defence? New Evidence from the Close of the Brezhnev Era." *Warwick Economic Research Papers* (2003): 1–39. Although the numbers are notoriously hard to calculate, Harrison has estimated that roughly 20% of Soviet Gross Domestic Product (GDP) went to defense. Economist Konstantin Sonin marks it even higher, as much as 40%. By comparison, outside the Korean and Vietnam Wars, in 1985 US defense spending was only 5.7% of GDP.

Epilogue

Someday this epilogue, or distinct parts of it, will become final chapters all their own. Future history will create new books, new chapters, and new conclusions. Such is the fast-paced and vanguard nature of the Space Age. Its achievements and attractions always draw us forward. I hope it will become a long book of many chapters. For now, we will draw a circle back to where we started, to our interpreted past rather than the inscrutable present, to history rather than to publicity. Current newsworthy topics like space elevators or space sails, nanosatellites and nuclear propulsion, belong more to the future than the past, to chapters yet written.

Thus far, we have centered our historical survey on the Cold War in space. Its rivalries drove almost all of the major spaceflight achievements of the last 60 years. Competition was the rule even well before the Cold War, as we discovered in the contest between the rocket pioneers for national priority, or in the pursuit of the world's first ballistic missiles, just before and during the Second World War. Competition remains the rule even today, although in new and potentially more promising ways.

Because of the complexities and costs of rocket science and spacecraft engineering, only three nations have sent human beings into space so far: the US and the USSR (Russian Federation), as well as the People's Republic of China.

By far, the US continued to invest most of its national treasury and political capital in spaceflight exploration. In 2014, for example, the investment of the US in its civil space budget amounted to $16.5 billion, nearly as much as the other spacefaring nations combined: the European Space Agency at $5.6 billion, Russia at $4.4 billion, Japan at $3.8 billion, China at $3 billion, and India at $1.5 billion. These numbers have grown steadily and proportionately ever since, though also with a significant measure of increased Chinese spending. For the US, NASA remains a rather modest investment, usually receiving just a few billion dollars a year more than the Department of Transportation, quite an appropriate measure of what has become expected and routine.

On the one hand, the US may seem singularly foolish over the years for having relinquished so many of its premier spaceflight programs, squandering its investments in Gemini, Apollo, and the Space Shuttle, dispensing with them for the better or the newer. This was the rule of the project system, the cycle of innovation and obsolescence. Still, these superb spacefaring systems achieved thresholds of rendezvous and docking, lunar exploration, and orbital architectures. Their legacies live on, namely in NASA's Space Launch System, which marries the best of the Apollo program, as, for example, its liquid hydrogen and liquid oxygen engines, and the best of the Space Shuttle program, as in its main engines and solid rocket boosters. The Artemis Accords (2020), encompassing a broad and democratic international coalition, and with plans for a crewed mission to the Moon and Mars, also extend the international collaborations of the Space Shuttle and ISS. The success of NASA's Artemis rocket in December 2022, when the uncrewed Orion spacecraft achieved a circumlunar trajectory, offered a perfect moment to remember another lunar event, the last of the Apollo series 50 years before, Apollo 17 in December 1972.

On the other hand, the USSR and its successor state, the Russian Federation, seem doubly wise with their long-term investments in the tenacious Soyuz booster and spacecraft, and the Proton launch system (mostly for satellites). Between 1957 and 1998, for example, of their 4,655 successful space launches, 66% were Russian and only 34% were American. Even at the collapse of the USSR, between 1990 and 1994, the Russians launched twice as many space missions as the US. Most dramatically, between 2011 and 2020, by government contract, Soyuz became America's sole crewed launch platform for space. NASA paid the Russian Federation $753 million for the initial two-year contract (at about $63 million a seat). Here was an era of international cooperation in space recalling Apollo–Soyuz, the Shuttle–Mir, and the building of the ISS.

As we have surveyed all along, these questions are less about which country was better or worse. They are more questions about differences. The political, cultural, and technological differences that have sometimes divided but have also united the enterprise of human spaceflight. The history of the Space Age shows how each country's unique approaches and achievements can contribute to the common good. One image may represent that singular human factor best: Leonardo da Vinci's Vitruvian man, our point of departure in Chapter 1. We have seen it repeated and adapted several times, as in the metallized human form on the front cover of Walter Hohmann's *The Attainability of Heavenly Bodies*; and in the space patch of Voskhod 2, with cosmonaut Aleksei Leonov in his spacewalk, flying horizontal.

By 2011, Vitruvian man appeared across cultures as a universal standard for the human dream of spaceflight. An exhibit in the central hall of the Museum of Cosmonautics in Moscow presented a larger-than-life statue of Yuri Gagarin in the Vitruvian pose, against a background of the Earth in a stained-glass mosaic. The aerospace industrial firm, Hamilton Sundstrand, stitched the image into a space patch to honor the Space Shuttle astronauts who built the ISS. The Belarusian rock group, Liapis Trubetskoi, featured one of its lead members in a Vitruvian pose, for the front cover of its album, *Funny Pictures / Veselye Kartinki* (2011). The lines of one of its songs, "Cosmonauts / *Kosmonavty*," compared us to new little planets among the large ones, on a path to immortality, as Fedorov and Tsiolkovsky once promised.

Gravity Breaches

We earthlings can actually experience the vicarious frill of spaceflight right here on earth. The National Aerospace Research and Training Center in Southampton, Pennsylvania, part of the Environmental Tectonics Corporation, has offered the ATFS-400–25 human centrifuge to train candidates for commercial human spaceflight, taking them up to 15 g-forces for sub-orbital, orbital, and advanced spaceflight modes. The centrifuge will "simulate any phase of air or spaceflight with acceleration, vibration, fully controllable axes with continuous 360° pitch and roll, and real-world displays and controls." Looking for more of a challenge? Try a weightless flight on a modified Boeing 727. The Zero Gravity Corporation will take you to an altitude of 32,000 feet. Then the plane will do a series of aerobatic maneuvers known as parabolas to achieve weightlessness at intervals, turning you into a parabola all your own, as its logo attests. Beginning in 2017, MIT students created the Telemetron musical instrument for some of the flights, a transparent 12-sided box, with chimes and gyroscopes that "emit their telemetry as they spin and collide," randomly creating their own sound compositions, a "poetics of zero gravity" and perhaps even a foundation for a new space culture. How about astronaut training at the US Space and Rocket Center's Adult Space Academy in Huntsville, Alabama? It has offered weekend programs and a ride on the one-sixth gravity device, a chair suspended from the ceiling that gives you a very bumpy ride (you can pretend you are jumping over moon craters). They also have a "Multi-Axis Trainer" which twists and turns, allowing you to feel what astronauts experience on re-entry.

The People's Republic of China entered the Space Age rather late, if with a kind of vengeance. Its Communist Party-state and military forces launched human beings into orbit and built a small space station of their own beginning in 2003. China first adapted its spacecraft processes and machinery from the Russian space program through the 1990s, improving the Soyuz models for its own program. Yang Liwei became the first Chinese "space traveler" (*taikonaut*), orbiting the Earth in the "Divine Vessel" (Shenzou 5). The Chinese have followed upon this success with spacewalks, three-person crews, and the assembly of several space laboratories and stations in orbit, as well as orbiters and landers to explore the Moon and Mars. These feats followed along the well-trodden paths of prestige-driven Russian and American spacefaring achievements. China has entered spaceflight history.

China's own space firsts defined a new threshold in the history of the Space Age: another break away from the locked pairing of the US and USSR. These national Chinese firsts also illustrate how countries on earth continue to see outer space as a realm of status and prestige, largely still bankrolled from their own treasuries; how they mark space as an extension of their own political and military realms, in China's case the Belt and Road Initiatives, and its incursions into the South China Sea. China's closed and secretive space programs are inseparable from its geopolitical ambitions here on earth, and from its party-state authoritarian regime. This is a regime that has denied the sovereignty of the peoples of Tibet and of the Uyghurs of East Turkestan, that has extinguished democracy in Hong Kong, that threatens independent and democratic Taiwan, and that has leveraged the military power of its client state, North Korea, against the US. The interwoven realms of rockets and rimlands still matter.

Besides the established precedents of Europe and Japan in raising satellites and interplanetary probes, exploratory spaceflight is becoming ever more international. India, Israel, Ukraine, South Korea, North Korea, and Iran have all proven their satellite and orbital launch capabilities. Spaceflight has diversified, especially among the democratic states. India has cooperated with the Russian Federation, but especially with NASA and US enterprises and universities, to launch space probes, including the Chandrayaan lunar probes, landers and rovers after 2008, one of which was first to land in the southern polar region; and after 2014 the Mars Orbiter Mission to study the Martian atmosphere and planetary features. India has developed its satellite capacities with regular launches and state funding. Another free-market, democratic state, South Korea is joining the pace with its expertise in robotics and advanced technologies.

These kinds of spaceflight events are becoming less and less like spectacular historical achievements, more like simpler news events. The Space Age is transforming into something akin to the Industrial Age of old, something that surrounds us rather than surprises us. Someday spaceflight might just disappear as a normal part of our lives. We will simply stop talking about it. The general public knows this truth. People may have short attention spans. They may be gullible for media hype. But there is also a common, everyday wisdom in how people lose interest in spaceflight, as in the cases of both the Apollo and Shuttle programs. People stopped attending launches or watching missions on TV. They went on with their lives.

The initiatives that we have covered always returned us back to Earth. Satellite capabilities, as the Sputnik event made manifest, also mean launch systems such as ballistic missiles, the delivery systems for nuclear weapons. After 2006, the US allies in India and South Krea developed IRBM weapons. So did China's client state North Korea, with the potential for its Hwasong-17 ICBM to launch a nuclear warhead against the US. Russia's new client state, Iran, is developing longer-range ballistic missiles as well. The proliferation of nuclear weapons and their delivery systems, along with a tenuous global regime of threat and deterrence, is becoming the rule.

In answer to the ballistic missile reach of rogue states like North Korea and Iran, as well as those of its geostrategic rivals like the Russian Federation and China, the US retooled its Strategic Defense Initiative into a variety of Ballistic Missile Defense (BMD) networks, parts of which have been adapted by both its allies and enemies. In 2019, the US Air Force also achieved a signal first when it successfully destroyed an ICBM in flight with an interceptor (using the Ground-based Midcourse Defense). It has redoubled its efforts to perfect military surveillance and anti satellite capabilities in space. In an epilogue to the Space Shuttle, the US Air Force has flown its highly secretive reusable military shuttle, the drone X-37, to serve these purposes; and has morphed into the new Space Force (2020), by order of President Donald Trump.

The US aerospace industries have also transformed in response to the new battle lines of cyberspace and outer space. Two mergers in 2023 offer examples, when the RTX Corporation formed out of Raytheon (Raytheon Intelligence and Space, and Raytheon Missiles and Defense), Pratt and Whitney, and Collins Aerospace; and when the L3Harris corporation, with its specialty in defense information technology, purchased the great Aerojet Rocketdyne company (for $4.7 billion). Information systems have become as big and rich, if not more so, than the old aerospace models.

Satellites have become a new battlefield of geopolitical competition, one not yet covered by international agreements. The US and USSR tested both ground- and air-based anti-satellite weapons throughout the Cold War, with recent tests of anti-satellite missiles (and R&D of new directed-energy weapons) by the US in 2008 and by Russia after 2021. China has also tested anti-satellite weapons since 2007, and is preparing a variety of military space platforms, indicative of the geopolitical imperatives at the center of all its spaceflight initiatives. This too is nothing new in history. As with its other series of firsts, Chinas has simply been joining a military race for superiority already long in play. Both India and Israel have joined this race for advantage.

The Russo–Ukraine War (2014–2024) reveals the continuing relevance of geopolitics and war, the sometimes open and violent battle for rockets and for rimlands, for global power and for aerospace reach. The Russian Federation invaded Ukraine to accomplish a broad package of goals. Among these was to destroy a rival, in a largely Russian-speaking, democratic "non-Russia" at its borders, as the Kremlin rationalized. To guarantee access to the Black Sea and its littorals, as well as break the rising alliance of democratic states reaching from the Baltic to the Black Seas (the "Intermarium"). To dominate Ukraine, a recent imperial possession rich in intellectual talent, natural resources (grain and coal), and in energy and industrial sites. And also, to seize Ukraine's significant engineering R&D and spaceflight assets at Dnipro, just as they were forming new partnerships with the US and Europe. It is worth remembering just how central a role Ukraine played in Russian spaceflight achievements, as in the Ukrainian backgrounds of Sergei Korolev and Valentin Glushko, and the design bureau and plants of Mikahil Yangel at Dnipro. There, and at places like Kharkiv and Zaporizhzhia, Ukraine's defense and aerospace assets accounted for up to 30% of total Soviet production. The Russian invasion was a matter of resentment and irredentism at their loss.

The spaceflight dimensions of the Russian invasion are manifest. On the one side, US satellite surveillance and SpaceX Starlink satellites have given Ukraine some leverage in military targeting and communications. US Patriot Defense Radars and Missiles have defended Ukrainian cities from Russian terror bombings. NASA's Fire Information for Resource Management System (FIRMS) has offered open-access observation of the heat signatures of fighting on the front line. On the other side, Putin has threatened the West, brandishing his new missiles (like the Zircon and Sarmat), warning that further support for Ukraine will lead to nuclear war. From the new Vostochnyi Cosmodrome in Siberia, against a backdrop of Russian spacecraft assets, Putin and his deputy Dmitrii Medvedev made these threats, and hosted their allies in the terror war against Ukraine, the dictators Alexander Lukashenko of Belarus and Kim Jong Un of North Korea. Space still mattered. Only in this case to leverage the symbols of spaceflight so as to destroy a neighboring people.

For the Russian Federation, space power was inseparable from the war. Dmitry Rogozin, after serving as director of *Roskosmos* (2018–2022), appeared in the Russian-occupied region of the Donbas in 2022, heading a military-political occupation initiative (his call sign was "Cosmos") before he suffered life-threatening injuries. In a rank violation of its peaceful, international spirit, in July 2022 Russian cosmonauts on the ISS even performed a propaganda stunt in support of the war, justifying the occupation and destruction of the Ukrainian Luhansk region. The Putin regime has also promoted a new aerospace brutalism and monumentalism in art and design, as in the poster, "Starships" (2019) by artist Alexei Belyaev-Gintovt, the image of cyborg humans conquering the planet and outer space, red stars and Russian imperial pride leading the way. For all these dreams, the Russian space program is in a state of stagnation. The country's economy remains decidedly earthbound, or better yet underground, propped up by the exports of extracted crude oil and refined petroleum, and by the import of high-technology products from the West and China.

From the very start of the full invasion in January of 2022, the Russian Aerospace Forces have launched the country's cruise and ballistic missiles in a campaign of terror bombings, targeting Ukrainian cities and civilians. Some of these missiles were the very ones Ukraine transferred to Russia after the Budapest Memorandum (1994) when it relinquished its nuclear weapons and their delivery systems after Russia promised to "refrain from the threat or use of force against the territorial integrity or political independence of Ukraine." The Russian anti-war singer, Vladi, caught the irony and tragedy in his song, "How the F--- is this Possible / *Kak eto bliad' vozmozhno*" (2022), raging against the Russian Federation's use of its aerospace resources, the heritage of Baikonur, as terror weapons against Ukrainian cities and people.

The war has reshaped and corrupted domestic politics in the Russian Federation. The Putin regime has transformed into a neo-Stalinist, retro-Soviet state, surviving on censorship and propaganda, beatings and arrests, persecution and imprisonments; propped up by the Federal Protective Service, the Federal Security Service, and a variety of internal security and police forces. Young people call them the "cosmonauts," for their protective helmets and armored bubble gear. They look the part. The rock singer Ivan Alekseev (Noise MC) recognized them in his anti-war song, "Land of Rains / *Strana dozhdei*" (2022). How could any normal person, he asked, unarmed and afraid, possibly fight against them?

Rocket Metaphors

In the US, rockets have been a favorite metaphor in popular culture over the twentieth century, featured in science-fiction novels and films, and even in consumer brands. The Nike "swoosh" (1971) may have actually started the trend, back in the Apollo era, as an emblem designed to express "movement," the sweep of parabolic speed. In recent years, especially since the retirement of the Space Shuttle in 2011, rockets have become a dominant symbol in American advertising campaigns. The rocket became a metaphor for us, as in its simple pairings with Girl Scouts or middle school students, to signify aspiration, launch, and success. Rocket Math promised that. Entrepreneurs adapted the image of the rocket into an app, a link to their discrete services. We call them "icons." The rocket as internet icon. The rocket became the small emblem to launch your new web page via Bluehost, or launch your materials to storage with vCloud, or to send your files to your Android phone via Firefox, or to educate yourself with daily livestreams via Adobe Live. Digital media seems to be the mind's dream of becoming like a rocket.

The rocket became a metaphor for accessibility, ease of use, and most of all speed. Television commercials, by Verizon's DSL (Digital Subscriber Live) and Xfinity mobile's 4G LTE, made these appeals. Forget the Moon or Mars. Buy their access and bring outer space down to earth, make all things weightless, within reach. All you need is to swipe your finger, or press a key, and hear the ping or ding, to find a date, or buy a car, or get delivery, or rent a ride, or photoshop your image, or "buy anything anywhere." We have become, as Rachel Plotnick has argued, the masters of the "digital command."

On a lighter note, the rocket has also helped launch the Mountain Dew Kick Start drink, an explosion of flavors; and a new variety of cannabis seed, called Rocket Fuel. Musicians and artists have celebrated this new trend. As, for example, the Electronic Dance Music and DJ sounds of Lindstrom, in *I Feel Space* (2005), with a cosmic beat of both ascent and drift. Or the artwork of Benjamon Edwards's *Immersion* (2004), his canvass of a hyper-settled earth landscape ascending into an aerial architecture. Or Julie Mehretu's, *Stadia II* (2004), of all the earth's things rising into a whirlwind of air and space, the shrinking Earth below. Hugo Rami Perez's, *Uncertain Future* (2021), even launches our cities and highways by parabolic trajectory into Earth orbit. Civilization becomes the rocket: space bound.

Both China and the Russian Federation, with their few allies in North Korea, Cuba, and Iran, pose some formidable competition to the US. Yet the US has matched China's breakthrough into space with an even more transformative development: "NewSpace," a term coined by the Space Frontier Foundation in 2006 to promote the private commercialization of outer space, especially in reusability and lower payload coasts. The Reagan administration first empowered privatization, in the Commercial Space Launch Act (1984) and in the platforms of the National Space Policy program. A uniquely American phenomenon, the internet mogul, helped do the rest.

Paul Allen, a co-founder of Microsoft, was one of the first. In 2004, he helped to finance Burt Rutan's "SpaceShipOne," built by Scaled Composites. The investment paid off. Rutan won the $10 million Ansari X Prize when his sub-orbital space plane, raised to altitude by a mother ship, launched by rocket spaceward, flying above 328,000 feet into microgravity, twice in ten days. Recognizing the new potential of these projects, the US Congress enacted further legislation in 2004 to encourage NASA "to promote the development of the emerging human space flight industry." This meant co-operating with a variety of new commercial ventures, all dating from around the turn of the century and directed by internet pioneers. Among them were Allen's Vulcan Aerospace, Richard Branson's Virgin Galactic, Jeff Bezos's Blue Origin, and Elon Musk's SpaceX. Together they emphasized reliability, reusability, and profitability. Each was a private company, not a country. Spaceflight, in essence, need no longer depend wholly on either government funding or its national-political imperatives. Free markets mattered. Thanks to these new enterprises, since 2019 the US has accounted for over half of all spacecraft launches worldwide and double the numbers of Russia and China combined.

These commercial ventures are currently focused either on sub-orbital flights, with brief stints in microgravity for space tourism; or on near-term launches of cargo and human beings into Earth orbit, especially in order to beat the rather expensive payload costs of the Shuttle, Soyuz, and Proton, among others. Each has had its share of successes and failures, including rocket explosions and one fatality, the death of co-pilot Michael Alsbury in a 2014 crash. Virgin Galactic initially charged $250,000 for its winged sub-orbital tourist flights, based on the SpaceShipOne model, and achieving its first successful crewed launch in 2021. XCOR Aerospace, its prime competitor, had hoped to beat this price with its own reusable shuttle-like launch vehicle (for a mere $95,000), though the project failed by 2017. Bridging these models, and competing for the lucrative orbital space launch market, was Allen's Stratolaunch Systems, designed to launch a rocket from a high altitude, raised by a six-engine, dual-bodied, jet aircraft, all to free the system from weather problems and fixed launch sites. But Stratolaunch failed too. Bezos's Blue Origins took a more measured, paced approach that includes both reusable sub orbital flights (with its first crewed launch in 2021) and orbital flights. It has also been the most secretive, maybe not surprising given some of the monopoly practices of its founder at Amazon.com.

The most promising of these new spaceflight companies has been SpaceX, Space Exploration Technologies Corporation, established at Hawthorne California, first situated in the hanger of an old factory that once made Boeing 747s. Elon Musk, an immigrant from South Africa, founded the company in 2002 from the sale of his PayPal to eBay (for $1.5 billion), and with an initial investment of $500 million of his own money. Among SpaceX's successes since 2010 have been the launch of the Falcon 9 rocket, the orbiting and return of its reusable Dragon spacecraft, and the vertical landing of its first stage Falcon 9 back to earth (for re-use) by both land and sea. Space X has launched satellites (including the remarkable Starlink network), has resupplied the ISS with cargo, and has even sent the cremated remains of hundreds of people into space, including those of Gordon Cooper, one of the original Mercury Seven astronauts, and those of James Doohan, the actor who played Scotty on the Star Trek series. The company Celestis helped pioneer this business of sending human remains into space in 1994. The USSR laid the cremated remains of its heroes in the bricks of the Kremlin Wall. The US launches its into space.

In May of 2020, Space X achieved a milestone when it sent two NASA astronauts, Bob Behnken and Doug Hurley, to the ISS aboard a Dragon capsule. The pause in US-sponsored spaceflights, from its own soil, was over. Over 10 million people watched the event on a variety of media platforms, breaking NASA internet viewership records. SpaceX crewed launch schedules had begun, by 2023 publicized and promoted on the old platform Twitter, purchased by Musk and renamed "X." The formal marriage of a spaceflight and media company was finally complete. Falcon and Starship may be becoming the franchises Apollo never did.

These commercial successes are not as radically new as they may seem. Once SpaceX proved the reliability of its techniques and machinery, NASA welcomed the new venture into its established "project system" of aerospace contractors. SpaceX now competes with such outfits as Northrop Grumman and Boeing for billions of dollars in government and military contracts for the resupply of the ISS and for the development of continuing orbital and exploratory platforms. There is a difference. SpaceX spurns the old model of "cost plus contracting," with its emphasis on higher profits, overhead, and bureaucracy. In its place: the streamlined and quick efficiency of fixed-price contracts, and a culture of ingenuity and innovation, for reusability and lower payload costs.

Spaceflight reached a threshold year in 2021. For the first time, there were 19 people in space. Russia launched the Japanese billionaire and space tourist (and creator of an online shopping service), Yusaku Maezawa, along with his filmmaker, aboard the Soyuz to the ISS, for around $80 million. China launched two crewed missions, Shenzou 12 and 13, and continued building its own space station. SpaceX flew two four-person international crews for NASA to the ISS (for about $55 million a seat). It flew its own four-person civilian crew to orbit (a first), and that mission actually raised $243 million in charity funding for the St. Jude Children's Research Hospital. Virgin Galactic flew two SpaceShipTwo flights into sub-orbit for the first time, including its founder Branson. The cost: $450,000 a seat. Blue Origin made its first three crewed launches into sub-orbit, which included Bezos himself and *Star Trek* actor, William Shatner. Unlike Bezos's Amazon, Blue Origin did not reveal its prices.

All these new spacefaring countries and firms have contributed to the remarkable expansion of a global space industry (mostly satellite communications), with nearly $500 billion in annual spending as of 2023. The US is the major player in state and private terms, with the highest revenues (approach 44% globally). The assets of the global industry are mostly in space, as satellites, but also include a growing number of spaceports around the world, in places like the space centers at Tanegashima (Japan), Satish Dhawan, Andhra Pradesh (India), and Naro, South Jeolla (South Korea); or new spaceports at Wallops Island (Virginia), and at Corn Ranch and Boca Chica (Texas). This textbook thus wraps up not that far from where we began, grounded on earth, surveying the peaceful exploration of space by rocket power.

The whole story so far has comprised a series of dramatic crossroads, moments of either creation or destruction, conflict or cooperation, intense achievements or frustrating pauses. This history of the Space Age has been one of firsts and seconds, starts and stops, gaps and breakthroughs, windfalls and shortfalls, liftoffs and spinoffs, tipping points and breaking points. More are sure to come.

Although we may need to take more care with our predictions. In 1985, NASA promised a US Moon base by 2010. In 1997, it predicted a first crewed mission to Mars by 2009. In 2016, Elon Musk predicted that SpaceX would send human beings to Mars by 2025. The future is not always as sure as it seems.

Transhumanism

We may end up solving the dilemmas of human physiology and psychology in spaceflight right here on earth, in our own self-evolution. We look for other Earths and for life beyond, but too often ignore how we are already transforming ourselves with a dizzying array of new technologies. Forget the cyborgs for space. We are already bound to become cyborgs right here. Thanks to artificial intelligence (AI), quantum computing, synthetic biology, neural implants and genetic engineering. We call all this transhumanism, the movement to seize control of human evolution for total progress and perfection, even immortality. Julian Huxley helped coin the term in 1957, Sputnik's year, emboldened by the "scientific and technical explorations" of the Space Age, seeing a new "cosmic self-awareness" and hoping for a "threshold of a new kind of existence."

But the promise of a new humanity was not just in space. Feminist scholars have argued that we have never been, nor ought to be, traditional or stable human men and women. Donna Haraway launched the "Manifesto for Cyborgs" in 1985, advocating for a new conception of the human being, to mix the male and female, animal and human, and human and machine. Katherine Hayles concurred, surveying the history and literature of these identities since the Enlightenment, finding all the fault lines and deceptions in the liberal conceptions of human freedom, personality, and dignity. Nothing, they argued, is really that sure anymore.

Recently branded as "Humanity Plus" (H+), transhumanism's advocates (among them Nick Bostrom, Hans Moravec, and Marvin Minsky) sell their trendy ideas on web pages, in books, and in journals like *Futures*, *The New Bioethics*, and the *Journal of Posthuman Studies*. Roger Launius, the former Chief Historian at NASA, and once director of the Space History Office at the National Air and Space Museum, has promoted the transhumanist option, predicting that humanity is bound to lose itself, become a mostly machine being readied for space travel.

In the best case, perhaps we will create AI-smart robots for distant space travel, where the human body is too frail, where robots are robust and replaceable, and can see and hear, touch and gather for us. Martin Rees, Donald Goldsmith, and Louis Friedman have advocated for this. In the worst case, maybe we will become those very robots, as cyborgs. In the very worst case, we will become the slaves of the higher overlords of AI intelligence. Amid all of these possibilities, what will be left of our human bodies and minds, the mortal coils that give meaning and purpose to our short lives, and that give promise to the generations that follow? Will outer space become more a realm of slavery than freedom? Critics like Bill Joy and David Deudney warn that it will. People at the heart of AI or in the know, from Eric Schmidt to Sam Altman, warn that it might extinguish human beings altogether. None of this is happening by surprise or by force. Some of these changes are happening in subtle, everyday ways, as average people make choices about pharmaceuticals, body modifications, and family planning. As Juan Enriquez and Steve Gullans have argued, we are already "evolving ourselves."

We began this textbook survey with a question. When did the Space Age begin? We found multiple answers, matching with the inventors and innovators at the start, people like Konstantin Tsiolkovsky and Robert Esnault-Pelterie, Robert Goddard and Hermann Oberth. We also found some rather problematic origins, as with the rise of the Stalinist rocket state, along with all its persecutions and terror; with the Nazi rocket teams under Dr. Wernher von Braun, driven by military conquest and slave labor; or with the missile race between the US and USSR that threatened the globe with nuclear annihilation.

As we conclude, let us pose the corollary question. When did the Space Age end? That is, if indeed it has. It might seem an odd question to ask. For some historical perspective, let us turn our attention to events here on earth. Perhaps the Space Age ended with the collapse of the USSR in 1991, brought on in part by Ronald Reagan's dramatic military spending, which included SDI. Maybe it ended with the terrorist attacks of September 2001, the catalyst for the wars in Afghanistan and Iraq that have consumed the many billions of dollars that we might have otherwise spent on human missions to Mars. Or perhaps the Space Age ended with the last flight of the Space Shuttle Atlantis in July of 2011, after 30 years and several hundreds of billions of dollars in investments, a Space Shuttle program that had grown far too expensive to continue. The lesson of these endings is that humanity's ventures into space have always been stubbornly grounded in our military, or political, or economic imperatives back on earth.

None of these terminal points seems quite right. For in spaceflight, as in many human endeavors, every end is also some kind of beginning. In the case of spaceflight, the historical tension draws us into the future, toward our far, vertical horizons. Space, in this sense, is overdetermined. It becomes the singular hope of the "long term." Through science fiction and popular science, our news and blogs, we may place too many dreams in it, stake our progress upon it, calibrate our reach for eternity and infinity through it. Space remains one of our most perfect utopias, the ultimate beyond, our highest calling and greatest good, a frontier of insatiable frontiers. That is why we have begun to send out probes to the planets, why we have briefly explored the Moon, why we plan still more.

The human race seems programmed for travel and migration, predisposed to move about the continents and planet, and ultimately to other planets.

However inspiring, we still ought to take caution with the long term, with the utopian vision that counts the end greater than means, subsumes the present and people to far future goals, to some sweeping, all-embracing, total transformation, to an optimism that time does not corrupt. In his *Thoughts*, Blaise Pascal warned us about the "eternal silence of these infinite spaces." C.S. Lewis, in his *Perelandra*, was wary of the allure of the "sweet poison of the false infinite." As David Malouf has argued more recently, we ought to take better care of ourselves, elude the emblems and ideologies of our big planet and unreachable other ones, the fetishes that so easily estrange and confuse us, detach us from our personhoods, our situations, our friendships. The band Beach House captured some of this theme in their "Space Song" (2015), a tribute to Sputnik's beeps and the human longing for dreams and distances, but also a recognition of human loss and grief, and of finding our bearings once again.

We saw the long term vision in Konstantin Tsiolkovsky, whose cosmist vision justified eugenics and euthanasia. In the labors of Wernher von Braun and his Peenemünde teams who built terror weapons in a Faustian bargain to get to the moon. We see it now in the transhumanist visions of publicists who foretell the end of our humanity, without a trace of irony or shame.

Transhumanism is nothing new. We have already traversed its origins in the mental landscapes of the Space Age. We have already moved beyond the human, beyond the earth, in the very prefixes that we assign to our new technologies and our new selves. We have become "super" (above), as in superman, supersonic, supercarriers, superbomb, and superhighways. We have become "ex" (away), as in exobiology, extraterrestrial, extremophile, extrasensory; and "X" (experimental) as in X-planes or X-Men. We have become "cyber" (virtual), as in cybernetics, cyberspace, cyberwar, and cybersecurity. We are now becoming ever more "hyper" (excess), as in hyperreality, hyperconnectivity, hypersonics, and hyperpowers. All that remains is our rising preference for the "meta" (beyond), the metaverse as a steppingstone, or alternative, to the universe itself.

By the very reaches of outer space, the Space Age will never end. Someday soon we might send our fellow human beings into outer space knowing that they might not return. They will set off to colonize new planets and moons as a matter of course, if still dependent on our home planet, Mother Earth. Based on all the evidence in human history so far, it is bound to happen. Even then, we will still be creating new Earths: new worlds built upon our very own biology and evolution, our own sustainable and improvable technologies; and we might hope, upon our very own free and integral bodies and minds, on our very own languages and civilizations. We await that Rubicon, that failsafe, that point of no return. Until then, and we should hope for long, the spacefaring Earth and its humanity will remain … indivisible.

Select Bibliography

Anderson, Chad. *The Space Economy: Capitalize on the Greatest Business Opportunity of Our Lifetime*. New York: Wiley, 2023.

Billings, Linda. "How Shall We Live in Space? Culture, Law and Ethics in Spacefaring Society," *Space Policy* 22/4 (November 2006): 249–255.

Blatt, Talia. "Anti-Satellite Weapons and the Emerging Space Arms Race," *Harvard International Review* 41/3 (2020): 29–34.

Bostrom, Nick. *Superintelligence: Paths, Dangers, Strategies*. New York: Oxford University Press, 2016.

Bouchey, Michael. "Unequal Partners: Canadian, Japanese, and American Space Programs," *Quest. The History of Spaceflight Quarterly* 2/22 (2015): 26–34.

Chiu, Joanna. *China Unbound: A New World Disorder*. Toronto: House of Anansi, 2021.

Chua, Amy. *Day of Empire: How Hyperpowers Rise to Global Dominance – and Why They Fall*. New York: Knopf, 2009.

Clegg, Brian. *Final Frontier: The Pioneering Science and Technology of Exploring the Universe*. New York: St. Martin's Press, 2014.

Cohen, Erik, and Samuel Spector. "Transhumanism and Cosmic Travel," *Tourism Recreation Research*, 45/2 (2020): 176–184.

Correll, Randall, and Simon P. Worden. "The Demise of US Spacepower: Not with a Bang but a Whimper," *Astropolitics* 3/3 (2005): 233–264.

Crawford, Kate. *Atlas of AI: Power, Politics, and the Planetary Costs of Artificial Intelligence*. New Haven: Yale University Press, 2021.

Crotts, Arlin. *The New Moon: Water, Exploration, and Future Habitation*. New York: Cambridge University Press, 2014.

Enriquez, Juan, and Steve Gullans. *Evolving Ourselves: How Unnatural Selection and Nonrandom Mutation are Changing Life on Earth*. New York: Penguin, 2015.

Evans, Ben. *The Twenty-First Century in Space*. New York: Springer Praxis, 2014.

Friedman, Louis. *Human Spaceflight: From Mars to the Stars*. Tucson: University of Arizona Press, 2015.

Geppert, Alexander C.T., ed. *Militarizing Outer Space: Astroculture, Dystopia and the Cold War*. New York: Palgrave, 2021.

Goldsmith, Donald, and Martin Rees. *The End of Astronauts: Why Robots Are the Future of Exploration*. Cambridge: Harvard University Press, 2022.

Guthrie, Julian. *How to Make a Spaceship: A Band of Renegades, an Epic Race, and the Birth of Private Spaceflight*. New York: Penguin, 2016.

Hayles, Katherine. *How We Became Posthuman: Virtual Bodies in Cybernetics, Literature and Informatics*. Chicago: University of Chicago Press, 1999.

Haraway, Donna. *Simians, Cyborgs and Women: The Reinvention of Nature*. New York: Routledge, 1991.

Harvey, Brian. *China in Space: The Great Leap Forward*. New York: Springer Praxis, 2013.

Huxley, Julian. "Transhumanism," in *New Bottles for New Wine*. London: Chatto & Windus, 1957: 13–17.

Impey, Chris. *Beyond: Our Future in Space*. New York: Norton, 2015.

Joy, Bill. "Why the Future Doesn't Need Us," *Ethics & Medicine* 17/1 (2001): 13–36 [originally in *Wired* magazine].

Johnson-Freese, Joan. *Space Warfare in the 21st Century: Arming the Heavens*. New York: Routledge, 2017.
Kaku, Michio. *The Future of Humanity: Terraforming Mars, Interstellar Travel, Immortality, and Our Destiny Beyond Earth*. New York: Doubleday, 2018.
Kurzweil, Ray. *The Singularity Is Nearer: When We Merge with AI*. New York: Viking, 2024.
Launius, Roger, and Howard McCurdy. *Robots in Space: Technology, Evolution, and Interplanetary Travel*. Baltimore: Johns Hopkins University Press, 2008.
Lewis, C.S. *Perelandra: A Novel*. New York: Macmillan, 1965
Luzin, Pavel. "The Future and Present of Roskosmos," *Riddle Russia* (1 March 2024).
Malouf, David. *The Happy Life: The Search for Contentment in the Modern World*. New York: Pantheon, 2013.
Minsky, Marvin. *The Emotion Machine: Commonsense Thinking, Artificial Intelligence, and the Future of the Human Mind*. New York: Simon & Schuster, 2006.
Moravec, Hans. *Robot: Mere Machine to Transcendent Mind*. New York: Oxford University Press, 2000.
Plotnick, Rachel. *Power Button: A History of Pleasure*. Cambridge: MIT Press, 2018.
Pyle, Rod. *Innovation the NASA Way: Harnessing the Power of Your Organization for Breakthrough Success*. New York: McGraw-Hill, 2014.
Rosa, Hartmut. *Social Acceleration: A New Theory of Modernity*. Jonathan Trejo-Mathys, trans. New York: Columbia University Press, 2013.
Soluri, Michael. *Infinite Worlds: The People and Places of Space Exploration*. New York: Simon & Schuster, 2014.
Tirosh-Samuelson, Hava. "In Pursuit of Perfection: The Misguided Transhumanist Vision," *Theology and Science*, 16/2 (2018): 200–222.
Weinzierl, Matthew. "Space, the Final Economic Frontier," *The Journal of Economic Perspectives* 32/2 (2018): 173–192.
Zubrin, Robert. *The Case for Space: How the Revolution in Spaceflight Opens Up a Future of Limitless Possibility*. Amherst: Prometheus Books, 2019.

For more works that remedy the excesses of transhumanism, that ennoble a more complete human person in culture, and steward in nature, see Simone Weil, *Gravity and Grace*. New York: Routledge, 2002 [1942]; Charles Taylor, *Sources of the Self*. Cambridge: Harvard University Press, 1989; Roger Scruton, *Soul of the World*. Princeton: Princeton University Press, 2016.

A Note on Non-English Sources

Besides the English-language sources listed for each chapter, I also have used, or recommend, a series of books in other languages. Here is a selection.

Arkhipova, Aleksandra, and Anna Kirziuk. *Opasnye sovetskie veshchi: Gorodskie legendy i strakhi v SSSR*. Moscow: Novoe Literaturnoe Obozrenie, 2020.

Buinovskii, Eduard. *Povsednevnaia zhizn' pervykh rossiiskikh raketchikov i kosmonavtov*. Moskva: Molodaia gvardiia, 2004.

Brandau, Daniel. *Raketenträume: Raumfahrt- und Technikenthusiasmus in Deutschland, 1923–1963*. Paderborn: Schöningh, 2019.

Demydovets, Maryna, and Oswald Ansgar. *Architektur für die russische Raumfahrt: vom Konstruktivismus zur Kosmonautik: Pläne, Projekte und Bauten*. Berlin: DOM Publishers, 2013.

Dominoni, Annalisa, and Benedetto Quaquaro. *Le città dell'universo: Come sarà abitare nello spazio*. Milan: Il saggiatore, 2023.

Fortikov, Ivan Petrovich. "Biosfera," *Priroda* 9 (1936): 27–32.

Golovanov, Iaroslav. *Korolev: Fakty i mify*. Moscow: Nauka, 1994.

Langemak, G.E., and V. P. Glushko, *Rakety*. Moscow: NKTP, 1935.

Kamanin, Nikolai P. *Skrytyi kosmos*, in four volumes. Moscow: Infortekst and Novosti kosmonavtiki, 1995–2001.

Korolev, Sergei P. *Raketnyi polet v stratosfere*. Moscow: Gosvoenizd, 1934.

Lapirov-Skoblo, M.Ia. "O puteshestviiakh v mezhplanetnye prostranstva," *Molodaia gvardiia* 5 (1924): 75.

Lapirov-Skoblo, M.Ia "Puteshestviia v mezhplanetnye prostranstva," *Pravda* 86 (15 April 1924), 6.

Lapirov-Skoblo, M.Ia., and A.E. Fersman, ed. *Uspekhi i dostizheniia sovremennoi nauki i tekhniki*. Moscow: Rabotnik prosveshcheniia, 1926.

Liapunov, Boris. *Otkrytie mira*. Moscow: Molodaia gvardiia, 1954.

Matveev, Oleg, et al., ed. *Nekotoryi opyt rossiiskoi kosmicheskoi deiatel'nosti (1991–2015 gg.)*. Moscow: Izd. Vitiuk, 2016.

Mayakovsky, Vladimir. "Protestuiu," *Krasnaia niva* 3 (1924): 72.

Nuti, Leopoldo. *La sfida nucleare: La politica estera italiana e le armi nucleari, 1945–1991*. Bologna: Il Mulino, 2007.

Pervushin, Anton. *Krasnyi kosmos: Zvezdnye korabli sovetskoi imperii*. Moscow: Eksmo, 2007.

Polianski, Igor. J., et al., ed. *Die Spur des Sputnik: Kulturhistorische Expeditionen ins kosmische Zeitalter*. Frankfurt am Main: Campus Verlag, 2009.

Ponomareva, Valentina. *Zhenskoe litso kosmosa*. Moscow: Gelios, 2002.

Portmann, Adolf. "Die Erde als Heimat des Lebes," in *Eranos Jahrbuch 1953*, volume 22, *Mensch und Erde*, Olga Fröbe-Kapteyn, ed. Zürich: Rhein-Verlag, 1954: 473–494.

Rakhmanin, Valentin F., and L.E. Sternin, ed., *Odnazhdyi i navsegda: Dokumenty i liudi o sozdatele raketnykh dvigatelei*. Moscow: Mashinostroenie, 1998.

Sandvoss, Ernst. *Vom Homo sapiens zum Homo spaciens: Eine Sinnperspektive der Menschheitsentwicklung*. Berlin: Logos Verlag, 2002.

Schwartz, Matthias. *Die Erfindung des Kosmos. Zur sowjetischen Science Fiction und populärwissenschaftlichen Publizistik vom Sputnikflug bis zum Ende der Tauwetterzeit*. Berlin: Peter Lang, 2003.

Semenov, Iu. P. ed., *Raketno-kosmicheskaia korporatsiia "Energiia."* Korolev: Energiia, 1996.

Shternfel'd, Ari A. *Initiation à la cosmonautique / Vvedenie v kosmonavtiku*. Moscow: 1937.

Sonin, Konstantin. "Malenkaia tragediia 2.0," *The Moscow Times* (3 September 2023).

Tarasenko, Maxim. *Voennye aspekty sovetskoi kosmonavtiki*. Moscow: Nikol, 1992.

Troitsky, Evgeny Sergeevich. *Russkaia ideia: Slaviansksia kosmizm in stantsiia "Mir."* Kaluga: Oblizdat, 2000.

Zakharchenko, Vasilii. *Puteshestvie v zavtra*. Moscow and Leningrad: Detlit, 1952.

Main Index

Art Index

Fiction and Poetry Index

Film, TV and Radio Index

Music Index

Kolibri 360 GmbH
Ferdinand-Jühlke-Straße 7
99095 Erfurt, Deutschland
produktsicherheit@kolibri360.de